Green Energy and Technology

For other titles published in this series, go to
http://www.springer.com/series/8059

Takeshi Yao
Editor

Zero-Carbon Energy Kyoto 2009

Proceedings of the First International Symposium of Global COE Program "Energy Science in the Age of Global Warming—Toward CO_2 Zero-emission Energy System"

Springer

Editor
Takeshi Yao
Program Leader
Professor of the Graduate School of Energy Science
Kyoto University
Steering Committee of GCOE Unit for Energy Science Education
Yoshida-honmachi, Sakyo-ku
Kyoto 606-8501, Japan
gcoe-office@energy.kyoto-u.ac.jp

ISSN 1865-3529 e-ISSN 1865-3537
ISBN 978-4-431-54078-6 e-ISBN 978-4-431-99779-5
DOI 10.1007/978-4-431-99779-5
Springer Tokyo Berlin Heidelberg New York

Softcover reprint of the hardcover 1st edition 2010

Cover design: WMXDesign, Heidelberg, Germany

Printed on acid-free paper

Springer is part of Springer Science+Business Media (www.springer.com)

Preface

Securing energy and conservation of the environment are the most important issues for the sustainable development of human beings. Until now, people have relied heavily on fossil fuels for their energy requirements and have released large amounts of greenhouse gases such as carbon dioxide (Here, we have abbreviated all greenhouse gases including carbon dioxide to "CO2."). Emissions of CO2 have been regarded as the main factor in climate change in recent years, and how to control them is becoming a pressing issue in the world. The energy problem cannot simply be labeled a technological one, as it is also deeply involved with social and economic issues. It is necessary to establish "Low carbon Energy science" as an interdisciplinary field integrating social science and human science with the natural sciences.

From 2008, four departments of Kyoto University, Japan — the Graduate School of Energy Science, the Institute of Advanced Energy, the Department of Nuclear Engineering, and the Research Reactor Institute—have joined forces, and with the participation of the Institute of Economic Research, have been engaged in a program entitled "Energy Science in the Age of Global Warming — Toward a CO2 Zero-Emission Energy System" for a Global Center of Excellence (COE) Program of the Ministry of Education, Culture, Sports, Science and Technology of Japan, with the support of university faculty members. This program aims to establish an international education and research platform to foster educators, researchers, and policy makers who can develop technologies and propose policies for establishing a scenario toward a CO2 zero-emission society no longer dependent on fossil fuels by the year 2100.

In the course of implementing the Global COE, we placed the GCOE Unit for Energy Science Education at its center, and we are proceeding from the Scenario Planning Group and the Advanced Research Cluster to Evaluation, forming mutual associations as we progress. The Scenario Planning Group is setting out a CO2 zero-emission technology roadmap and establishing a CO2 zero-emission scenario. They will also conduct analyses from the standpoints of social values and human behavior. The Advanced Research Cluster, as an education platform based on research, promotes socio-economic study of energy, study of new technologies for renewable energies, and research for advanced nuclear energy by following the roadmap established by the Scenario Planning Group. Evaluation is conducted by exchanging ideas among advisors inside and outside the university, including those from abroad, to gather feedback on the scenario, education, and research.

For education, which is the central activity of the Global COE, we have established the GCOE Unit for Energy Science Education and have selected students from the doctoral course, and are fostering these human resources. The students, on their own initiative, are planning and conducting interdisciplinary group research combining social and human science with natural science, working toward CO2 zero emission. The students will acquire the ability to survey the whole energy system through participation in scenario planning and interaction with researchers from other fields, and will apply that experience to their own research. This approach is expected to become a major feature of human resources cultivation. We will strive to foster young researchers who will be able to employ their skills and knowledge with a broad international perspective and expertise in their field of study in order to respond to the needs of society in terms of various energy and environmental problems. Those new researchers also will become leaders in the twenty-first century, full of vitality and creativity and working toward harmony between the environment and mankind.

We held the First International Symposium of the Global COE titled "Zero-Carbon Energy, Kyoto 2009" on August 20–21, 2009, at Kyoto University Clock Tower in parallel with the First International Summer School on Energy Science for Young Generations (ISSES-YGN) on August 20–22, 2009, at Kyoto University Clock Tower and Kyodai Kaikan. There were many important lectures by invited speakers and members of the Global COE, with interesting presentations by students at the GCOE Unit for Energy Science Education. This book is a compilation of the lectures and presentations. We hope that it will provide the impetus for the establishment of Low carbon Energy science.

Takeshi Yao
Program Leader
Global COE "Energy Science in the Age of Global Warming
– Toward a CO_2 Zero-Emission Energy System"

Contents

(ii) Renewable Energy Research and CO_2 Reduction Research

(iii) Advanced Nuclear Energy Research

Part I
Plenary and Invited Papers

What Can We Learn from Photosynthesis About How to Convert Solar Energy into Fuels?

Richard J. Cogdell, Katsunori Nakagawa, Masaharu Kondo, Mamoru Nango, and Hideki Hashimoto

Abstract We briefly review the need for construction of novel systems for the production of clean renewable fuels to replace oil and gas. Then the case is made that if it will be possible to gain a sufficient understanding of photosynthesis that it should be possible to use this information to produce "artificial leaves". These artificial leaves will be designed to convert solar energy into dense portable fuel.

Keywords Solar fuels • Photosynthesis • Artificial leaf • Global warming

1 Introduction

Currently in the developed world we get our energy mainly from fossil fuels. In fact approximately 70–80% of our current energy needs are met by burning coal, oil and gas. Unfortunately oil and gas supplies are predicted to be largely exhausted by the end of this century. Also we have a major problem caused by the increasing rates at which we currently consume fossil fuels, namely global warming caused by elevated levels of CO_2 in the atmosphere. As a result of these two imperatives there is an urgent need to develop new, clean, scalable, and renewable sources of fuels. Providing for our requirements for electricity is not such a fundamental problem. There are many clean and renewable energy sources that can be used to produce electricity, e.g. wind, solar energy, hydro, thermal, etc. The main challenge is to

R.J. Cogdell (✉)
University of Glasgow, Glasgow G12 8TA, Scotland, UK
e-mail: R.Cogdell@bio.gla.ac.uk

K. Nakagawa, M. Kondo, and M. Nango
Nagoya Institute of Technology, Gokiso-cho, Showa-ku, Nagoya, 466-8555, Japan

K. Nakagawa, M. Kondo, M. Nango and H. Hashimoto
CREST/JST, Saitama, Japan

H. Hashimoto
Osaka City University, 3-3-138 Sugimoto, Sumiyoshi-ku, Osaka 558-8585, Japan

T. Yao (ed.), *Zero-Carbon Energy Kyoto 2009*, Green Energy and Technology,
DOI 10.1007/978-4-431-99779-5_1,

find ways of producing clean sources for the production of dense portable fuel. If the aeroplanes are to be kept in the sky and the ships are to be kept at sea, then a fuel equivalent to gasoline will be required.

One possible abundant energy source that in principle could be harnessed to produce fuel is the sun. More than enough solar energy reaches the surface of earth each hour to satisfy all our current energy requirement for one year. How can we use this plentiful supply of solar energy to produce fuels? There is already a process that takes place on our planet that converts solar energy into fuel. This process is photosynthesis.

2 What Is Photosynthesis?

Photosynthesis is the process whereby plants, algae, and some bacteria are able to use solar energy to convert atmospheric carbon dioxide into sugar (a fuel). Indeed all the fossil fuels that man is currently so greedily consuming represent photosynthetic activity that occurred in the past millennia. If we could fully understand photosynthesis would it be possible to use this knowledge to produce robust, efficient artificial systems to convert solar energy into fuels. Although, at present we do not have all the detailed information that is needed in order to produce such systems it is possible from a consideration of the essence of photosynthesis to start a long the path towards succeeding in this aim.

Photosynthesis can be divided into four key partial reactions [1]. These are light-harvesting (light-concentration), using this concentrated light-energy to separate charge across a membrane, accumulation of positive charges on one side of this membrane in order to extract electrons from water (water splitting) and accumulation of the negative charges on the other side of this membrane in order to do catalysis to produce a fuel (e.g. the conversion of carbon dioxide to carbohydrate) (Fig. 1).

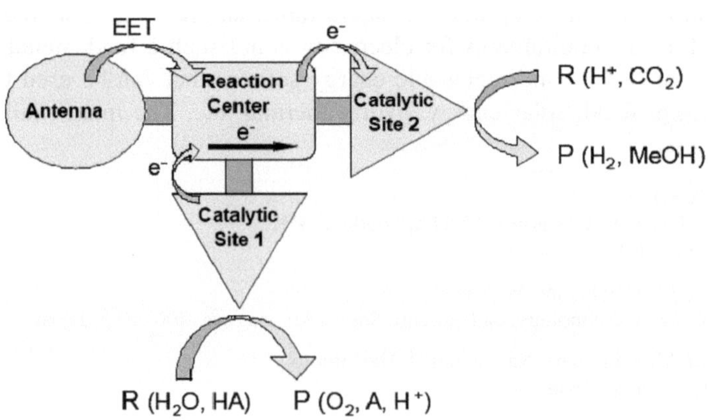

Fig. 1 A representation of the four key partial reactions of photosynthesis

3 Concept of the Artificial Leaf

Jim Barber from Imperial College in London has championed the idea of an artificial leaf. This is an elegant concept that really clearly illustrates the idea of using the photosynthetic blueprint in order to design ways of using solar energy to produce fuels. We will now consider each of the four key partial reactions of photosynthesis, outlined above, in order to assess where current research is along path to producing fuels from solar energy. We will describe both current approaches and where we think the major bottlenecks are.

Solar energy, though an abundant energy source, is a diffuse low-density energy source. This means that relatively large surface areas are required to harvest and concentrate this energy before it can be used to make fuel. Photosynthesis achieves this through its light-harvesting pigment-protein complexes. We are now in the fortunate position of having several high-resolution X-ray crystal structures of light-harvesting complexes from a variety of different photosynthetic organisms. It is possible therefore to ask whether there are some key common design features that can be found in these structures (Fig. 2).

Remarkably the structures of antenna complexes from different species are found to be highly variable. Initially it might be thought that this is a very disappointing result. Why should these structures be so variable? The answer is that the

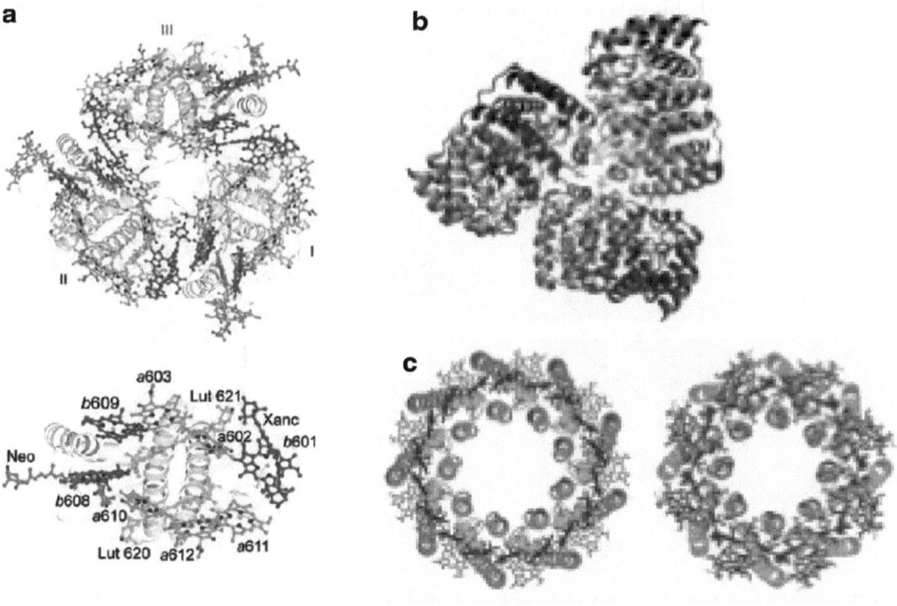

Fig. 2 Examples of the X-ray crystal structures of different light-harvesting complexes. (**a**) LHCII from higher plants [2], (**b**) peridinin-chlorophyll *a* protein from dinoflagellates [3], and (**c**) LH2 complex from two different species of purple bacteria [4, 5]

physics of energy-transfer is very tolerant. So long as the light-absorbing pigments are arranged close enough singlet–singlet energy-transfer will remain efficient even when the positioning of these pigments is quite variable. This design tolerance rather than being disappointing is encouraging to researchers who are trying to construct artificial systems designed to replicate biological light-harvesting. It means that there will be many different potentially successful ways of constructing light-harvesting modules [6].

Photosynthesis uses reaction centres to drive a transmembrane charge separation process powered by light energy provided by the light-harvesting system. Again there are several high-resolution X-ray crystal structures of photosynthetic reaction centres from several different species of photosynthetic organisms. In this case the basic structure of these reaction centres and organization of the redox centres within them is very highly conserved [6] (Fig. 3).

In contrast to energy-transfer the structural constraints on electron transfer are much more stringent. Even though this is true several artificial analogues of the basic reaction centre structure have been synthesized and shown to successfully separate change upon illumination. It appears therefore that it is not too difficult to construct artificial systems that can successfully mimic both light-harvesting and charge-separation.

At present the major bottlenecks both conceptually and practically are where the one electron redox reactions characteristic of a basic reaction centre interface with the chemical reactions that require either multiple positive or negative charges. The reaction centre of photosystem II also houses the water splitting apparatus.

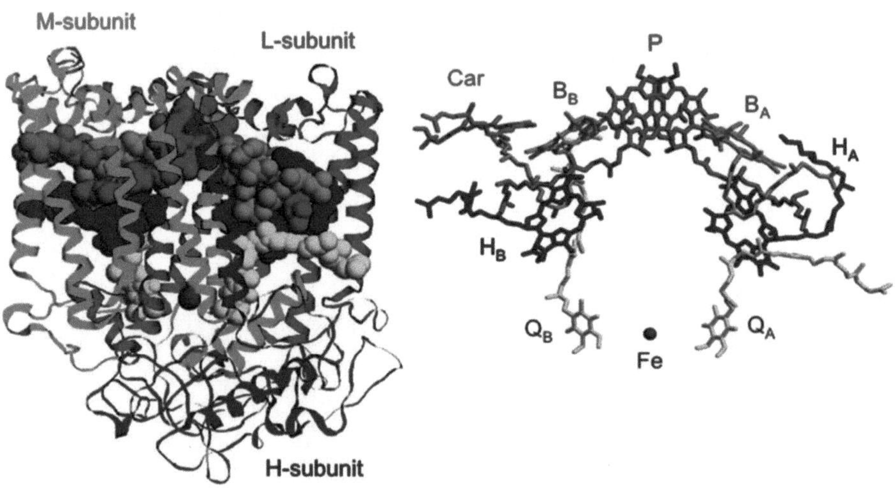

Fig. 3 Structure of the purple bacterial reaction centre and a view of the organization of the reaction centre redox carriers with the protein subunits removed (PDB: 1RGN). P is the special pair of bacteriochlorophyll molecules that go oxidized upon illumination. B is a monomeric bacteriochlorophyll. H is a bacteriopheophytin molecule. Q is a quinone. Charge is separated down the A branch and the negative charges are accumulated by Q_B

Unfortunately the X-ray crystal structure in this area only reveals an outline of the catalytic centre that is capable of storing four positive charges and then using them in a concerted reaction to split water into oxygen, protons and electrons (Fig. 4).

What is needed is structures of this catalytic centre in each of its individual redox states together with a detailed understanding of how the protein in the region of manganese centre participates in the reaction mechanism. There have been numerous attempts to mimic the structure of the water splitting centre with just the manganese/calcium/oxygen atoms [e.g. 8]. None of these metal complexes are able to reproduce the catalytic power of the natural system. We expect that this will remain to be the case until models include the function of the protein (a smart matrix) as well as just the ion centre. A similar barrier to progress exists on the side of the negative charges. There are however enzymes that are able to store negative charges in order to do a catalytic reaction required to produce fuels. The simplest

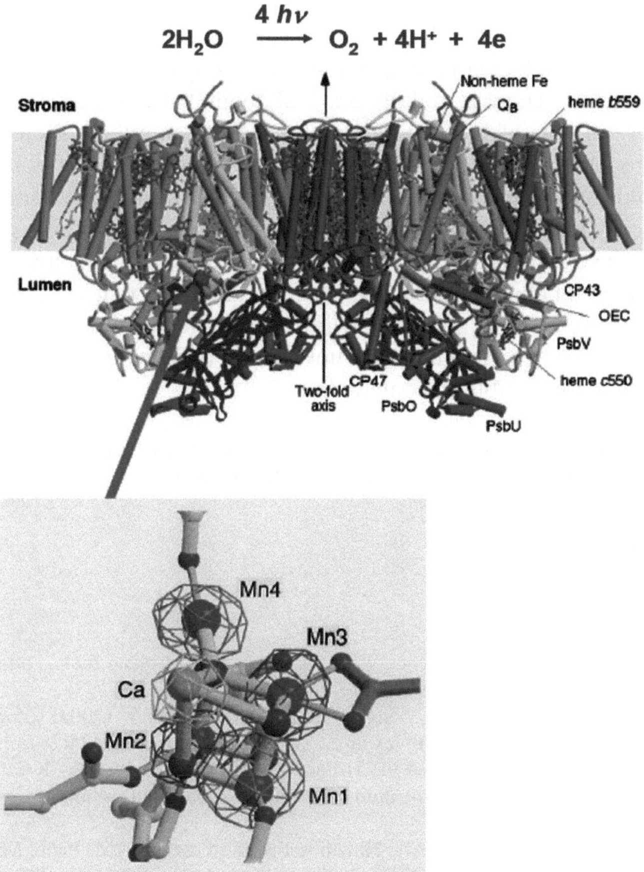

Fig. 4 The overall structure of photosystem II together with the picture of the water splitting centre [7]

example of such enzyme is hydrogenase [9]. This enzyme is able to reduce protons to produce hydrogen. Although hydrogen is not very dense fuel, there are many situations in which it could usefully substitute for denser carbon based fuels. Hydrogenase does provide a useful model system with which to develop the methodology to couple the reaction centre to an output capable of producing fuel. Unfortunately most hydrogenases are very oxygen sensitive. They are inactivated in the presence of oxygen. This is a major problem, since on the other side of the reaction centre oxygen is going to be produced. Recently hydrogenases have been found in anaerobic purple photosynthetic bacteria that are much less oxygen sensitive. It is to be hoped that when the details of the origin of this oxygen insensitivity are understood that hydrogenases resistant to oxygen can be incorporated into devices designed to be artificial leaves.

4 Outlook for the Future

Photosynthesis is a subject to the same laws of chemistry and physics as every other process on earth is. There is nothing magical about photosynthesis. It has however evolved over millions of years to the point where it can rather efficiently use solar energy to produce fuels. We believe that the production of an artificial leaf is possible and holds out the prospect of producing practical scalable systems for converting solar energy into fuel. Moreover, the need for such a system is so great that now is the time to invest in the research that is required to realize this dream. We see this as one of the grand challenges facing mankind and hope that enough of the really talented next generation of scientists will dedicate themselves to solving this challenge. Like all grand challenges it will not be easy but the result of not facing up to this challenge is impossible to contemplate.

Acknowledgements RJC acknowledges the support of the EPSRC. RJC and HH thank HFSP for support. HH and MN thank Nissan Science Foundation for support.

References

1. ESF report on "Harnessing solar energy for the production of clean fuels". http://ssnmr.leidenuniv.nl/files/ssnmr/CleanSolarFuels.pdf
2. Liu Z, Yan H, Wang K, Kuang T, Zhang J, Gui L, An X, Chang W (2004) Crystal structure of spinach major light-harvesting complex at 2.72 Å resolution. Nature 428:287–292
3. Hofmann E, Wrench PM, Sharples FP, Hiller RG, Welte W, Diederichs K (1996) Structural basis of light harvesting by Carotenoids: peridinin-chlorophyll-protein from *Amphidinium carterae*. Science 272:1788–1791
4. McDermott G, Prince SM, Freer AA, Hawthornthwaite-Lawless AM, Papiz MZ, Cogdell RJ, Isaacs NW (1995) Crystal structure of an integral membrane light-harvesting complex from photosynthetic bacteria. Nature 374:517–521

5. Koepke J, Hu X, Muenke C, Schulten K, Michel H (1996) The crystal structure of the light-harvesting complex II (B800–850) from *Rhodospirillum molischianum*. Structure 4:581–597
6. Moser CC, Page CC, Cogdell RJ, Barber J, Wraight CA, Dutton PL (2003) Length, time, and energy scales of photosystems. Advances in protein chemistry. Academic, New York, pp. 71–109
7. Ferreira KN, Iverson TM, Maghlaoui K, Barber J, Iwata S (2004) Architecture of the photosynthetic oxygen-evolving center. Science 303:1831–1838
8. Sproviero EM, Gascon JA, McEvoy JP, Brudvig GW, Batista VS (2006) Characterization of synthetic oxomanganese complexes and the inorganic core of the O_2-evolving complex in photosystem II: evaluation and the DFT/B3LYP level of theory. J Inorg Biochem 100:786–800
9. Vignais P, Billoud B (2007) Occurrence, classification, and biological function of hydrogenases: an overview. Chem Rev 107:4206–4272

Renaissance of Nuclear Energy in the USA: Opportunities, Challenges and Future Research Needs

Masahiro Kawaji and Sanjoy Banerjee

Abstract The future of nuclear energy is an important issue for many countries intending to reduce their dependence on fossil fuels and achieve the reduction targets for green house gas (GHG) emissions. As of June, 2008, there were 439 operating nuclear reactors with a total generating capacity of 372 GWe and 42 power reactors under construction in 15 countries. In the USA, a total of 104 nuclear reactors currently produce 20% of the electricity and account for at least 70% of all GHG-free electricity generation. Their performance has been improving steadily over the past 20 years and has now reached 90% capacity factor. The Energy Policy Act of 2005 authorized future nuclear R&D and provided incentives for construction of new nuclear plants. As a result, there are now 17 COL applications for construction of as many as 26 new reactors in the USA. This paper summarizes some of the opportunities, challenges and future research needs for achieving and sustaining nuclear renaissance in the USA.

Keywords Nuclear energy • Nuclear reactors • Nuclear power • LWR • PWR • BWR

1 Introduction

The future of nuclear energy is an important issue for many countries in the world aiming to reduce both their dependence on fossil fuels and green house gas (GHG) emissions. As of June, 2008, there were 439 operating nuclear reactors with a total generating capacity of 372 GWe and 42 power reactors were under construction in 15 countries. Today, the nuclear power accounts for approximately 17% of worldwide electricity generation. In 2004, the United States, France and Japan together accounted for ~56% of the nuclear electricity generation capacity as shown in Fig. 1, and their share is expected to decrease slightly to ~50% in 2020 as other countries, especially

M. Kawaji (✉) and S. Banerjee
The Energy Institute, City University of New York, New York, USA
e-mail: mkawaji@ccny.cuny.edu

T. Yao (ed.), *Zero-Carbon Energy Kyoto 2009*, Green Energy and Technology,
DOI 10.1007/978-4-431-99779-5_2, © Springer 2010

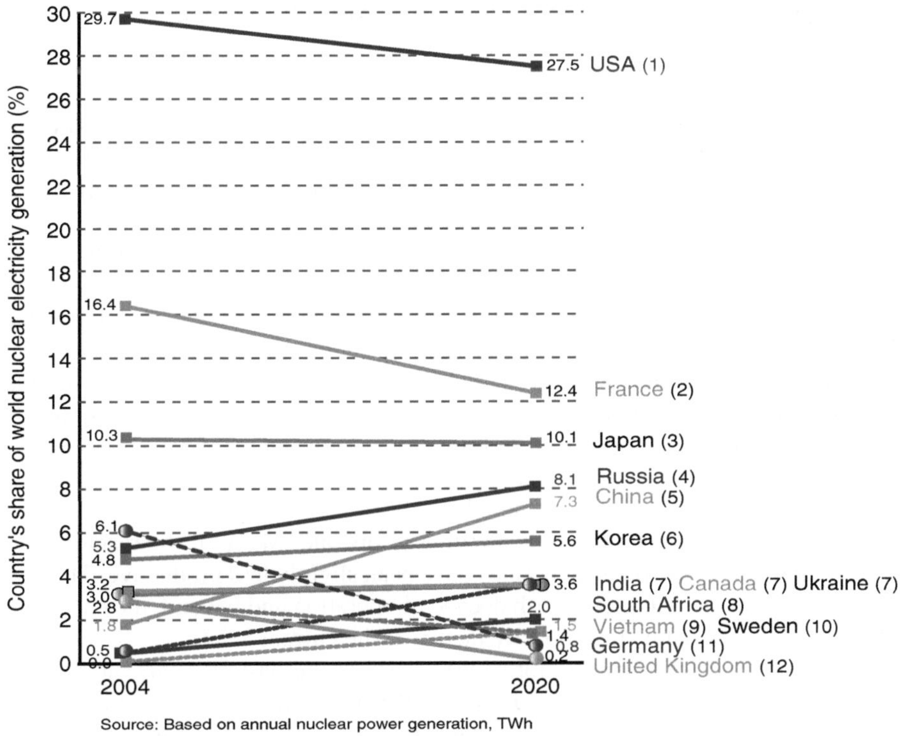

Fig. 1 Share of world nuclear electricity generation [1]

China, Russia and India plan to expand their nuclear energy generation [1]. European countries, on the other hand, have reduced their use of nuclear power in recent years but countries such as United Kingdom and Italy have decided to deploy more nuclear power in the future.

In the USA, 85% of all the energy consumed comes from fossil fuels: oil, natural gas, and coal [2]. The rest is provided by nuclear and hydro. The renewable energy sources such as solar, wind and biomass contribute very little at the present time. In electricity generation, the fuels used in US power plants are coal (48.5%), natural gas (21.3%), nuclear (19.6%), hydro (5.9%), wind (1.3%), petroleum (1.1%), wood (0.4%), waste (0.4%), geothermal (0.4%) and solar/PV (<0.1%).

As developing countries increase their electricity use and plug-in hybrid and electric vehicles are commercialized in the near future, the global consumption of electricity is expected to keep growing at a rapid pace. If GHG emission is to be significantly reduced in the next 20–40 years, the role of nuclear energy in the global energy supply needs to be expanded significantly since the renewable energy sources could take many years to become a significant source of non-fossil energy.

2 Opportunities and Challenges for Nuclear Energy in the USA

Nuclear and hydroelectric power accounts for most of non-CO_2 emitting source of electricity in the USA as shown in Fig. 2a. The nuclear share has been steady at about 20% over the past 20 years, and now accounts for at least 70% of all GHG-free electricity generation (Nuclear Energy Institute website: http://www.nei.org/resourcesandstats/documentlibrary/reliableandaffordableenergy/graphicsandcharts/uselectricitygenerationfuelshares/) (Fig. 2b).

A total of 104 nuclear reactors are currently operating in the USA and produce 800 billion kilowatt hours of electricity per year. Some nuclear power plants have increased their power output so that additional electricity equivalent to 16 new units has been added to the grid through power uprates. The existing nuclear power plants have also improved their operating performance significantly in the past 20 years as evident from their capacity factor data [3], reaching 90% as shown in Fig. 3.

In 2009, the USA officially entered the license renewal era, as two reactors passed the 40-year mark and are continuing to operate. Many of the reactors that soon enter their license renewal periods have been slightly less productive. Thus, as more reactors move into their fourth decade of operation and beyond, the challenge now is to continue achieving capacity factors at the current level.

In spite of their excellent performance record in the past 20 years, no nuclear power plants have been built and nuclear R&D was severely curtailed after the last of the Gen-II reactors went online in the early 1990s. A decade later, the Energy Policy Act of 2005 authorized future nuclear R&D and provided incentives for construction of new nuclear plants. As of June, 2009, 17 Construction and Operating License (COL) applications for as many as 26 new reactors have been docketed by the US Nuclear Regulatory Commission [4]. The planned sites and reactor types to be built are shown in Fig. 4. The first COL licenses are expected to be granted in July 2011 as shown in Table 1.

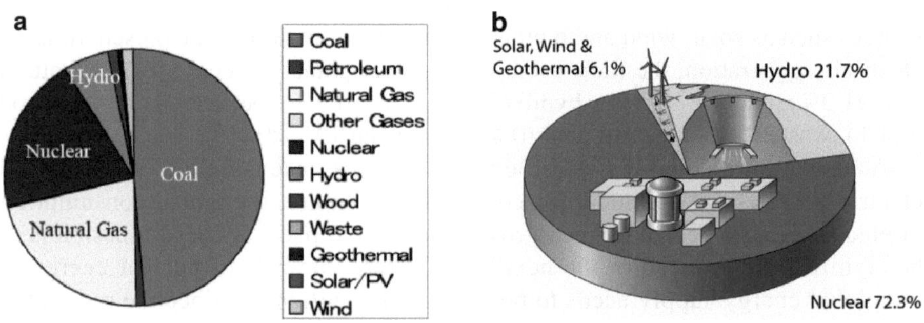

Fig. 2 Power plant fuels used (**a**) and emission-free electricity generation (**b**) in the USA (2008)

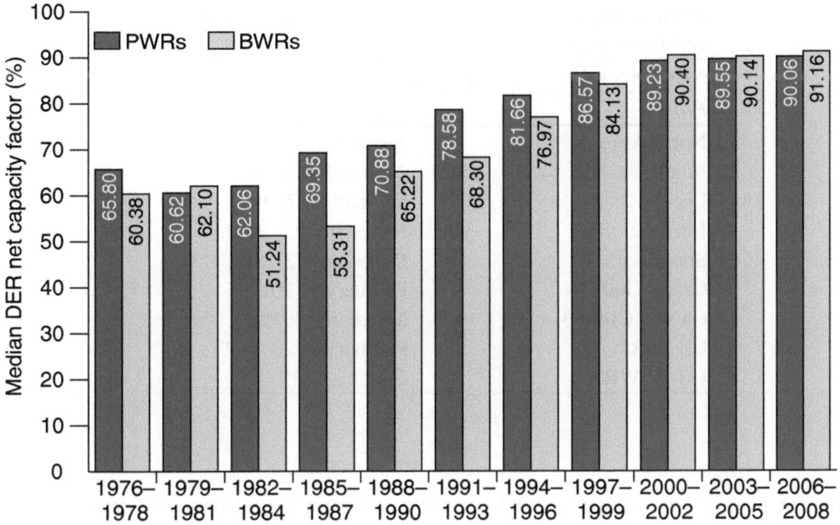

Fig. 3 Nuclear reactor capacity factors in the USA

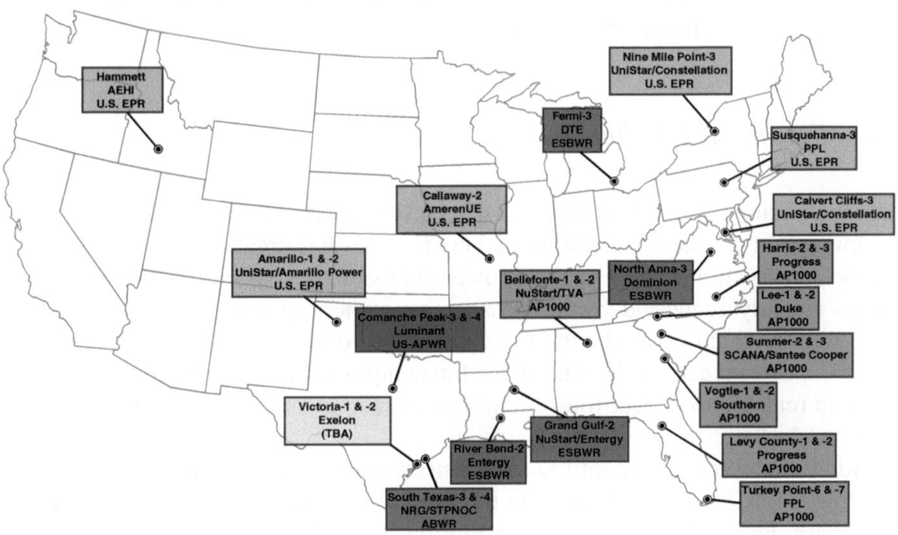

Fig. 4 COL applications announced for new nuclear reactors in the USA as of January, 2009 [4]

There are five new designs of advanced reactors which will be built in the future. Some designs have already been certified by the US Nuclear Regulatory Commission (NRC), and others are currently under review as summarized in Table 2.

Table 1 Expected dates for COL issuance and design certification [4]

NRC target dates for COL issuance and design certification	
Project	*Date*
North Anna-3	July 13, 2011
Vogtle-3, -4	August 24, 2011
Lee-1, -2	Seplemher27, 2011
Harris-2, -3	November 22, 2011
Summer-2, -3	December 13, 2011
Grand Gulf-3	February 1, 2012
Calvert Cliffs-3	March 16, 2012
US EPR	February 5, 2012
US-APWR	June 25, 2012

These new reactors possess the following advanced features and improvements over the existing reactors: standardized designs; easier to operate; faster and cheaper to build, operate and maintain; simpler and safer; latest technology; less equipment and components; passive safety systems (AP1000 and ESBWR); simplified operations and maintenance. Their power output ranges from 1,100 to 1,700 MWe, and some are already operating (ABWR in Japan) or under construction (US-EPR in Europe, AP1000 in China).

2.1 Financial Challenges

Although many COL applications have been submitted to NRC and are undergoing review, recent economic downturn and credit crisis has created financial obstacles for the construction of new nuclear power plants in the near future. The loan guarantees recently requested by the prospective owners of new reactors amount to $122 billion which is far above the original DOE offer of $18.5 billion. However, under the current economic conditions, financing is difficult to obtain and it would be more realistic to expect four to eight new reactors entering service in the 2018 time frame [5].

Additional financial incentives for construction of new nuclear power plants can be provided by local governments in the form of rate recovery during the construction phase. For example, in 2007, the Florida Public Service Commission adopted new rules that will let investor-owned utility companies recover some of the costs of the new plants before they begin operation (http://www.floridapsc.com/home/news/?id=459). The partial recovery of the planning and construction costs of a new nuclear plant before it begins operation would allow the companies to recoup those costs earlier and will encourage more investment in the facilities while lessening the chance for "rate shock" that could occur if the company waited to recoup all its construction costs when the plant began operation.

Table 2 Advanced reactor designs

	Reactor type	Vendor	Design certification	Reactor power (MWt)	Electric output (MWe)	Design life (years)
AP1000	Advanced pressurized water reactor (passive design)	Toshiba-Westinghouse	Certified *by* NRC – 2005 Revision under review – expected in 2011	3,400	1,117	60
ESBWR	Boiling water reactor (passive design)	GE-Hitachi	Under review – expected in 2010	4,500	1,560	60
ABWR	Advanced boiling water reactor	Hitachi, GE-Hitachi, Toshiba	Certified by NRC – 1997	3,926	1,350	60
US-EPR	Pressurized water reactor	AREVA	Under review – expected in 2012	4,300	1,600	60
US-APWR	Advanced pressurized water reactor	Mitsubishi	Under review – expected in 2012	4,451	1,700	60

3 Future Needs

In the next quarter century, aggressive investments in new plants will be needed along with an ongoing effort to up-rate existing plants; extend operating licenses from 40 to 60 years; and license and construct Gen-III nuclear plants. At the same time, the US government needs to control nuclear materials, assure nuclear nonproliferation abroad, and conduct ultimate management of used nuclear fuel.

Nuclear R&D requirements identified by the US Department of Energy (DOE) include understanding how materials age in a harsh reactor environment over many decades of service; developing a sustainable fuel cycle consisting of fuel recycling, advanced reactors, robust waste forms, and a geologic repository; and developing very high-temperature gas reactors (VHTRs) for process industry applications as one of the main Gen-IV reactor designs to be developed by the USA [6].

From a licensing perspective, the thermal-hydraulic performance of nuclear systems during normal operation and accident conditions continues to be central to reactor safety evaluation [7]. This is because many of the current generation of light water reactors have either been granted or are seeking increases in power outputs, requiring better estimates of safety margins. As well, several reactor designs with new, and sometimes passive, safety features require improved understanding for design certification and construction. Generic safety issues have also arisen, e.g., with regard to potential blockage of screens or strainers by debris generated, during blowdown in postulated loss-of-coolant accidents, which may impact long term coolant recirculation and core cooling. An overview of Light Water Reactor Thermalhydraulics and Safety issues is summarized in Table 3.

3.1 Nuclear Engineering Education

To perform the required research, development, and deployment of new reactor technologies in the future, renewed investments into the human capital and infrastructure capabilities as well as expansion of international collaboration will be required [6]. The expansion of nuclear power for electricity generation would lead to an increased demand for skilled labor at all levels. It is expected that each new reactor will require between 1,400 and 1,800 workers for construction with peak employment of up to 2,300 workers. Once built, these potential power plants would require tens of thousands of permanent, full-time workers to operate the plants and additional supplemental labor for maintenance and outages.

American industry faces increased competition for skilled talent and the nuclear industry is not an exception. In addition, the nuclear industry is also challenged by an aging work force, with nearly 50% of workers aged 47 or older who will be eligible to retire during the next 10 years. Along with plans for industry growth, the expected attrition of a large portion of the industry's total work force has prompted an unprecedented recruitment effort throughout the industry. Still, recruitment of skilled workers remains a significant challenge for the nuclear industry.

Table 3 LWR thermalhydraulics/safety issues overview [7]

	Current			Next generation			
Issues	BWRs EPU	PWRs EPU	Generic areas	AP1000	ESBWR	USAPWR	EPR
CHF (new fuel designs, experiments & correlations)	X	X	X	X	X	X	X
Neutronic-thermalhydraulic instability/ATWS (coupling analysis tools)	X				X		
LOCA (best estimate & uncertainties modeling)		X		X		X	X
CCFL/reflux condensation (experiments & models)			Gas in safety injection lines	Gas in safety injection lines	Gas in safety injection lines	Refluxing	Refluxing
Sump screen/strainer blockage	?	X	X	?	?	?	?
Containment overpressure credit (model accuracy)	X	X	X				
Steam dryer failure (experiments & models)	X				?		
Specific safety features (behavior)				ADS systems	Containment noncondensable distribution	Accumulator, delayed injection secondary depressurization	Secondary depressurization safety injection

Table 4 Numbers of nuclear engineering and health physics degrees granted in the USA

	Nuclear engineering degrees[a]			Health physics degrees[b]			
Year	B.S.	M.S.	Ph.D.	Year	B.S.	M.S.	Ph.D.
2008	454	260	127	2008	73	108	8
2007	413	227	89	2007	79	91	28
2006	346	214	70	2006	71	90	12
2005	268	171	74	2005	78	77	14
2004	219	154	75	2004	54	64	14
2003	166	132	78	2003	56	73	25
2002	195	130	67	2002	41	76	20
2001	120	145	80	2001	37	71	23
2000	159	133	74	2000	33	79	24
				1999	55	115	22

[a]Survey of 31 universities with nuclear engineering programs
[b]Survey of 26 universities granting health physics degrees

The US NRC has estimated that the nuclear industry as a whole will need an influx of 90,000 new workers within 10 years. Fortunately, increasing public recognition of the value of nuclear energy as a clean, reliable electricity source is leading more young people to identify nuclear energy as a career path. The number of nuclear engineering programs at US institutions dropped from about 50 programs in 1990 to fewer than 30 in the late 1990s, but bounced back to more than 30 programs currently. A recent Department of Energy study also found that enrollments in undergraduate nuclear energy programs have grown to more than 1,900 in the 2006–2007 academic year, compared to fewer than 500 eight years ago. Graduate enrollments also have jumped to more than 1,100 in the 2006–2007 year vs. just 220 in 1998–1999. The numbers of undergraduate and graduate degrees awarded in nuclear engineering and health physics programs between 2000 and 2008 are shown in Table 4 [8].

4 Summary

Nuclear power is an important source of emission-free electricity that can contribute to reduced dependence on fossil fuels and mitigation of global warming effects around the world. In the USA, aggressive investments in new plants will be needed in the next quarter century, along with an ongoing effort to uprate existing plants, extend operating licenses from 40 to 60 years, and license and construct Gen-III nuclear plants. To date, a total of 17 Construction and Operating License (COL) applications have been submitted for construction of five types of advanced reactors, reflective of the opportunities for nuclear renaissance in the USA. At the same time, there are challenges to be faced in controlling nuclear materials, assuring nuclear nonproliferation abroad, and conducting ultimate management of used nuclear fuel. To perform the required R&D and new reactor deployment, renewed investments into the human capital and infrastructure capabilities will be required.

References

1. *Nuclear News* (2009) American Nuclear Society, January, 2009, p. 44
2. Nuclear Energy Agency (2008) Nuclear energy outlook 2008. OECD, Washington, DC, ISBN: 9789264054103
3. *Nuclear News* (2009) American Nuclear Society, May, 2009, p. 33
4. *Nuclear News* (2009) American Nuclear Society, January, 2009, pp. 35–36
5. *Nuclear News* (2009) American Nuclear Society, June, 2009, p. 28
6. US Department of Energy (2008) Required assets for a nuclear energy applied R&D program. Draft Report by Idaho National Laboratory (September, 2008)
7. Banerjee S, Abdullahi Z (2009) Thermal and hydraulic issues related to light water reactors. A Keynote paper to be presented at the 13th International Topical Meeting on Nuclear Reactor Thermal Hydraulics (NURETH-13), Sept. 27 – Oct. 2, 2009, Kanazawa, Japan
8. *Nuclear News* (2009) American Nuclear Society, July, 2009, pp. 93–94

Eco-Friendly Production of Biodiesel by Utilizing Solid Base Catalysis of Calcium Oxide for Reaction to Convert Vegetable Oil into Its Methyl Esters

Masato Kouzu

Abstract Since biodiesel is commonly produced by converting vegetable oil into its methyl esters with the help of catalysis of alkali-hydroxide dissolved in methanol, it is necessary to eliminate the homogeneous catalyst from the crude biodiesel by washing with a large amount of water. With a view of studying the eco-friendly production without discharging wastewater, we investigated the solid base catalysis of calcium oxide. Primarily, transesterification of soybean oil at reflux methanol under atmospheric pressure was carried out on a glass batch reactor, for testing calcium oxide as compared with the other solid base such as calcium hydroxide magnesium oxide, and alumina supported potassium. Calcium oxide was superior in the catalytic activity and the reusability to the other solid bases. Additionally, the interesting facts on the solid base catalysis were found through the primary test. Based on these results, the practical catalyst was manufactured as an experiment. For testing the practical catalyst, rapeseed oil was transesterified on the laboratory scale pilot plant.

Keywords Biodiesel • Solid base catalyst • Calcium oxide • Lime stone

1 Introduction

Biodiesel is an eco-friendly alternative to fossil diesel fuel, because the raw material is vegetable oil that is one of the renewable energy resources. Additionally, diesel engine fueled by biodiesel is sure to reduce toxic emissions like SO_x, unburned hydrocarbons, and soot particles [1]. In 2008, the total of biodiesel produced globally seems to reach ten-million tons.

Commonly, biodiesel is produced by converting vegetable oil into its methyl esters with the help of catalysis of alkali-hydroxide dissolved in methanol. The homogeneous catalyst brings about the very fast conversion of vegetable oil:

M. Kouzu (✉)
Research Center of Fine Particle Science and Technology, Doshisha University, Kyoto, Japan
e-mail: mkozu@mail.doshisha.ac.jp

T. Yao (ed.), *Zero-Carbon Energy Kyoto 2009*, Green Energy and Technology, DOI 10.1007/978-4-431-99779-5_3, © Springer 2010

soybean oil was turned into biodiesel after 1 h of the base-catalyzed reaction carried out at 333 K under atmospheric pressure in the presence of sodium hydroxide dissolved in methanol [2]. However, a massive amount of wastewater is discharged from the production process, because it is necessary to wash the homogeneous catalyst off the crude biodiesel. Also, emulsification of biodiesel occurs in the washing step with the result that it is very difficult to operate the production process.

For the problems mentioned above, various solutions have been proposed by many researchers. Saka et al. has investigated the non-catalytic reaction in supercritical methanol [3]. The reaction was performed at 623 K under 30 MPa as a result and rapeseed oil was converted into biodiesel after only 4 min. In their recent study, the non catalytic reaction was modified by pre-treating the feedstock oil with subcritical water [4]. On the other hand, the enzymatic reaction can convert vegetable oil into its methyl esters at room temperature [5]. In order to improve the cost efficiency of the biodiesel production, lipase-producing microbial cells immobilized within porous support material was used as the whole-cell biocatalyst [6]. Also, the heterogeneous catalytic reaction is useful in producing biodiesel without discharging wastewater. There was a research paper stressing that anion-exchange resin is a candidate for the solid base catalyst [7]. From the economical point of view, our interests were taken in utilizing the solid base catalysis of calcium oxide [8].

Calcium oxide is one of typical solid bases that can catalyze reactions making carbon–carbon bond, such as aldol addition, Michael addition, and Tishchenko reaction [9–11]. For these reactions, as well as the reaction to produce biodiesel, the solid base catalysis is effective in generating nucleophile [12]. Due to the molecular structure formed out of ionic crystal, oxygen anion functions as the basic active site. However, the catalytic activity was seriously reduced by contacting with CO_2 and H_2O contained in air.

In this paper, we investigated solid base catalysis of calcium oxide for the reaction to produce biodiesel, with a view of studying the eco-friendly production. Primarily, on a batch glass reactor, soybean oil was transesterified with refluxing methanol under atmospheric pressure, in order to test calcium oxide as compared with calcium hydroxide, magnesium oxide, strontium oxide, anion-exchange resin, and alumina supported potassium. Mechanism on the heterogeneous catalytic reaction was discussed with relating to properties of the catalyst collected after the reaction. Furthermore, based on results of the primary test, the practical catalyst was manufactured as an experiment. In order to test the practical catalyst, the rapeseed oil transesterification was performed on the laboratory scale pilot plant.

2 Materials and Methods

Calcium oxide (CaO) was prepared by calcining lime stone powder at 1,173 K for 1.5 h in helium gas flow of 100 ml min^{-1}. Purity of the lime stone powder was 99.5%. Properties of CaO were shown in Table 1, in comparison with those

Table 1 Properties of solid base catalysts tested in the present study

	Surface area ($m^2\,g^{-1}$)[a]	Basic strength[b]
CaO	13	$15.0<H\_<18.4$
$Ca(OH)_2$	16	$9.3<H\_<15.0$
MgO	198	$15.0<H\_<18.4$
SrO	2	$15.0<H\_<18.4$
K/Al_2O_3	102	$15.0<H\_<18.4$
Resin[c]	–	–

[a] Calculated by BET method using data on nitrogen adsorption at 77 K
[b] Determined by indicator method
[c] Anion-exchange resin, a commercially available product

of the other solid bases tested in the present study. Calcium hydroxide ($Ca(OH)_2$) was made of CaO, by storing in humid nitrogen. For obtaining magnesium oxide (MgO) and strontium oxide (SrO), their carbonates were calcined at the prescribed temperature: 773 K for MgO and 1,323 K for SrO. For alumina supported potassium (K/Al_2O_3), its precursor was prepared by impregnating potassium nitrate onto alumina support. Then, calcination of the precursor was conducted at 773 K for transforming into the solid base catalyst. Anion-exchange resin was the commercially available product. Prior to the test, the resin was washed with methanol on a glass column and a tubular pump.

For the practical catalyst manufactured as an experiment, uneven form of the roughly crushed lime stone was the raw material. Size distribution of the raw material was regulated within the range of 1.0–1.7 mm by sieving.

Primarily, on a glass batch reactor, soybean oil was transesterified at reflux of methanol in the presence of the solid base tested. Soybean oil was of edible grade: Acid value <0.1 mg-KOH g^{-1} and water content <0.01%. For methanol, water content was below 0.1%. After 100 ml of soybean oil was mixed with 50 ml of methanol in the glass batch reactor, 14 mmol of the catalyst was stirred into mixture of the reactants. Then, the glass batch reactor was heated on a mantle heater, and reflux of methanol was continued for 2 h. At the interval of 0.5 h, a small amount of the transesterified oil was withdrawn for analysis to determine the yield of fatty acid methyl esters (FAME). The analysis was carried out on Agilent 6890 gas chromatograph which was equipped with a cool-on column reactor, a stainless steel capillary column and a flame ionization detector.

Following the transesterifying operation, whole of the product containing the catalyst was collected. The catalyst was separated from the product by filtration, in order to examine the properties by X-ray diffraction (XRD) and scanning electron microscopy (SEM). In advance of these instrumental analyses, the separated catalyst was washed several times with methanol.

The practical catalyst, prepared on the basis of results of the primary test, was employed for the rapeseed oil transesterification on a laboratory scale pilot plant. As shown in Fig. 1, the laboratory scale pilot plant was characterized by the batch unit consisting of a circulating stream passing through a column reactor. Into the column reactor, 20 ml of the practical catalyst was packed. First, 270 ml of rapeseed

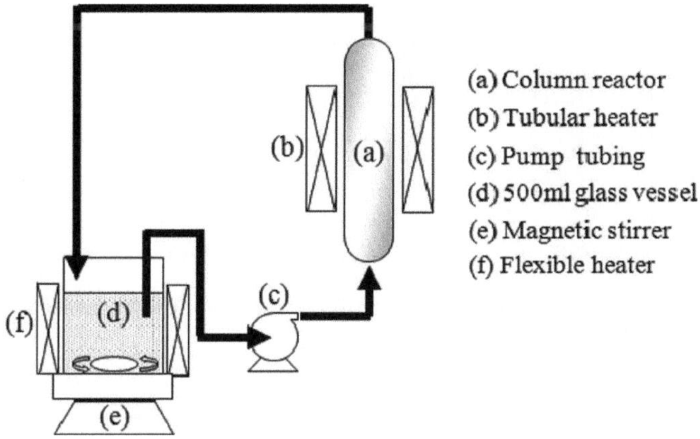

Fig. 1 Schematic flow diagram of a laboratory scale pilot plant to test practical catalyst manufactured as an experiment

oil was emulsified with 210 ml of methanol in the glass vessel at the temperature of 333 K under atmospheric pressure. Then, the emulsified reactants were pumped into the column reactor at the feeding rate of 50 ml min^{-1}. The circulation was continued with keeping temperatures of the column reactor and the glass vessel at 333 K till rapeseed oil was converted into biodiesel.

3 Results and Discussion

3.1 Catalytic Activity

Figure 2 shows the yields of FAME produced by transestrifying soybean oil for 1 h in the presence of the solid base tested. CaO was employed for the reaction as a result and the yield of FAME reached 93%. For $Ca(OH)_2$, MgO, and anion-exchange resin, the FAME yield was 13%, 4%, and 14%, respectively. Therefore, it was evident that CaO was very active in the reaction as compared to the other solid bases mentioned above. However, relationship between the catalytic activity and the basic strength among these solid bases was not appreciable.

As shown in Fig. 3, there were the solid bases matching CaO for the catalytic activity: SrO and K/Al_2O_3. But, SrO could not reused for the reaction successively repeated, due to the serious leaching of the catalyst. Although K/Al_2O_3 could be collected after the reaction, the yield of FAME produced in the presence of the reused catalyst seriously decreased. The deactivation of the catalyst was caused by leaching of the active phase from the catalyst. On the other hand, CaO seemed to be reused without deactivating. From these results, we concluded that calcium

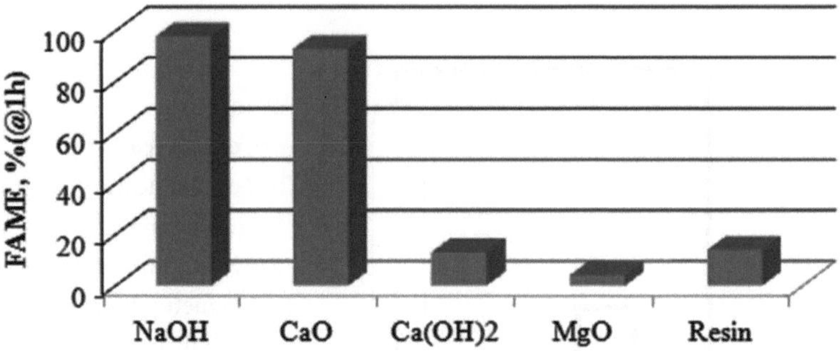

Fig. 2 Yield of FAME produced by transesterifying soybean oil with methanol for 1 h in the presence of the solid base tested

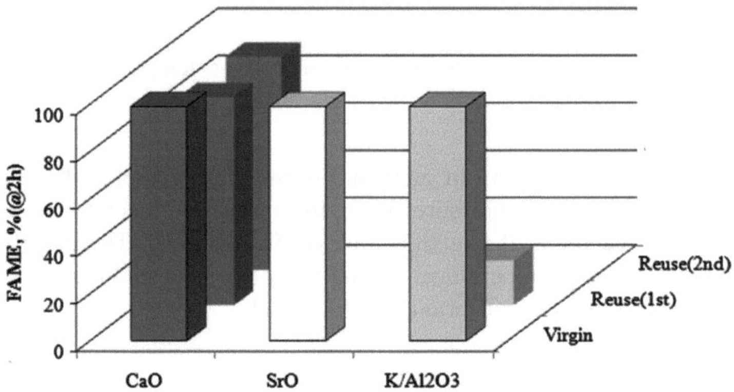

Fig. 3 Variation in yield of FAME with successive repetition of reaction, with reusing the solid base catalyst

oxide is the solid base catalyst useful in producing biodiesel. Also, we were very interested in the good reusability of CaO, because researchers in the field of solid base catalyst know that calcium oxide is deactivated by contacting with a slight CO_2 and moisture contained in air [12]. Since it was difficult to guard the catalyst perfectly from air-exposure during the test to evaluate the reusability, it was expected that the reused CaO was deactivated to some extent. Accordingly, it was very important to understand the mechanism on the reaction catalyzed by CaO.

3.2 Mechanism on the Catalytic Reaction

From our interest in the reusability of CaO, a mechanism on the catalytic reaction was investigated by appreciating change in properties of the catalyst during the soybean oil transesterification. Figure 4 show change in shape of the catalyst

Fig. 4 SEM images of catalyst

particles, observed by SEM. Fresh CaO, obtained after calcination of the raw material, was formed into particles of sub-micron size. After the reaction, the catalyst consisted of particles of two types. One was shaped a large angular rock, and the other looked like a cluster of thin plates. The drastic change of the shape was likely to reflect the chemical conversion of the catalyst.

Figure 5 compares fresh CaO and reused one using their XRD patterns. Only a diffraction pattern of calcium oxide was drawn for fresh CaO, while the diffraction pattern of reused CaO differed obviously from that of fresh one. From the diffraction pattern, it was evident that calcium oxide was turned into calcium diglyceroxide, $Ca(C_3H_7O_3)_2$. Also, Fig. 5 show that the chemical conversion of CaO was not obvious at 15 min. of the reaction time. When the FAME yield reached 70% at 30 min. of the reaction time, the major compound of the catalyst was calcium diglyceroxide. These results indicated calcium oxide was combined with glycerol, and not methanol, under the transesterifying condition.

Since calcium diglyceroxide seemed to be the catalytically active phase of the reused CaO, the reference sample of calcium diglyceroxide (Ca-Gly) was prepared by immersing CaO in glycerol blended with methanol. The soybean oil transesterification was carried out in the presence of Ca-Gly as a result and the catalyst was as active as the reused CaO: The yield of FAME produced after 2 h was close to 90% for Ca-Gly. Furthermore, data from the reaction indicated that Ca-Gly was the tolerant catalyst to air-exposure.

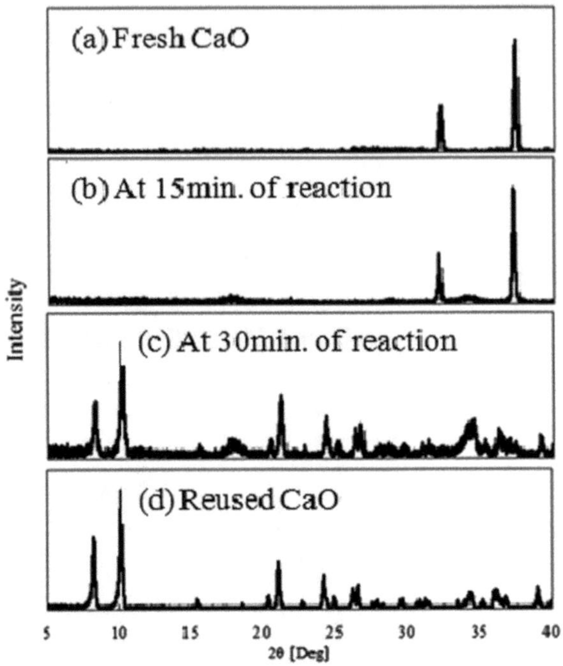

Fig. 5 XRD patterns of catalyst

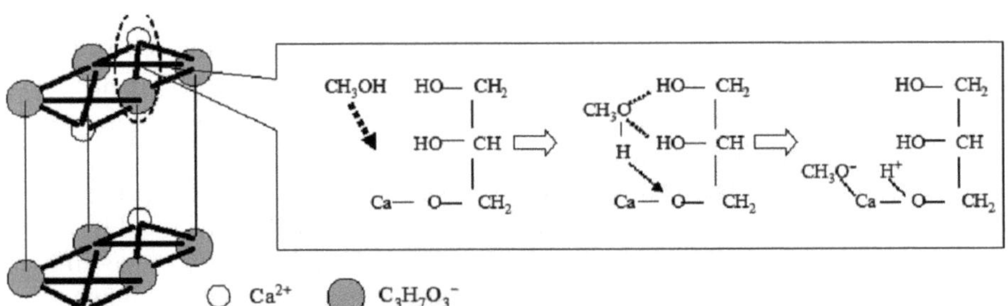

Fig. 6 Mechanism on conversion of methanol into nucleophile by solid base catalysis of calcium diglyceroxide

A mechanism on the reaction catalyzed by calcium diglyceroxide was illustrated with Fig. 6, assuming that oxygen neighboring calcium cation functioned as the basic active site [13]. When methanol got access to the basic active sites, an attractive intermolecular force causing hydrogen bond was possibly generated between OH groups in the glyceroxide anion and oxygen in methanol. The accessibility was enhanced by the attractive intermolecular force with the result that abstraction of proton for converting methanol into the nucleophile was promoted. Also, the basic site was weaker for Ca-Gly ($9.3<H\_<15.0$) than for CaO ($15.0<H\_<18.4$).

Probably, this was the reason why Ca-Gly was not deactivated in air. Although the catalyst adsorbed CO_2 and moisture, they were possibly desorbed from the catalyst under the reacting condition due to weakness of the basic property.

3.3 *Performance of the Practical Catalyst*

Since data from test on the glass batch reactor indicated that the solid base catalysis of calcium oxide was appropriate to the reaction producing biodiesel, we manufactured the practical catalyst as an experiment. For the practical use, the proper shape matching with the column reactor and the sufficient mechanical strength are required. In order to meet the requirement, the roughly crushed lime stone was selected as the raw material. The manufactured practical catalyst was tested by employing it for the rapeseed oil transesterification performed on the laboratory scale pilot plant.

Figure 7 shows variation in the reaction efficiency with successive repetition of the operation to transesterify rapeseed oil. The transesterifying operation was repeated 17 times with reusing the catalyst. Also, the FAME yield measured after 2 h of the reaction reached 96.5% successively till the number of the repetition times was over 10. Thereafter, the reaction efficiency gradually deceased. In our future work, it is necessary to elucidate the reason that the practical catalyst was deactivated with repetition of the transesterifying operation.

4 Conclusion

In order to study the eco-friendly production of biodiesel, solid base catalysis of calcium oxide for a chemical reaction to convert vegetable oil into its methyl esters was investigated. Data from the test carried out on a batch glass reactor indicated

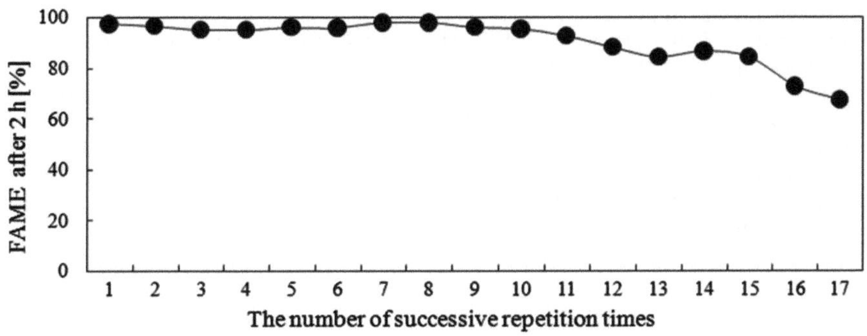

Fig. 7 Variation in reaction efficiency with successive repetition of operation transesterifying rapeseed oil on a laboratory scale pilot plant

that calcium oxide was superior in the catalytic activity and the reusability to the other solid bases: calcium hydroxide, magnesium oxide, strontium oxide, anion-exchange resin and alumina supported potassium. Interestingly, calcium oxide was converted into calcium diglyceroxide during the reaction. Thereafter, calcium diglyceroxide functioned as the solid base catalyst. Furthermore, the transformation of the active phase made the catalyst tolerant to air-exposure. When the practical catalyst prepared on the basis of the results mentioned above was tested on the laboratory scale pilot plant, the operation to transesterify rapeseed oil was successively repeated 17 times. However, the reaction efficiency gradually decreased on and after the 11th operation.

References

1. Ma F, Hanna MA (1999) Biodiesel production: a review. Bioresour Technol 70:1–15
2. Freedman B, Pryde EH, Mounts TL (1984) Variables affecting the yield of fatty esters from transesterified vegetable oil. J Am Oil Chem Soc 61:1638–1643
3. Saka S, Kusdiana D (2001) Biodiesel fuel from rapeseed oil as prepared in supercritical methanol. Fuel 80:225–231
4. Saka S (2005) Biodiesel fuel production by supercritical methanol technology. J Jpn Inst Energy 84:413–419
5. Shimada Y, Watanabe Y, Samukawa T, Sugihara A, Noda H, Fukuda H, Tominaga Y (1999) Conversion of vegetable oil to biodiesel using immobilized *Candida antarctica* lipase. J Am Oil Chem Soc 76:789–793
6. Oda M, Kaieda M, Hama S, Yamaji H, Kondo A, Izumoto E, Fukuda H (2005) Facilitatory effect of immobilized lipase-producing *Rhizopus oryzae* cells on acryl migration in biodiesel-fuel production. Biochem Eng J 23:45–51
7. Shibasaki-Kitagawa N, Honda H, Kuribayashi H, Toda T, Fukumura T, Yonemoto T (2007) Biodiesel production using anionic ion-exchange resin as heterogeneous catalyst. Bioresour Technol 98:415–421
8. Kouzu M, Umemoto M, Kasuno T, Tajika M, Aihara Y, Sugimoto Y, Hidaka J (2006) Biodiesel production from soybean oil using calcium oxide as a heterogeneous catalyst. J Jpn Inst Energy 85:135–141
9. Zhang G, Hattori H, Tanabe K (1988) Aldol addition of aceton, catalyzed by solid base catalysts: magnesium oxide, calcium oxide, strontium oxide, barium oxide, lanthanum oxide and zirconium oxide. Appl Catal 36:189–197
10. Kabashima H, Tsuji H, Hattori H (1997) Michael addition of methyl crotonate over solid base catalysts. Appl Catal A Gen 165:319–325
11. Seki T, Kabashima H, Akutsu K, Tachikawa H, Hattori H (2001) Mixed Tishchenko reaction over solid base catalysts. J Catal 204:393–401
12. Hattori H (2004) Solid base catalysts: generation, characterization, and catalytic behavior of basic sites. J Jpn Petrol Inst 47:67–81
13. Gryglewicz S (1999) Rapeseed oil methyl esters preparation using heterogeneous catalysts. Bioresour Technol 70:249–253

Part II
Contributed Papers

γ-Ferric Oxide/Carbon Composite Synthesized by Aqueous Solution Method as a Cathode for Lithium-Ion Batteries

Mitsuhiro Hibino and Takeshi Yao

Abstract Iron oxide is one of the most promising materials as an electrode of lithium-ion batteries due to its low toxicity and low cost. In this study, composites of nano-sized γ-ferric oxide (γ-Fe_2O_3) and carbon material is synthesized by aqueous solution method, which is generally conducted at low cost. Acetylene black (AB) or ketjen black (KB) is adopted for the carbon component in the composite. These composites (γ-Fe_2O_3/AB and γ-Fe_2O_3/KB) are examined as cathodes of lithium-ion batteries and they exhibit high coulombic efficiency and high cycle performance. Furthermore the γ-Fe_2O_3/KB composite is found to allow rapid discharge and charge.

Keywords Lithium-ion battery • Ferric oxide • Cathode material • Composite material

1 Introduction

For effective use of new energy and various electric vehicle systems such as HEV, P-HEV and pure EV, there is a growing need for electric energy storage with high power density as well as high energy density. The secondary battery and electric double layer capacitor (EDLC) are expected for the electric energy storage now and also in future. Among many kinds of electric energy storage devices, the lithium-ion battery is expected for the above mentioned vehicle systems because of its high energy density. The present battery has reached the allowable level for HEV from performance aspect. However its price should be reduced below half of the present one. The P-HEV and EV require further performance-growing as well as cost-reducing. Thus, component materials for the future lithium-ion battery should

M. Hibino (✉) and T. Yao
Graduate School of Energy Science, Kyoto University, Yoshida, Sakyo-ku, Kyoto, 606-8501, Japan
e-mail: hibino@energy.kyoto-u.ac.jp

T. Yao (ed.), *Zero-Carbon Energy Kyoto 2009*, Green Energy and Technology, DOI 10.1007/978-4-431-99779-5_4, © Springer 2010

be made from inexpensive raw materials in low-cost manners. Light environmental load throughout the manufacturing process is also crucial.

A cathode of lithium-ion battery is, in general, required to react with lithium ions at high electrode potential. For this requirement, transition metal oxides are often used. In terms of light-weight, the fourth period transition metal oxides are usually employed. For low toxicity and low cost, iron based materials are eagerly anticipated. We focus iron oxides, which is a simple binary system of iron and oxygen. There are some iron oxides having various valence states of iron ions and various crystal structures. Since iron ions undergo reduction by lithium intercalation reaction during discharging process, iron ions in the higher valence state before discharging allow more lithium ions reacting with cathode[1] and it leads to high energy density battery. Therefore we have selected ferric oxide, Fe_2O_3, which is stable and includes only iron ions in the trivalent state. The stable phase of Fe_2O_3 is alpha form, so called hematite. In contrast, gamma phase, which is the metastable phase, is called maghemite. These materials have been reported for cathode materials of lithium-ion batteries already [1–8] and we also have reported electrode property of γ-Fe_2O_3 [8]. Before Komaba's study [3], Fe_2O_3 were considered to be difficult to use for cathode material due to its big problems of poor reversibility and cycle durability. Recently their cathode performance has been improved by using nano-sized particles [3–8]. In particular, we aim to use γ-Fe_2O_3 for the cathode of the lithium-ion battery for rapid discharge and charge.

Our strategy for preparation of rapidly dischargeable and chargeable electrode materials is as follows: A number of lithium ions have to move from the anode to cathode (during discharging), and from the cathode to the anode (during charging) in short time for such rapid discharge and charge. Then diffusion length of lithium ion is limited to short distance. Adequately small particles permit lithium to reach all parts of particles even if such a short diffusion length. For preparation of an active material in the small particle, aqueous solution method, which can be, in general, conducted at low cost, is appropriate. On the other hand, since the rapid discharge and charge means fast electrochemical reactions, high electronic conductivity is necessary. It is effectively achieved by combination of an active material with conducting additives such as graphitic carbon materials. We attempted to prepare a composite including carbon material. From the above viewpoints, in order to obtain a γ-Fe_2O_3/carbon composite with favorably contacting condition between particles, carbon material as conducting additive is introduced during the synthetic stage of γ-Fe_2O_3 small particles by aqueous solution method. In this study, we adopted acetylene black (AB) and ketjen black (KB) for the carbon component in the composite, which are representative materials for conducting additives to electrodes. These composites were examined as cathodes of lithium-ion batteries.

[1] The number of lithium ions reacting with electrode materials is expressed as the term "specific capacity" using a practical unit of charge amount per electrode weight, "$mA\,h\,g^{-1}$". The use of electrode material with high specific capacity is essential for high energy density battery.

2 Experimental

Tetrahydrated iron (II) chloride was introduced into a buffer solution of pH 6.2. The buffer solution was prepared beforehand from potassium acetate and acetic acid. Acetylene black (AB) or ketjen black (KB) was put into the solution, and the dispersion was stirred for 3 h at 25°C. By this treatment, mixture of γ-FeOOH and the carbon (AB or KB) was precipitated in the solution. The precipitate was suction-filtrated using a 0.1 μm PTFE membrane filter and then dried at 70°C in air. After that, γ-FeOOH was transformed into γ-Fe_2O_3 by heating at 200°C in a vacuum for 72 h and then γ-Fe_2O_3/carbon composite was obtained. The weight ratio of γ-Fe_2O_3 in the composite was determined by means of Inductively Coupled Plasma – Atomic Emission Spectroscopy (ICP-AES) and thermogravimetry (TG).

The electrochemical property was measured by using three electrodes glass cell. For the working electrode, the γ-Fe_2O_3/carbon composite was mixed with another KB and PTFE, at the weight ratio of 8:1:1, and pressed on Ni mesh. For both the counter electrode and the reference electrode, lithium metal pressed on Ni mesh was used. The electrolyte was 1 mol dm^{-3} $LiClO_4$ in a mixture of EC and DME (1:1, v/v). In this study, current density and specific capacity were expressed on the basis of the weight of the γ-Fe_2O_3/carbon composite.

From the X-ray diffraction (XRD) analysis, iron oxide was identified and the cell parameter was determined. To investigate the structural change induced by the discharge, ex-situ XRD was performed. After the required electrochemical conditions were attained, the working electrode was detached from the cell and set in a sealed holder in an argon gas system. This holder was fixed to the diffractometer for the XRD analysis.

3 Results and Discussion

3.1 Characterization

XRD profiles are illustrated in Fig. 1a for the composite including AB (γ-Fe_2O_3/AB) and Fig. 1b for the composite including KB (γ-Fe_2O_3/KB). The peaks of γ-Fe_2O_3 based on JCPDS No.39-1346 are also shown. Each profile is composed of

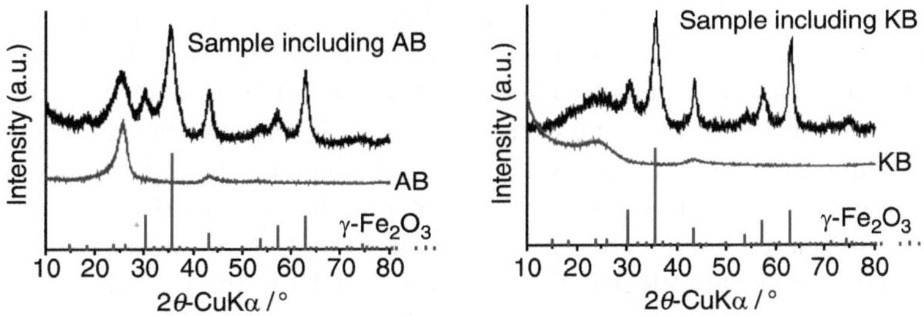

Fig. 1 XRD profiles for the samples obtained using AB (**a**) and KB (**b**). The profiles of pristine carbons and peaks of γ-Fe_2O_3 according to JCPDS No. 39-1346 are also shown

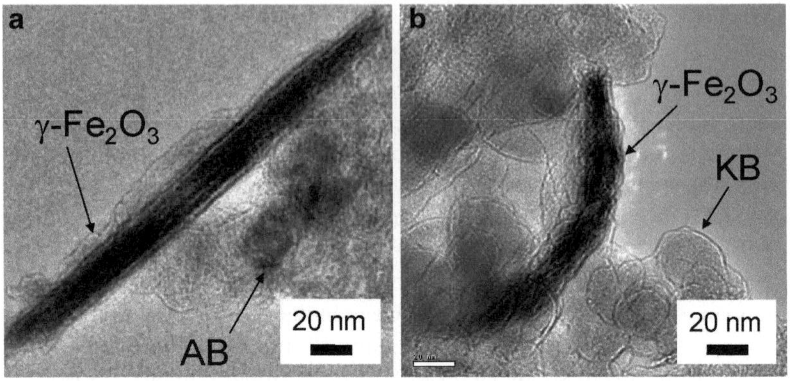

Fig. 2 TEM photographs for the composite of (**a**) γ-Fe_2O_3/AB and (**b**) γ-Fe_2O_3/KB

the peaks of each carbon and γ-Fe_2O_3. No peak other than those of γ-Fe_2O_3 and the carbon was found in the profile. Thus, the composite of γ-Fe_2O_3/AB or γ-Fe_2O_3/KB was successfully prepared. The weight ratio of γ-Fe_2O_3 determined by ICP-AES and TG was 47.4wt% in γ-Fe_2O_3/AB and 38.5wt% in γ-Fe_2O_3/KB.

In TEM photographs for γ-Fe_2O_3/AB (Fig. 2a) and γ-Fe_2O_3/KB (Fig. 2b), acicular particles with typical sizes of 20 nm × 200 nm were observed at the surfaces of the AB or KB particles. They were found to be dispersed instead of being aggregated.

3.2 *Electrochemical Measurements*

Discharge-charge profiles at current density of 0.4 $A\,g^{-1}$ for the composites are demonstrated in Fig. 3a for γ-Fe_2O_3/AB and Fig. 3b for γ-Fe_2O_3/KB. Both of them showed reversible discharge and charge. Specific capacities at the 2nd cycle are 52 and 83 $mA\,h\,g^{-1}$ for γ-Fe_2O_3/AB and γ-Fe_2O_3/KB, respectively.

Discharge and charge capacity for cycle repetition is plotted in Fig. 4. Both composites exhibited high retention rate. The specific capacity of γ-Fe_2O_3/KB was higher than that of γ-Fe_2O_3/AB. The KB is known to be able to raise conductivity of materials more effectively than other conducting additives at the same weight ratio. Therefore, higher conductivity could allow γ-Fe_2O_3 to react electrochemically with more lithium ions. For γ-Fe_2O_3/KB, the specific capacity in the 1st charge is markedly higher than that in the 1st discharge. As seen in Fig. 3, in the second and later cycles, discharge curve started at the potential higher than the initial potential of the 1st discharge curve. This indicates that γ-Fe_2O_3, AB, and KB are charged non-faradaically at their electric double layers (EDL) in the first charge process. An additional charge due to charge at the EDL was responsible for the observation that the specific capacity in the first charge was above that in the first discharge.

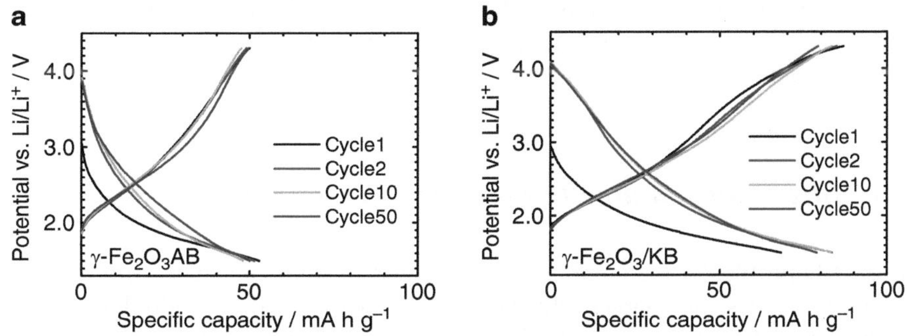

Fig. 3 Discharge-charge profiles at current density 0.4 $A g^{-1}$ between 1.5 and 4.3 V for of γ-Fe_2O_3/AB (**a**) and γ-Fe_2O_3/KB (**b**)

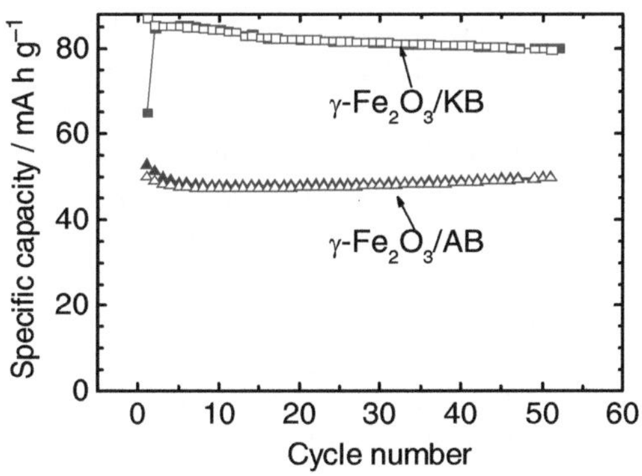

Fig. 4 Cycle performances at current density 0.4 $A g^{-1}$ between 1.5 and 4.3 V of g-Fe_2O_3/AB and γ-Fe_2O_3/KB composite. The *closed symbols* are for discharge and *open ones* for charge

In particular, KB has high specific surface area (1,420 $m^2 g^{-1}$)[2] and its electric double layer capacitance was expected to be nonnegligibly observed. Since γ-Fe_2O_3 was in the form of small particles, as seen in TEM photographs (Fig. 2), it also could have high surface area. Then the total capacity was the sum of capacity due to lithium insertion in γ-Fe_2O_3 and capacity due to EDL of γ-Fe_2O_3 and carbon.

For development of electrode materials, it is significant to have knowledge about the number of lithium ion reacting with γ-Fe_2O_3. We can estimate it, if the contribution of capacitance due to EDL to total capacity is obtained. We assumed that the capacity discharged in the second or later cycles above the initial potential was attributed to capacitance due to the EDL. By means of obtaining the EDL contribution from data of Fig. 3 and subtracting it from total capacity, we estimated the

[2] This value was BET surface area calculated from nitrogen adsorption-desorption isotherm of KB as purchased. The surface area of AB was described as 133 $m^2 g^{-1}$ in a catalog of supplier.

number of lithium ions reacting with γ-Fe_2O_3 to be 0.52 and 0.85 for γ-Fe_2O_3/AB and γ-Fe_2O_3/KB, respectively. These values will be discussed in connection with lattice parameter variation later.

3.3 Structural Change Induced by Li Insertion

The XRD profiles of the composites discharged to 1.5 V are shown in Fig. 5. For both composites, no marked change between before and after discharging was found. The lattice parameter of γ-Fe_2O_3 in the γ-Fe_2O_3/AB composite was almost unchanged: 0.8350(6) nm (before discharge) and 0.8352(6) nm (after discharge). In contrast, γ-Fe_2O_3 in the γ-Fe_2O_3/KB composite underwent enlargement from 0.8343(8) to 0.8378(3) nm. This increase was probably caused by enlargement of ionic radius through reduction of iron ions from Fe^{3+} to Fe^{2+}. The number of lithium ions reacting with γ-Fe_2O_3 was larger for γ-Fe_2O_3/KB than for γ-Fe_2O_3/AB, as described in the preceding section. Thereby γ-Fe_2O_3/KB after discharging was in the higher reduction level than γ-Fe_2O_3/AB. However, the relationship between the number of lithium ion reacting with γ-Fe_2O_3 and the lattice parameter was not proportional. Such unproportionality implied that the site of lithium ions changed depending on the lithium insertion level or/and that the iron ions moves to other site due to its radius change by reduction. Assuming lattice parameter of 0.835 nm and oxygen parameter of 0.38, the size of various sites of Fd3m are 8a: 49 pm, 8b: 35 pm, 16c: 74 pm and 16d: 66 pm on the basis of Shannon's ionic radii [9]. The 8a and the 8b sites are tetrahedrally coordinated by four oxygen atoms and the 16c and the 16d octahedrally six oxygen atoms. The radii of Fe^{3+} in the high spin (HS) state are 49 and 65 pm for coordination number (CN) 4 and 6, respectively. Hence both the 8a and the 16d sites are suitable in size for Fe^{3+}. On the other hand, Fe^{2+} in HS state has radii of 63 pm (CN 4) and 78 pm (CN 6). The size of Fe^{2+} is too big

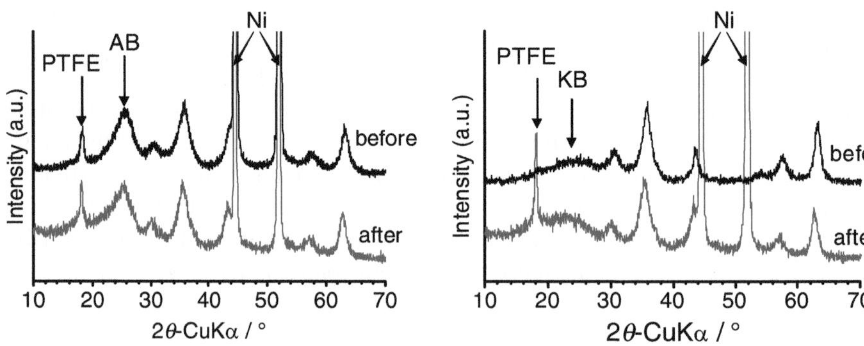

Fig. 5 XRD profiles of composites before and after discharging to 1.5 V at current density 0.4 A g^{-1} for (**a**) γ-Fe_2O_3/AB and (**b**) γ-Fe_2O_3/KB

for the 8a but appropriate for the empty 16c. Thus the migration of the iron ions from the 8a to the 16c is the likely result and was confirmed at the high Li insertion level [1, 2]. The ionic radii of Li^+ are 59 pm (CN 4) and 76 pm (CN 6). Thus, the 16c site (74 pm) seems to accommodate both Fe^{2+} (78 pm) and Li^+ (76 pm) with lattice and oxygen parameters almost unchanged. If the 16c site is occupied fully by Fe^{2+} and Li^+, chemical formula of the lithium inserted iron oxide is expressed by $(\ )^{8a}(Li_{0.5}{}^{+}Fe_{0.5}{}^{2+})_2{}^{16c}(Fe_{0.833}{}^{3+})_2{}^{16d}O_4$ on the basis of the space group Fd3m,[3] where $(\ )^{8a}$ represents an empty 8a site. Then the number of lithium ion on the basis of the chemical formula Fe_2O_3 is 0.75. The observation that the lattice parameter was unchanged at Li/Fe_2O_3 of 0.52 and enlarged at Li/Fe_2O_3 of 0.85 is consistent with the hypothesis that the 16c site is used by Fe^{2+} and Li^+ ions below Li/Fe_2O_3 of 0.75. Above Li/Fe_2O_3 of 0.75, lithium ions probably begin to enter the 8a sites. Alternatively, if lattice parameter increases, as observed really in γ-Fe_2O_3/KB, some Fe^{2+} ions may reside in the 8a site without moving to the 16c site. Then Li^+ ions can occupy the 16c site more than Li/Fe_2O_3 of 0.75. Establishing the sites of Li^+ and Fe^{2+} quantitatively along with site occupancy requires future research.

3.4 Rapid Discharge and Charge Property

Figure 6a shows discharge capacities at the 2nd cycle between 1.5 and 4.3 V vs. Li/Li^+ at various current densities. Though in this stage no data above current density 4 A g^{-1} has been obtained, the low decreasing rate observed in the figure implies

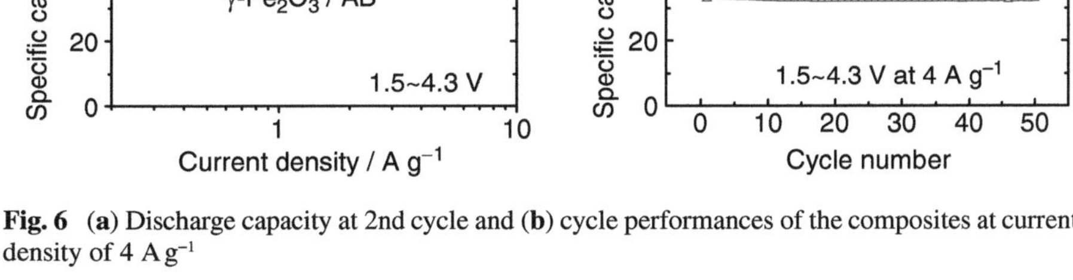

Fig. 6 (**a**) Discharge capacity at 2nd cycle and (**b**) cycle performances of the composites at current density of 4 A g^{-1}

[3] γ-Fe_2O_3 can be expressed by $(Fe^{3+})^{8a}(\ )^{16c}(Fe_{0.833}{}^{3+})_2{}^{16d}O_4$ on the basis of the space group Fd3m.

that the γ-Fe_2O_3/KB composite allows higher current density than 4 A g^{-1}. The specific capacity of the γ-Fe_2O_3/KB composite was 80 mA h g^{-1} at a current density of 4 A g^{-1}: the capacity 80 mA h g^{-1} could be discharged rapidly in 1.2 min. The cycle performances at current density of 4 A g^{-1} is shown in Fig. 6b. The composites exhibited high retention rate of specific capacity; the ratio of discharge capacity of the 50th cycle to that of the 5th cycle was 97.8% for γ-Fe_2O_3/KB composite. These results indicate that the γ-Fe_2O_3/KB composite is a promising cathode material of rapidly discharging and charging lithium-ion batteries.

References

1. Thackeray MM, David WIF, Goodenough JB (1982) Mater Res Bull 17:785
2. Pernet M, Strobel P, Bonnet B, Bordet P, Chabre Y (1993) Solid State Ionics 66:259–265
3. Komaba S, Suzuki K, Kumagai N (2002) Electrochemistry 70:506
4. Larcher D, Masquelier C, Bonnin D, Chabre Y, Masson V, Leriche J-B, Tarascon J-M (2003) J Electrochem Soc 150:A133
5. Kanzaki S, Inada T, Matsumura T, Sonoyama N, Yamada A, Takano M, Kanno R (2005) J Power Sources 146:323
6. Kanzaki S, Yamada A, Kanno R (2007) J Power Sources 165:403
7. Quintin M, Devosa O, Delville MH, Campet G (2006) Electrochim Acta 51:6426
8. Hibino M, Terashima J, Yao T (2007) J Electrochem Soc 154:A1107
9. Shannon RD (1976) Acta Cryst A32:751

Morphology Control of TiO_2-Based Nanomaterials for Sustainable Energy Applications

Yoshikazu Suzuki

Abstract TiO_2-based nanomaterials (including layered titanates, $H_2Ti_nO_{2n+1}$, which transform into TiO_2 at high temperatures) have attracted much attention because of their fascinating characteristics relating to renewable energy and environmental applications. Recently, one-dimensional (1D) TiO_2 nanomaterials, such as nanotube, nanowire, nanorod, nanofiber, have been investigated for electrode materials in dye-sensitized solar cells (DSC) to improve light-to-electricity conversion efficiency. These 1D nanomaterials are also applied for lithium ion or hydrogen storage, photocatalysts, multifunctional filter materials and so on. In this paper, our recent studies on 1D TiO_2 nanomaterials, particularly on sustainable energy applications, are briefly overviewed. Some collaborative studies on TiO_2 nanomaterials with French groups, i.e., Mines ParisTech (from 2006) and Strasbourg University (from 2008), are introduced.

Keywords TiO_2 • Dye-sensitized solar cells (DSC) • One-dimensional nanomaterials

1 Introduction

TiO_2 powders are widely used for industrial applications, such as pigments, cosmetics, food ingredients, electronic devises, photocatalysts, catalyst supports, dye-sensitized solar cells, and so on. Phase-, size- and morphology-controls have been extensively studied in order to improve various functions of TiO_2 powders [1]. Since the innovative work by Kasuga et al. in 1998–1999 [2, 3], TiO_2-derived one-dimensional (1D) nanomaterials synthesized by the hydrothermal alkali treatment method, such as nanotubes [4–21], nanowires [22–26], and nanofibers [27–29], have attracted much attention. Their unique microstructure and promising energy-related applications have been focused: e.g., lithium ion batteries [24, 25, 30], hydrogen storage [31, 32], dye-sensitized solar cells (DSCs) [33–36], and photocatalytic hydrogen evolution [37, 38].

Y. Suzuki (✉)
Institute of Advanced Energy, Kyoto University, Kyoto, 611-0011, Japan
e-mail: suzuki@iae.kyoto-u.ac.jp

T. Yao (ed.), *Zero-Carbon Energy Kyoto 2009*, Green Energy and Technology,
DOI 10.1007/978-4-431-99779-5_5,

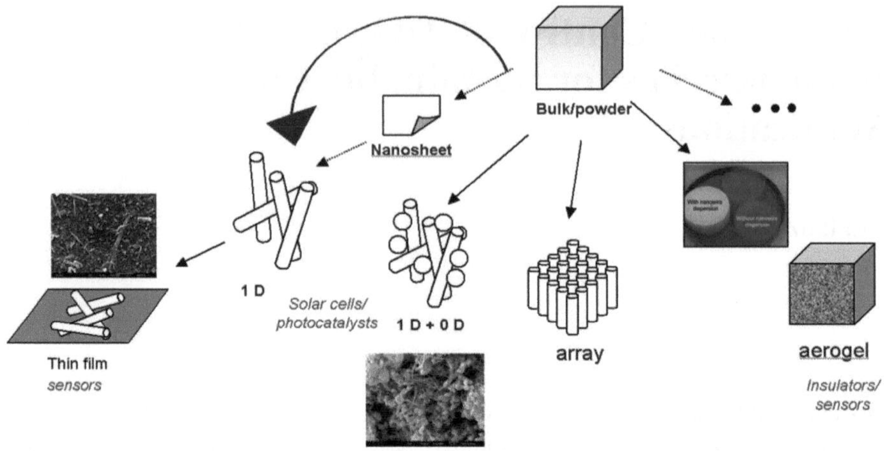

Fig. 1 Morphology control of TiO_2-based nanomaterials; Nanostructured TiO_2 with controlled morphology is useful to design various devices for renewable energy applications

Figure 1 schematically illustrates a variety of TiO_2-based nanomaterials developed by our group. In our previous article [35], our recent studies on 1D TiO_2-based DSC (around 2005–2007) were demonstrated. Thus, in this article, some collaborative studies on TiO_2 nanomaterials with French groups, i.e., Mines ParisTech (from 2006) and Strasbourg University (from 2008), are introduced [39–42].

2 TiO_2 Aerogel/TiO_2 Nanowire Composite

By combining high-surface area TiO_2 aerogel and TiO_2 nanowire, it is possible to prepare a new TiO_2-based nanocomposite with homogeneous 3D mesoporous structure reinforced with 1D nanodispersoid (Figs. 2 and 3) [39].

Through the preparation of such a nanocomposite structure, it is expected to harmonize high surface area, efficient charge transfer, and improved mechanical properties (in case of bulk, granular or film form). Although there were some nanowire-free regions in the composite, the composite possessed unique "nanowire-network" microstructure, where TiO_2 nanowires were embedded within aerogel matrix. Since some mechanical (toughening), optical (light scattering) and electrical (conducting path) effects are expected for this new TiO_2/TiO_2 composite powder, more homogeneous dispersion of nanowires and further characterization are required [39].

3 N-Doped TiO_2 by Plasma Processing for Visible-Light Sensitive Photocatalysts

From a viewpoint of microstructure control, maintaining fine nanostructure during post-treatments is also an important issue. Non-equilibrium (low-temperature) nitrogen DC-arc plasma treatment of a commercial TiO_2 anatase nanopowder

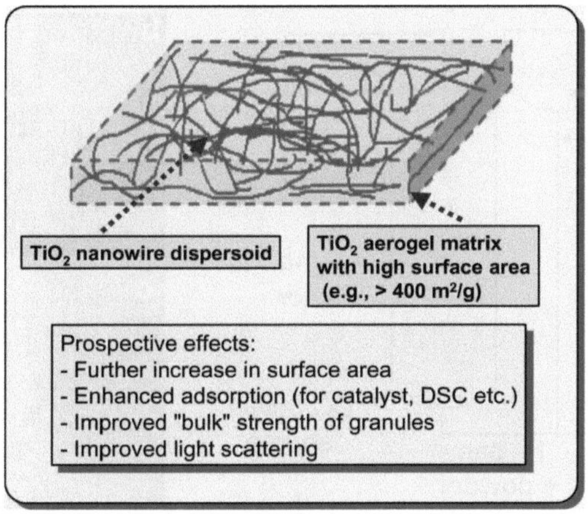

Fig. 2 Design concept of TiO_2 aerogel/TiO_2 nanowire composite [39]

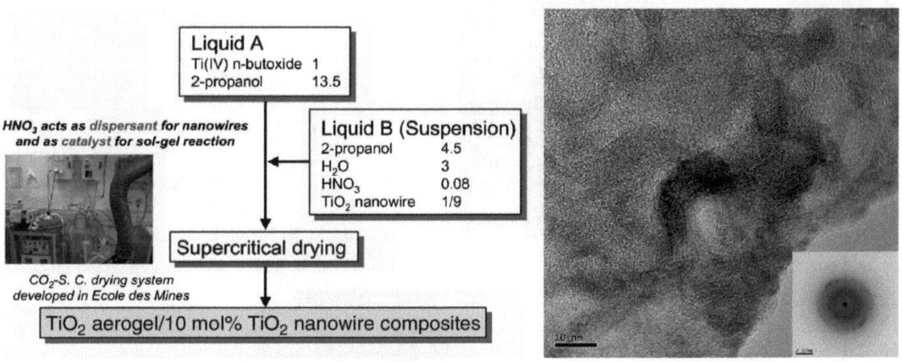

Fig. 3 (*Left*) Schematic diagram of the preparation of TiO_2 aerogel/TiO_2 nanowire composite, and (*right*) its microstructure (scale bar: 10 nm) [39]

was examined to obtain nitrogen-doped TiO_2, without changing the shape and size of primary nanoparticles [40]. By using a non-thermal discharge at low current (150 mA) and high voltage (1,200 V) using pure N_2 gas (Fig. 4), light yellowish-gray TiO_2 powder was successfully obtained within a short period of 5–10 min.

XPS and TEM-EELS studies confirmed the existence of doped nitrogen. Due to the relatively mild conditions (plasma power of 180 W), metastable anatase structure and fine crystallite size of TiO_2 (ca. 10 nm) were maintained after the plasma treatment (Fig. 5). The in-flight powder treatment system used in this study is promising for various type of powder treatment [40].

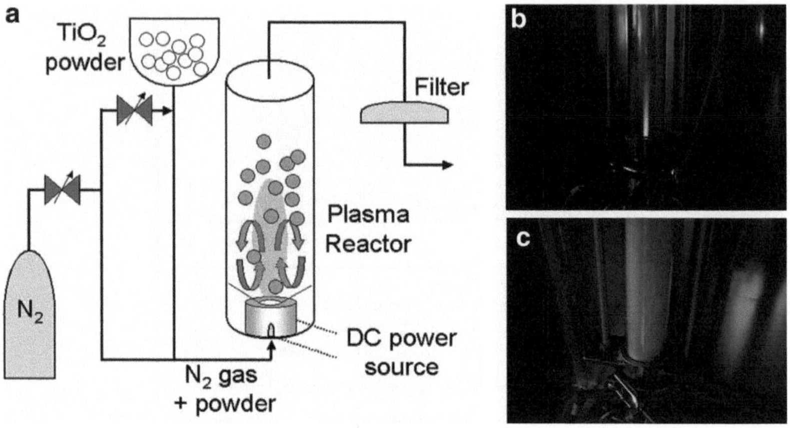

Fig. 4 Non-equilibrium nitrogen DC-arc plasma apparatus with a vertical type reactor: (**a**) Schematic illustration, (**b**) typical nitrogen plasma without powder feeding, and (**c**) nitrogen plasma treatment with powder feeding [40]

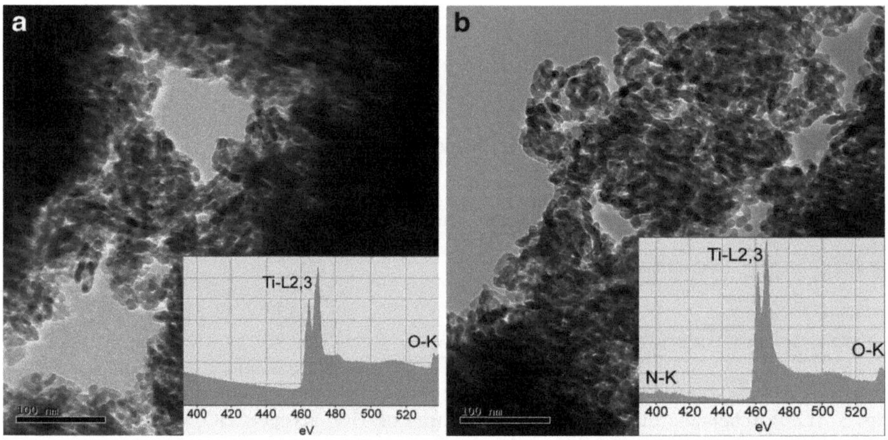

Fig. 5 TEM micrographs and EELS spectra (insert) of (**a**) starting TiO_2 anatase powder and (**b**) nitrogen-plasma treated powder. Primary particle size remained ca. 10 nm after plasma treatment, which indicates the plasma treatment was non-thermal [40]

4 Titanate Nanowire Thin-Film (Spray LbL Process)

Our recent study revealed that TiO_2 nanowires (obtained from the post-heat treatment of hydrogen titanate nanowires) showed much better hydrogen production ability due to their higher crystalinity [38]. In order to extend the application fields of these nanowires and other 1D nanomaterials, thin-film processing technologies must be established. Since the titanate and TiO_2 1D nanomaterials are temperature-sensitive, an ambient temperature process is favored. Aqueous solution processes have high potential to fit these demands. Spray Layer-by-Layer self-assembly method

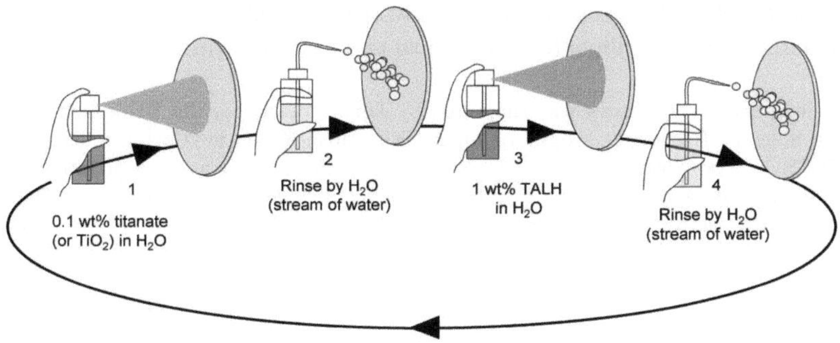

Fig. 6 Spray Layer-by-Layer process in this study (based on the work of Izquierdo et al. [41]). Intervals of each operation were 2 min (i.e., 8 min for one cycle) [42]

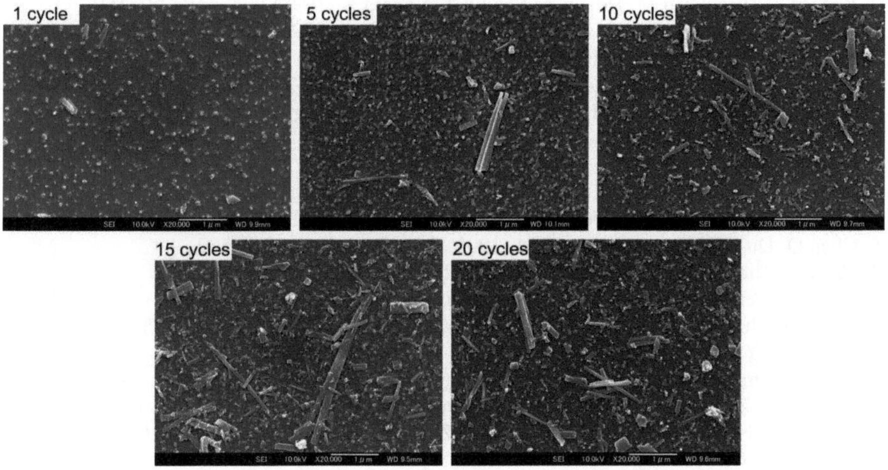

Fig. 7 Microstructure of titanate-nanowire thin films prepared by splay-LbL method (the scale bars are 1 μm) [42]

(or Spray LbL method) is a rapid and versatile coating process using aqueous solutions (or suspensions) with nanometer-scale controllability (Fig. 6) [41, 42].

Figure 7 shows titanate nanowire thin films, successfully obtained by the Spray LbL method within a short period [42]. The adhesion of titanate nanowires to the substrate under a rapid spray process was attributed to their linear morphology and good dispersion in the aqueous suspension.

5 Further Studies on Various TiO_2-Based Nanomaterials

Furthermore, various TiO_2-based nanomaterials with unique morphologies have been prepared in our group, such as iron-oxide coated titanate nanowire [43], TiO_2 rutile nanorod array [44], and TiO_2 nanoparticles by electrospray pyrolysis method

(Y Terada, S Tohno, Y Suzuki, unpublished work). The author wishes these TiO_2-based nanomaterials will contribute to solve global energy and environmental problems.

Acknowledgments This work was supported by Japan Society for the Promotion of Science (FY2006 Scientist Exchanges Program), and MEXT, Japan (Grant-in-Aid for Science Research No. 19685020 For Young Scientist: Category A). This study was also supported by the Collaboration Agreement between Strasbourg University and Kyoto University. The author acknowledge the active collaborations of Mines ParisTech groups and Strasbourg University groups.

References

1. Bavykin DV, Friedrich JM, Walsh FC (2006) Adv Mater 18:2807
2. Kasuga T, Hiramatsu M, Hoson A, Sekino T, Niihara K (1998) Langmuir 14:3160
3. Kasuga T, Hiramatsu M, Hoson A, Sekino T, Niihara K (1999) Adv Mater 11:1307
4. Seo D-S, Lee J-K, Kim H (2001) J Cryst Growth 229:428
5. Zhang QH, Gao LA, Sun J, Zheng S (2002) Chem Lett 31:226
6. Lin CH, Chien SH, Chao JH, Sheu CY, Cheng YC, Huang YJ, Tsai CH (2002) Catal Lett 80:153
7. Mao YB, Banerjee S, Wong SS (2003) Chem Comm 2003
8. Wang YQ, Hu GQ, Duan XF, Sun HL, Xue QK (2002) Chem Phys Lett 365:427
9. Du GH, Chen Q, Che RC, Yuan ZY, Peng LM (2001) Appl Phys Lett 79:3702
10. Chen Q, Du GH, Zhang S, Peng LM (2002) Acta Crystallogr B 58:587
11. Chen Q, Zhou WZ, Du GH, Peng LM (2002) Adv Mater 14:1208
12. Zhang S, Peng LM, Chen Q, Du GH, Dawson G, Zhou WZ (2003) Phys Rev Lett 91:256103
13. Yao BD, Chan YF, Zhang XY, Zhang WF, Yang ZY, Wang N (2003) Appl Phys Lett 82:281
14. Sun X, Li Y (2003) Chem Euro J 9:2229
15. Ma RZ, Bando Y, Sasaki T (2003) Chem Phys Lett 380:577
16. Yang J, Jin Z, Wang X, Li W, Zhang J, Zhang S, Guo X, Zhang Z (2003) Darton Trans 2003
17. Zhang S, Li W, Jin Z, Yang J, Zhang J, Du Z, Zhang Z (2004) J Solid State Chem 177:1365
18. Zhang M, Jin ZS, Zhang JW, Guo XY, Yang JJ, Li W, Wang XD, Zhang ZJ (2004) J Mol Catal A Chem 217:203
19. Suzuki Y, Yoshikawa S (2004) J Mater Res 19:982
20. Yoshida R, Suzuki Y, Yoshikawa S (2005) Mater Chem Phys 91:409
21. Nakahira A, Kato W, Tamai M, Isshiki T, Nishio K, Aritani H (2004) J Mater Sci 39:4239
22. Du GH, Chen Q, Han PD, Yu Y, Peng LM (2003) Phys Rev B 67:035323
23. Yin S, Fujishiro Y, Wu J, Aki M, Sato T (2003) J Mater Proc Tech 137:45–48
24. Armstrong AR, Armstrong G, Canales J, Bruce PG (2004) Angew Chem Int Ed 43:2286
25. Kavan L, Kalbac M, Zukalova M, Exnar I, Lorenzen V, Nesper R, Grätzel M (2004) Chem Mater 16:477
26. Yoshida R, Suzuki Y, Yoshikawa S (2005) J Solid State Chem 178:2179
27. Suzuki Y, Pavasupree S, Yoshikawa S, Kawahata R (2005) J Mater Res 20:1063
28. Pavasupree S, Suzuki Y, Yoshikawa S, Kawahata R (2005) J Solid State Chem 178:3110
29. Suzuki Y, Pavasupree S, Yoshikawa S, Kawahata R (2007) Physica Status Solidi (a) 204:1757
30. Zhou YK, Cao L, Zhang FB, He BL, Li HL (2003) J Electrochem Soc 150:A1246–A1249
31. Lim SH, Luo JZ, Zhong ZY, Ji W, Lin JY (2005) Inorg Chem 44:4124
32. Bavykin DV, Lapkin AA, Plucinski PK, Friedrich JM, Walsh FC (2005) J Phys Chem B 109:19422
33. Uchida S, Chiba R, Tomiha M, Masaki N, Shirai M (2002) Electrochem 70:418

34. Suzuki Y, Ngamsinlapasathian S, Yoshida R, Yoshikawa S (2006) Central Eur J Chem 4:476
35. Suzuki Y, Ngamsinlapasathian S, Asagoe K, Yoshikawa S (2007) J Jpn Soc Powder Powder Metall 54:202
36. Asagoe K, Ngamsinlapasathian S, Suzuki Y, Yoshikawa S (2007) Central Eur J Chem 5:605
37. Jitputti J, Pavasupree S, Suzuki Y, Yoshikawa S (2008) Jpn J Appl Phys 47:751
38. Jitputti J, Suzuki Y, Yoshikawa S (2008) Catal Comm 9:1265
39. Suzuki Y, Berger MH, D'elia D, Ilbizian P, Beauger C, Rigacci A, Hochepied JF, Achard P (2008) Nano 3:373
40. Suzuki Y, Gonzalez-Aguilar J, Traisnel N, Berger M-H, Repoux M, Fulcheri L (2009) J Nanosci Nanotech 9:256
41. Izquierdo A, Ono SS, Voegel J-C, Schaaf P, Decher G (2005) Langmuir 21:7558
42. Suzuki Y, Pichon BP, D'Elia D, Beauger C, Yoshikawa S (2009) J Ceram Soc Jpn 117:381
43. BP Pichon, Y Suzuki, M-H Berger (2009) J Jpn Soc Powder Powder Metall (43) 56:640
44. Y Hayami, Y Suzuki, T Sagawa, S Yoshikawa (2010) J Nanosci Nanotech (in press)

New Material Processing and Evaluation for TiO_2 by Microwave and Mid-Infrared Light Techniques

Taro Sonobe, Mahmoud Bakr, Kyohei Yoshida, Kan Hachiya, Toshiteru Kii, and Hideaki Ohgaki

Abstract We present a systematic study of material processing and evaluation for TiO_2 using microwave/far/mid-infrared light techniques. A mid-infrared free electron laser (MIR-FEL: 5–20 μm) facility (KU-FEL: Kyoto University Free Electron Laser) has been constructed in Institute of Advanced Energy Kyoto University, and first laser saturation at 13.2 μm was achieved in May 2008. Currently, we have started to develop a measurement system of phonon and electron interaction in metal oxides using MIR-FEL. In addition, we have also developed several new techniques to control the surface condition of TiO_2 using microwave irradiation.

Keywords Microwave processing • Mid-infrared light • Free electron laser

1 Introduction

The technology of electromagnetic wave and infrared light energy application will become greatly contribute to the field of the environmental, medical, and welfare fields. Recently, MW has become paying attention to be as a clean, highly effective energy source. Shortening arrival time up to the setting temperature and achieving the uniform heating can be realized by using the micro wave energy [1]. This makes a new application to not only a high-speed sintering ceramics technology but also the high purity synthesis of the exotic material and the chemical reaction. On the other hand, an infrared region light has a good resonance with molecular and lattice vibration, therefore, many techniques have been developed to analyze structural, optical, and transport properties while paying attention to the interaction with infrared lights [2, 3].

T. Sonobe (✉) and K. Hachiya
Graduate School of Energy Science, Kyoto University, Yoshida-honmachi, Sakyo-ku, Kyoto, 606-8501, Japan
e-mail: t.sonobe@iae.kyoto-u.ac.jp

M. Bakr, K. Yoshida, T. Kii, and H. Ohgaki
Institute of Advanced Energy, Kyoto University, Gokasho, Uji, Kyoto, 611-0011, Japan

T. Yao (ed.), *Zero-Carbon Energy Kyoto 2009*, Green Energy and Technology, DOI 10.1007/978-4-431-99779-5_6, © Springer 2010

At the Institute of Advanced Energy of Kyoto University, A mid-infrared free electron laser (MIR-FEL: 5–20 μm) facility (KU-FEL: Kyoto University Free Electron Laser) has been constructed and the first laser saturation at 13.2 μm was observed in May 2008 [4]. Currently, we started to develop a measurement system of phonon and electron interaction in metal oxides using MIR-FEL. In addition, we have also developed several new techniques to control the surface condition of metal oxides such as TiO_2 using microwave (MW) irradiation [5, 6].

2 Carbon-Modified TiO_2 by Microwave Irradiation Technique

Carbon-modified TiO_2 through microwave carbonization of ethanol using a domestic microwave oven was developed [5]. This process enabled to form the carbonaceous compounds on the surface of TiO_2 and created several new mid-gap bands into the original bandgap (Fig. 1). The carbon-modified TiO_2 showed remarkable visible-light absorption and photocatalytic activity compared with pure TiO_2 as seen in Fig. 2.

The electron excitation to the mid-gap bands has been enhanced by UV as well as visible light irradiation. In addition, the excessive irradiation of microwave can change the optical transition mechanism due to the degradation of carbonaceous compounds on the surface which results in the discrepancy of photocatalytic activity under UV and visible light irradiation. As for the 3 min irradiated samples of MW3, the direct absorption edge was located at 2.8 eV with the indirect absorption edge

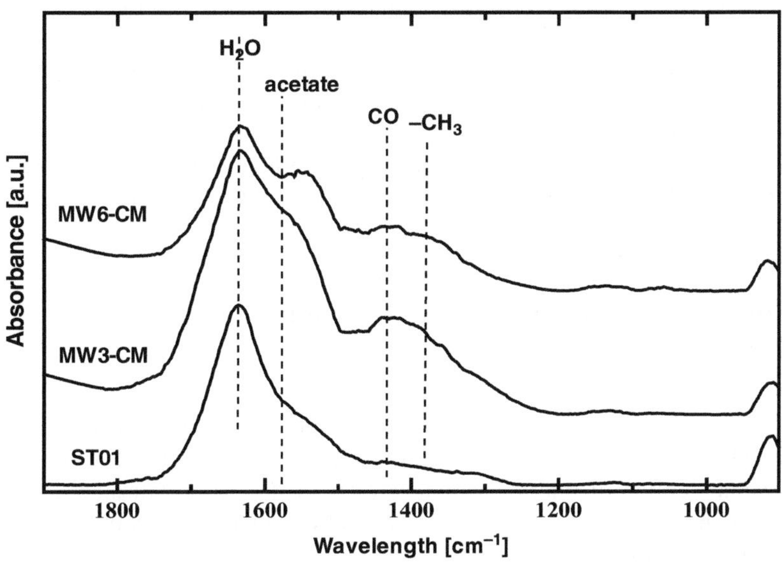

Fig. 1 ATR-FTIR spectra of untreated TiO_2 (ST01), and microwave treated TiO_2 for 3 min (MW3-CM) and 6 min (MW6-CM) [5]

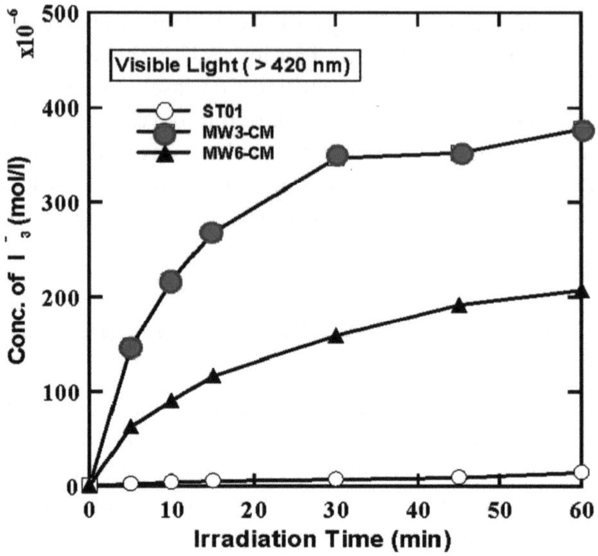

Fig. 2 Formation profile of I3- as a function of visible light (λ>420 nm) irradiation [5]

of another absorption lay at about 1.4 eV within the original bandgap of anatase TiO_2 at 3.2 eV, while the 6 min irradiated sample of MW6 showed two absorption edges of indirect transitions at 1.6 and 2.4 eV as seen Fig. 3. We assume that the above discrepancy of photocatalytic activity for MW6 is due to the creation of different gap states during microwave irradiation. This means that the straightforward photoexcitation of carriers to the gap states by visible light has limited possibility under indirect transitions, while carrier injection via bandgap excitation by UV light and carrier-trapping to the gap states becomes more favorable in turn. As for visible-light-active TiO_2, it seems likely that an introduction of the mid-gap bands with direct absorption as an absorption edge is effective to the enhanced carrier excitation for photocatalytic activity under visible light irradiation.

3 Surface Reduced TiO_2 by MW Atomic Oxygen Plasma Technique

Surface reduction of TiO_2 by microwave atomic oxygen plasma technique was developed, where atomic oxygen plasma is directly emitted from TiO_2 during an intensive absorption of microwave under vacuum [6]. After emission, a dark-colored solid specimen which has good electrical conductivity was obtained as seen Figs. 4 and 5 respectively.

The changes in the structural and optical properties of the sample surface were also investigated. The UV–vis–NIR absorption coefficient spectra showed a marked

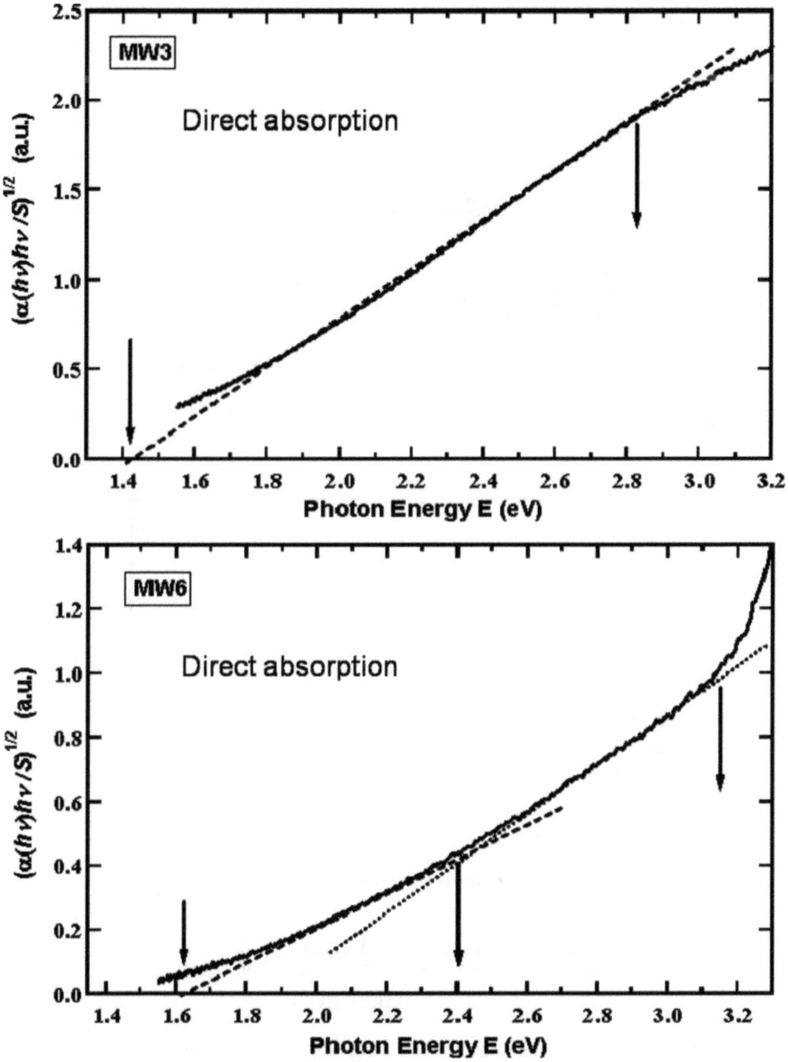

Fig. 3 The plot of $(\alpha h\nu/S)^{1/2}$ vs. photon energy for the absorption edge of MW3 and MW6 denoted by the *solid curve*. The *dotted line* is visual guides for $\alpha h\nu$ $(h\nu - E')^2$ and to evaluate the absorption threshold ($E' = E_g$ for bandgap absorption) [5]

absorption from 200 to 2,400 nm for the sample after plasma emission as seen in Fig. 6; the absorption significantly decreased and the band-gap was restored after annealing in air. The XRD patterns and Raman spectra also suggested the generation of oxygen vacancies after microwave plasma emission and its reforming to oxygenated TiO_2 after air annealing. Therefore, it was found that TiO_2 is directly reduced to the lower oxygenated condition through the emission of the oxygen atomic plasma by microwave irradiation under vacuum.

Fig. 4 Photographs of TiO_2 pellets after MW plasma emission [6]

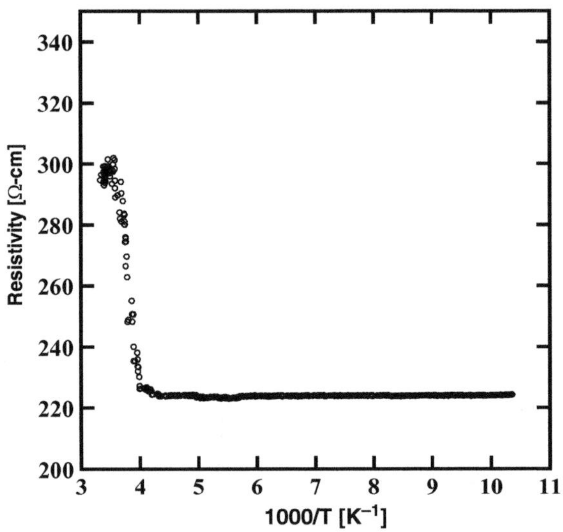

Fig. 5 Electrical resistivity as a function of 1,000/T for MW plasma after MW plasma emission [6]

4 Further Studies of the Development of Material Evaluation System by MIR-FEL

It is well known that an infrared region light has a good resonance with phonon in some solid compounds such as metal oxides. In particular, TiO_2 of wide bandgap semiconducting materials shows unique electrical and optical properties through coupling of phonon with electronic structures, resulting in photochemical phenomena with microwave irradiation [6]. Therefore, it is considered that the irradiation of MIR-

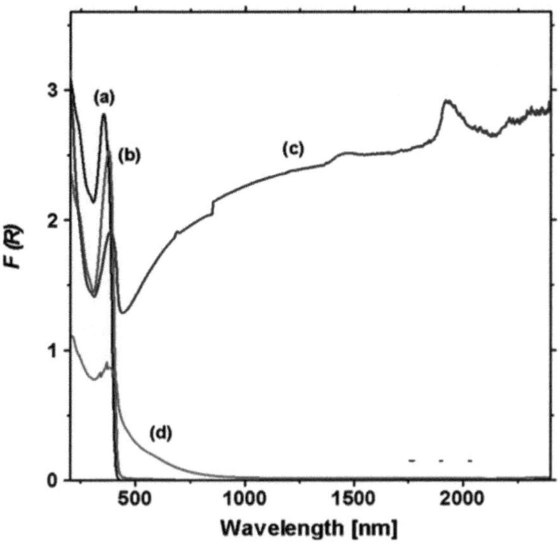

Fig. 6 UV–vis–NIR absorption coefficients obtained from diffuse reflectance spectra with Kubelka–Munk relation for TiO_2 pellets prepared under various conditions: (**a**) as-received, (**b**) sintered in air at 1,000°C, (**c**) MW plasma in vacuum, and (**d**) after annealing in air at 1,000°C [6]

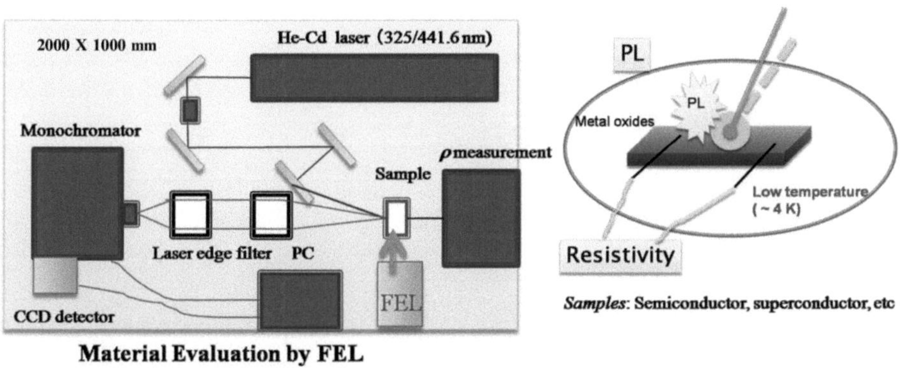

Fig. 7 Plan view of a in-situ PL and electric resistivity measurement system during FEL irradiation [7]

FEL on TiO_2 possibly give rise to the changes in both of properties such as temperature dependence of electric resistivity and photoluminescence at low temperature.

Currently, we started to develop a in-situ measurement system of PL spectra and electric resistivity during MIR-FEL irradiation for microwave modified TiO_2 [7]. Figure 7 shows the plan view of a in-situ PL and electric resistivity measurement system during FEL irradiation.

Acknowledgements The authors would like to express their gratitude to Associate Professor N. Shinohara and Assistant Professor T. Mitani of the Research Institute for Sustainable Humanosphere (RISH), Kyoto University, for their assistance with microwave. Gratitude is also forwarded to Dr. T. Hata of RISH, Kyoto University, for their assistance with the Raman spectroscopy.

References

1. Sato M, Matsubara A et al (2006) Proc. 5th Sohn Int. Symp. Advanced Processing of Metals and Materials, p 157
2. Xi X, Hwang J, Tanner DB et al (2009) Proc. 5th Int. WIRMS 2009 (in press)
3. P Hellwing, J Gross, et. al (2009) Proc. 5th Int. WIRMS 2009 (in press)
4. Ohgaki H et al (2008) Jpn J Appl Phys 47(10):8091
5. Sonobe T et al (2007) Jpn J Appl Phys 47:8456
6. Sonobe T et al (2009) Jpn J Appl Phys 48:116003
7. Sonobe T et al (2009) Proc. 5th Int. WIRMS 2009 (in press)

Construction of the Functional Biomolecules with the Ribonucleopeptide Complexes

Masatora Fukuda, Fong Fong Liew, Shun Nakano, and Takashi Morii

Abstract Development of an artificial photosynthesis complex composed of an assembly of functional biomolecules is desired for the establishment of the renewable energy system in the age of global warming. As the basic methodology for constructing such functional biomolecules, we have developed a stepwise functionalization strategy for RNA and peptide complex (ribonucleopeptide: RNP). Our strategy based on a framework of RNP enables to generate the fluorescent RNP sensors with a variety of binding and signal-transducing characteristics. Here we demonstrate the usefulness of this RNP strategy by the construction of fluorescent RNP sensors for the biologically active amines.

Keywords RNA • Peptide • Biosensor

1 Introduction

Greenhouse gases (written as CO_2) exhausted by the use of fossil resources cause serious environmental problems. In order to establish a CO_2 zero emission energy system that is free from the use of fossil fuels, it is necessary to develop new energy and environmental technologies. One of such energy and environmental technologies for the CO_2 zero emission energy system uses solar energy as the chemical energy, not only as the electrical energy.

The ultimate goal of our research is to create a new system that utilize the solar energy as the chemical energy by developing an artificial photosynthesis complex composed by an assembly of functional biomolecules. As the basic methodology

M. Fukuda, F.F. Liew, S. Nakano, and T. Morii
Graduate School of Energy Science, Kyoto University, Yoshidahonmachi, Sakyo-ku, Kyoto, 606-8501, Japan

T. Morii (✉)
Institute of Advanced Energy, Kyoto University, Gokasho, Uji, Kyoto, 611-0011, Japan
e-mail: t-morii@iae.kyoto-u.ac.jp

T. Yao (ed.), *Zero-Carbon Energy Kyoto 2009*, Green Energy and Technology,
DOI 10.1007/978-4-431-99779-5_7,

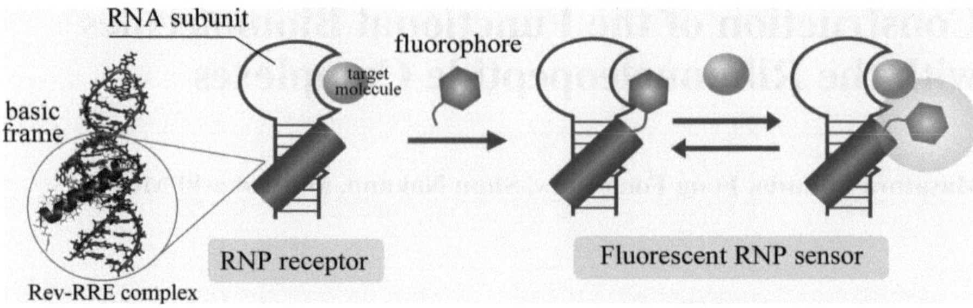

Fig. 1 Schematic illustration shows the stepwise fictionalization to construct ribonucleopeptide (RNP) receptors and fluorescent RNP sensors. The randomized sequences are introduced to RNA subunit of Rev-RRE complex [6] and then RNP receptors were selected by the in vitro selection method. Moreover, the peptide subunit of RNP receptors was modified by the fluorophore to convert RNP receptor to fluorescent RNP sensors that shows fluorescent intensity changing along with the ligand binding

for constructing a functional biomolecule, we have developed a strategy for constructing the functional biomolecular receptor from the noncovlent complex of RNA and peptide (ribonucleopeptide: RNP) [1–5]. Our strategy based on a framework of RNP also provides a wide variety of fluorescent biosensors with many functions, i.e., high signal-to-noise ratios, different emitting wavelengths and various concentration ranges for the ligand detection [3]. In the first step to construct RNP based fluorescent sensors, an RNA-derived RNP pool was prepared by a structure-based design of the Rev Responsive Element (RRE)-HIV Rev peptide complex [6] appended with a randomized nucleotides region as a ligand binding domain, adjacent to the RRE segment. In vitro selection method [7, 8] was applied to a randomized nucleotides region to afford a series of ATP-binding RNP receptors with high selectivity and affinity [3]. In the second step, ATP-binding RNP receptors were successfully converted to fluorescent sensors for ATP by the chemical modification of the N-terminal of the Rev peptide with a various kind of fluorophore. RNP-based fluorescent sensors are also applicable to the other class of small molecules, such as biologically active substances. We report here fluorescent RNP sensors for dopamine, which is one of the biologically active amines, by the RNP strategy (Fig. 1).

2 Results and Discussion

RNP receptors for dopamine were isolated from an RNP library by in vitro selection method as previously reported [3, 5]. In each round of selection, an RNP pool was incubated with a dopamine-immobilized agarose resin. After removal of unbound RNP fraction, a bound RNP fraction were recovered by specific elution with free dopamine. An ethanolamine-agarose resin was utilized in the negative selection step of dopamine, respectively. The bound RNA fractions were collected, reverse transcribed and applied to successive PCR amplification (RT-PCR) to generate new DNA pools for next round. DNA templates were transcribed again and the resulting RNA pools were subjected to the next round of selection. After several rounds of selection cycles, sequences of the selected RNA subunit were analyzed by DNA sequencing and dopamine-binding RNP receptors were collected.

In order to convert the dopamine-binding RNP receptor to fluorescent dopamine sensors, 7-methoxycoumarin-3-carboxylic acid were introduced to the N-terminal of the Rev peptide (7mC-Rev). From the structure of our basic frame, binding of RNP to the ligand would cause a local environmental perturbation around the fluorophore attached at N-terminal of the Rev peptide. Therefore, a fluorescent complex that composed RNA subunit of dopamine-binding RNP receptors and 7mC-Rev peptide was expected to change the emission properties along with dopamine binding. As expected, the RNP receptors with 7mC-Rev showed an increase in the fluorescence emission intensity upon addition of dopamine, as observed in the case of the ATP-binding RNP sensor reported previously [3]. Figure 2 shows the titration curve of the fluorescent RNP complex formed by 7mC-Rev peptide and DH05 RNA, which is one of the RNA subunit of dopamine-binding RNP receptors. Relative fluorescence intensity changes were measured with increasing concentrations of dopamine and a nonlinear regression analysis of the titration curve yielded a dissociation constant of 4 μM for the complex of DH05/7mC-Rev and dopamine (Fig. 2). Observed relative fluorescence intensity change (I/I_0) was 1.48 at the saturation of dopamine binding to DH05/7mC-Rev.

We next investigated the selectivity of the dopamine-binding RNP sensor by titration with various dopamine analogues. Substitutions of the hydroxyl group at position 3 (i.e., tyramine) decreased fluorescent response (Fig. 3). On the other hand, the complete deletion of the aliphatic chain (i.e., catechol) and removal of the

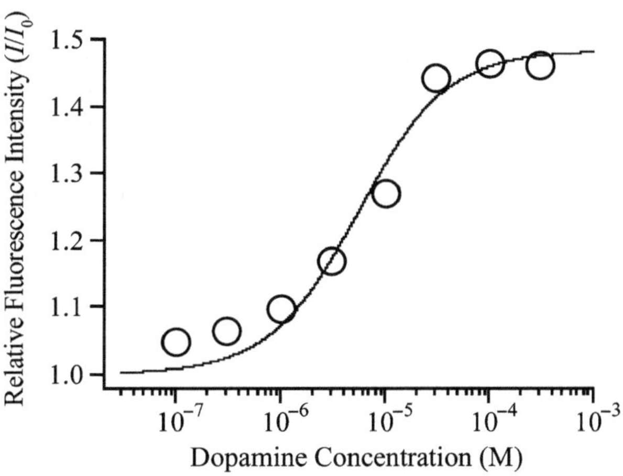

Fig. 2 Direct titration of a fluorescent RNP complex (1 μM) of the DH05 RNA subunit and 7mC-Rev (DH05/7mC-Rev) with dopamine (0.1, 0.3, 1, 3, 10, 30, 100, 300 μM)

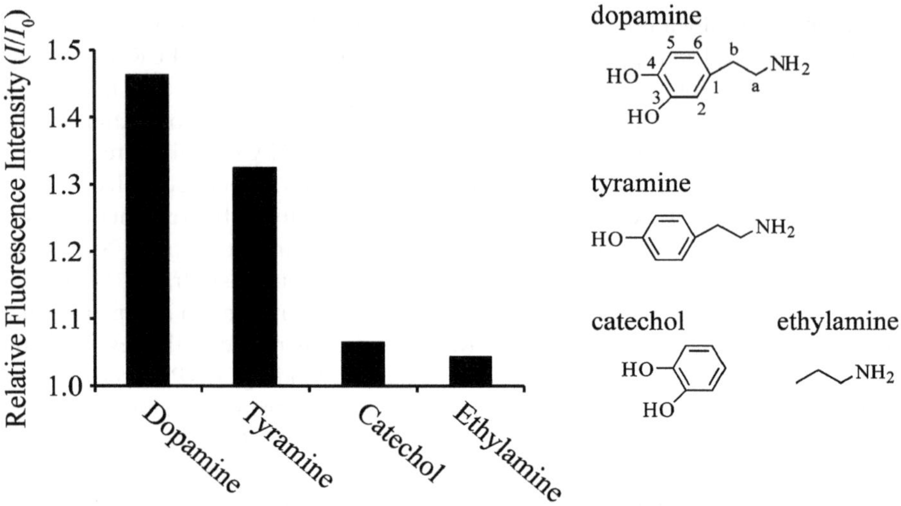

Fig. 3 Ligand specificity of fluorescent RNP sensor for dopamine and its derivatives. The bar graphs show Relative fluorescence intensity changes (I/I_0) of DH05/7mC-Rev in the presence of various dopamine derivatives (100 μM). The chemical structures of each ligand are indicated on the right

catechol ring (i.e., ethylamine) completely abolished the formation of a stable ligand-fluorescent RNP complex (Fig. 3). These results indicate that the existence of both the catechol ring and aliphatic chain with amino group are necessary for binding of this fluorescent RNP sensor.

3 Conclusions

In this study, we showed the construction of fluorescent RNP sensor for dopamine by utilizing the RNP strategy. Combination of the RNA subunit derived from the dopamine-binding RNP and the fluorophore-modified Rev peptides afforded fluorescent RNP sensors that showed distinct changes in the fluorescence emission intensities upon binding to dopamine. One of the dopamine sensors constructed in this study specifically recognized the catechol ring and the hydroxyl group at the position 3 of dopamine and both the catechol and ethylamine moiety are necessary for binding of fluorescent RNP sensor. Our results clearly showed that the strategy for stepwise functionalization of RNP is effectively applied to a fluorescent sensor for the biologically active amine.

References

1. Morii T, Hagihara M, Sato S, Makino K (2002) In vitro selection of ATP-binding receptors using a ribonucleopeptide complex. J Am Chem Soc 124:4617–4622
2. Sato S, Fukuda M, Hagihara M, Tanabe Y, Ohkubo K, Morii T (2005) Stepwise molding of a highly selective ribonucleopeptide receptor. J Am Chem Soc 127:30–31

3. Hagihara M, Fukuda M, Hasegawa T, Morii T (2006) A modular strategy for tailoring fluorescent biosensors from ribonucleopeptide complexes. J Am Chem Soc 128:12932–12940
4. Hasegawa T, Hagihara M, Fukuda M, Morii T (2007) Stepwise functionalization of ribonucleopeptides: optimization of the response of fluorescent ribonucleopeptide sensors for ATP. Nucleosides Nucleotides Nucleic Acids 26:1277–1281
5. Hasegawa T, Hagihara M, Fukuda M, Nakano S, Fujieda N, Morii T (2008) Context-dependent fluorescence detection of a phosphorylated tyrosine residue by a ribonucleopeptide. J Am Chem Soc 130:8804–8812
6. Battiste JL, Mao H, Rao NS, Tan R, Muhandiram DR, Kay LE, Frankel AD, Williamson JR (1996) Alpha helix-RNA major groove recognition in an HIV-1 rev peptide-RRE RNA complex. Science 273:1547–1551
7. Ellington AD, Szostak JW (1990) In vitro selection of RNA molecules that bind specific ligands. Nature 346:818–822
8. Tuerk C, Gold L (1990) Systematic evolution of ligands by exponential enrichment: RNA ligands to bacteriophage T4 DNA polymerase. Science 249:505–510

High-Pr Heat Transfer in Viscoelastic Drag-Reducing Turbulent Channel Flow

Yoshinobu Yamamoto, Tomoaki Kunugi, and Feng-Chen Li

Abstract In this study, direct numerical simulation (DNS) of heat transfer in a turbulent channel flow with Giesekus model was conducted at turbulent Reynolds number is 150, Prandtl number is 5 and Weissenberg number based on the friction velocity is 30. Temperature field was treated as a passive scalar and constant temperature difference condition was imposed at the wall. As the results, fundamental DNS database such as mean temperature, turbulent heat flux and turbulent Prandtl number were obtained. And heat transfer reduction rate exceeded drag reduction rate as well as previous DNS result by Yu and Kawaguchi (Int J Heat Mass Transf 48:4569–4578, 2005), despite of the Prandtl number distinction.

Keywords DNS • Heat transfer • Viscoelastic turbulent flow • High-Pr fluid

1 Introduction

To save a pumping power of a pipeline in heat transport systems such as district heating/cooling systems, application of the drag reduction effect is one of the most important schemes. Dramatic turbulent suppression effects were caused form mixing a small amount of polymers or surfactant additives with a liquid flow. It was well known as Toms effects [1] and many experimental and numerical studies have been conducted to understand the mechanism of the drag reduction. On the other hand, drag reduction effects on heat transfer were not understood well. Recently, Li et al. [2] reported that heat transfer coefficients in drag reducing flows with

---

Y. Yamamoto (✉) and T. Kunugi
Department of Nuclear Engineering, Kyoto University, Kyoto, Japan
e-mail: yyama@nucleng.kyoto-u.ac.jp

F.-C. Li
School of Energy Science and Engineering, Harbin Institute of Technology, Harbin, China

T. Yao (ed.), *Zero-Carbon Energy Kyoto 2009*, Green Energy and Technology,
DOI 10.1007/978-4-431-99779-5_8,

surfactant additives were also reduced. Yu and Kawaguchi [3] has conducted a direct numerical simulation (DNS) of heat transfer in a viscoelastic flow with Prandtl number (*Pr*) 0.71 and confirmed heat transfer reduction over drag reduction. However, Prandtl number of water flow covers more largely value from 3 to 7. In the high-*Pr* fluid ($Pr>2$), temperature length scale became smaller proportional to square of *Pr* and conductive sub-layer thickness was decreased proportional to cubic of *Pr*. These implies that analogy between drag reduction and heat transfer reduction is not always satisfied.

Therefore, in this study, DNS was conducted in order to investigate a high-*Pr* heat transfer in a viscoelastic drag reducing turbulent channel flow.

2 Numerical Condition and DNS Procedure

Objective flow field was a 2D fully developed turbulent channel flow and temperature field was treated as a passive scalar. In this study, Giesekus constitutive equation [4] was employed as a viscoelastic model and governing equations were continuity, incompressible Navier–Stokes, energy and Giesekus constitutive equations, respectively. Second order Finite Differencing Method (FDM) for velocity and temperature equations and MINMOD scheme [5] for Giesekus constitutive equation were adapted as the spatial discretization method. Time integration method was Crank–Nicolson scheme for Wall-normal viscous term, Euler Implicit scheme for Pressure terms and third order Runge–Kutta scheme for others, respectively. Constant mean pressure gradient was imposed as a driving force. Non-slip and periodic conditions were imposed for boundary conditions of velocity and the constant temperature at the top and bottom boundaries ($\theta_{top} > \theta_{bed}$), and periodic conditions were imposed for a passive scalar field. Numerical condition was tabled in Table 1. Time integration was conducted over 200,000 time steps after the fully-developed status on each case. The computational time per one step in CASE1 was about 1.2 s on Fujitsu HX600/16CPUs at ACCMS and IIMC, Kyoto University. Numerical results normalized by wall units based on the friction velocity, friction temperature, channel half height and kinetic viscosity, were shown by superscript +.

Table 1 Numerical condition

Case	Re_τ	Pr	We_τ	β	α	L_x, L_y, L_z	N_x, N_y, N_z
Viscoelastic	150	5	30	0.5	0.001	16*h*, 2*h*, 8*h*	160, 182, 160
Newton	150	5	–	–	–	16*h*, 2*h*, 8*h*	144, 182, 144

$Re_\tau = u_\tau h/\nu$ Turbulent Reynolds number; u_τ friction velocity; h channel half height; ν Kinetic viscosity; $Pr = D/\nu$ Prandtl number; D Thermal diffusivity; $We_\tau = \lambda\, u_\tau^2/\nu$ Weissenberg number; λ Relaxation time; β Ratio of solvent contribution to the total zero-shear viscosity; α Mobility factor; $L_x(N_x)$, $L_y(N_y)$, $L_z(N_z)$ Computational domain (or Grid Number) for stream(x), Wall-normal(y) and spanwise(z) direction, respectively

3 Results and Discussion

3.1 Drag Reduction

Mean velocity (U) profiles are plotted in Fig. 1a. In case of the viscoelastic flow ($We_\tau = 30$), mean velocity profile is largely up-shifted from the Newton flow case ($We_\tau = 0$), due to effects of the drag reduction. Figure 1b shows turbulent intensity profiles. In the drag reduction case, streamwise intensity (u_{rms}) is increased but wall-normal (v_{rms}) and spanwise (w_{rms}) intensities are decreased compared with the Newton flow case. The peak position of the streamwise intensity in the drag reduction case is shift to the channel-center side. These indicate that viscoelastic effects suppress streamwise vortices.

Figure 2a shows instantaneous streamwise turbulent velocity contours at near wall region. In both cases, typical wall-turbulence structures organized by high- and low-speed streaks are observed. In case of the drag-reduction, spanwise- and wall-normal fluctuations are suppressed, and streaks are shown as straight lines. Figure 2b shows velocity contours as same as Fig. 2a with velocity vector plots in end view. It can be seen that streaky structures are observed in not only near wall but also near half of the channel center regions and scale of wall-normal and spanwise vectors is decreased at near wall region, compared with the Newton flow. These visualization results are consistent with results of mean velocity and turbulent intensities and well accorded with previous DNS results [3, 6] despite of differences of numerical scheme and computational domain size.

3.2 Heat Transfer Reduction

Figure 3a shows mean temperature (Θ) profiles. In case of the viscoelastic flow ($We_\tau = 30$), mean velocity profile is largely up-shifted from the Newton flow case ($We_\tau = 0$) as well as mean velocity profiles as shown in Fig. 1a. Turbulent temperature

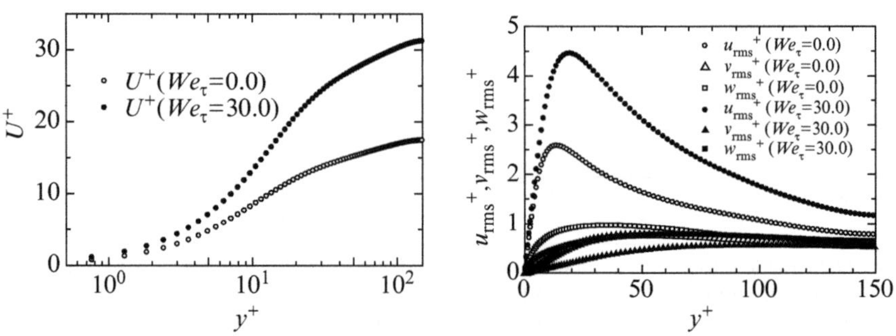

Fig. 1 Mean and statistics profiles of velocity, (**a**) mean velocity, (**b**) turbulent intensity

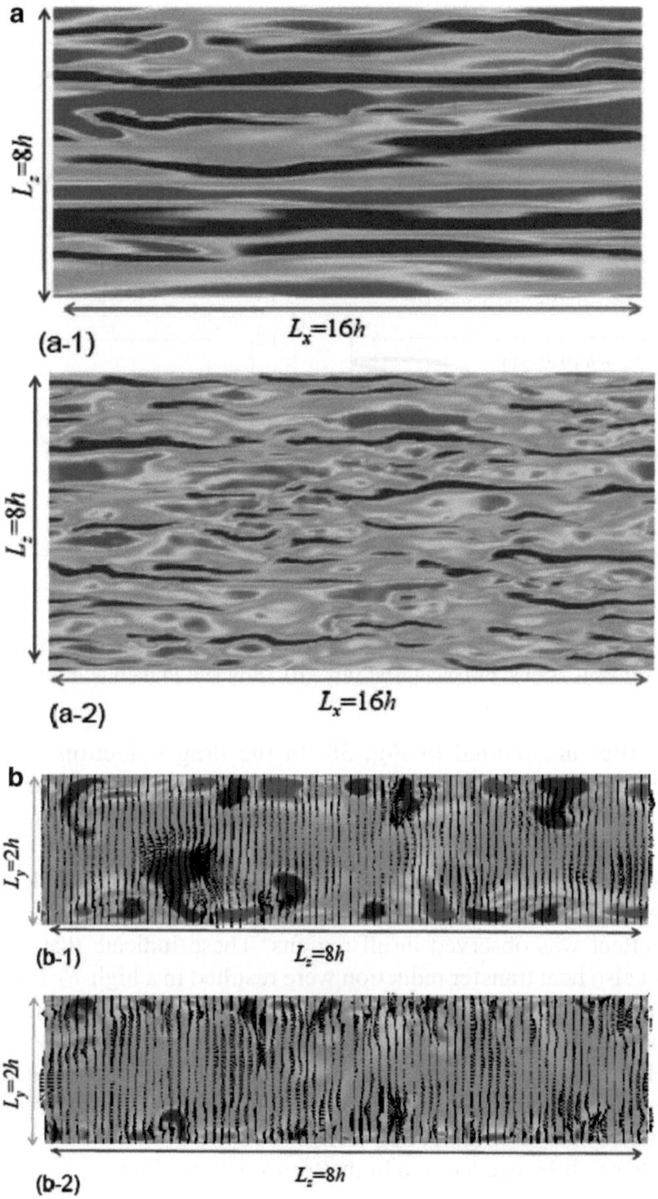

Fig. 2 Flow visualization, (**a**) Instantaneous streamwise turbulent velocity contours, –5.0(Blue) < u^+ < 5.0(Red), $y^+ = 15$, Top view, (**a-1**) Viscoelastic flow ($We_\tau = 30$), (**a-2**) Newton flow ($We_\tau = 0$). (**b**) Instantaneous streamwise turbulent velocity contours, –5.0(Blue) < u^+ < 5.0(Red) with turbulent velocity vector plots, end view, (**b-1**) Viscoelastic flow ($We_\tau = 30$), (**b-2**) Newton flow ($We_\tau = 0$)

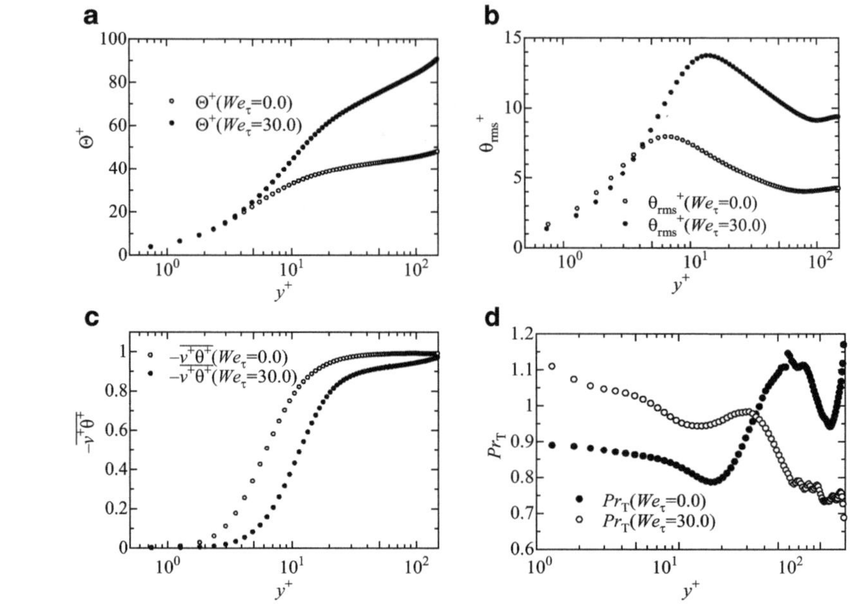

Fig. 3 Mean and statistic profiles of temperature, (**a**) Mean temperature, (**b**) Turbulent temperature intensity, (**c**) Wall-normal turbulent heat flux, (**d**) Turbulent Prandtl number

intensity profiles are plotted in Fig. 3b. In the drag reduction case, turbulent temperature intensity (θ_{rms}) is increased as well as the streamwise turbulent intensity and the peak position of the turbulent intensity is also shift to the channel-center side. Figure 3c shows wall-normal turbulent heat flux ($-v\theta$) profiles. Wall-normal turbulent heat flux is attributable mainly to turbulent diffusion. Compared with the Newton flow, the turbulent heat flux is decreased and the heat conduction effect was observed in all regions. These indicate that not only drag reduction but also heat transfer reduction were resulted in a high-*Pr* fluid. Turbulent Prandtl number is plotted in Fig. 3d. In the vicinity wall, the turbulent Prandtl number in the Newton flow is increased and one in the viscoelastic flow shows almost constant value (=0.9). Turbulent Prandtl profiles show opposite behavior from $y^+=30$ to the channel center. Turbulent Prandtl number is often used in a turbulent model simulation of an engineering design but some modification of turbulent Prandtl number will be needed in a high-*Pr* viscoelastic flow.

3.3 Comparison of Drag and Heat Transfer Reduction

As shown in Figs. 1b and 3b, the peak position of the streamwise turbulent intensity is $y^+=13.4(We_\tau=0)$, $19.2(We_\tau=30)$ and one of the turbulent temperature intensity is $y^+=6.4(We_\tau=0)$, $13.5(We_\tau=30)$, respectively. Rate of the peak position of the

Table 2 Friction drag coefficient and Nusselt number

Case	Re_m	Cf	Nu	DR%	HTR%
Viscoelastic	11,553	2.82×10^{-3}	8.3	62.1%	74.4%
Newton	4,560	8.66×10^{-3}	14.6	–	–

$Re_m = U_m 2h/\nu_{eff}$ Bulk Reynolds number, U_m Bulk velocity, ν_{eff} effective kinetic viscosity, $Cf = 2\nu_{eff} dU/dy_{wall}/U_m^2$ Friction drag coefficient, dU/dy_{wall} Mean velocity gradient at wall, $Nu = 2hd\Theta/dy_{wall}/(\theta_{top} - \theta_{bed})$ Nusselt number, $d\Theta/dy_{wall}$ Mean temperature gradient at wall, *DR%* Drag reduction rate, *HTR%* Heat transfer reduction rate [3]

streamwise turbulent and temperature turbulent intensity, is 2.1(=13.4/6.4) in the Newton flow and 1.4(=19.2/13.5) in the viscoelastic flow. This implies the promotion of the laminarization near wall region and the analogy between velocity and temperature fields in the viscoelastic flow, compared with the high-*Pr* Newton flow. Friction drag coefficient and Nusselt number are tabled in Table 2. Both of the friction drag coefficient and Nusselt number are decreased caused from viscoelastic effects. In the high-*Pr* flow, heat transfer ratio is over drag reduction ratio as well as DNS results with $Pr = 0.71$.

4 Conclusions

In this study, DNS was carried out to investigate a high-*Pr* heat transfer in a viscoelastic drag reducing turbulent channel flow. Obtained results are summarized as following:

(1) Drag reduction effects of mean velocity and turbulent intensities were good agreements with previous DNS [3, 6].
(2) Heat transfer reduction was confirmed in a high-*Pr* and viscoelastic fluid.
(3) In the viscoelastic flow, Turbulent Prandtl number estimated by the present DNS shows the distinct profile and some modification of turbulent Prandtl number might be needed in turbulent model analysis.
(4) Compared with the high-*Pr* Newton flow, promotion of the analogy between momentum transfer and heat transfer was observed in the viscoelastic flow.
(5) As well as previous DNS [3], heat transfer reduction rate exceeded drag reduction rate, despite of the Prandtl number distinction.

Acknowledgments This work was supported by Grant-in-aid for Young Scientist (B), MEXT KAKENHI (21760156) and present numerical simulation was supported partly by Collaborative Research Program for Young Scientists of ACCMS and IIMC, Kyoto University.

References

1. Toms BA (1948) Some observations on the flow of liner polymer solutions through straight tubes at large Reynolds numbers. In: Proc. 1st Int. Congress on Rheology, North Holland, Amsterdam, pp 135–141
2. Li FCh, Kawaguchi Y, Hishida K (2004) The influence of a drag-reducing surfactant on turbulent velocity and temperature field of a 2D channel flow. Exp Fluids 36(1):131–140
3. Yu B, Kawaguchi Y (2005) DNS of fully developed turbulent heat transfer of a viscoelastic drag-reducing flow. Int J Heat Mass Transf 48:4569–4578
4. Giesekus H (1982) A simple constitutive equation for polymer fluids based on the concept of deformation-dependent tensorial mobility. J Non-Newt Fluid Mech 11:69–109
5. Yu B, Kawaguchi Y (2004) Direct numerical simulation of the viscoelastic drag-reducing flow: a faithful finite-difference method. J Non-Newt Fluid Mech 116:431–466
6. Ishigaki T, Tukahara T, Kawaguchi Y, Yu B (2009) DNS study on viscoelastic effect in drag-reduced turbulent channel flow. Turbulent Shear Flow Phenomena 6(1):359–364

Current Status of Accelerator-Driven System with High-Energy Protons in Kyoto University Critical Assembly

Jae-Yong Lim, Cheol Ho Pyeon, Tsuyoshi Misawa, and Seiji Shiroya

Abstract At the Kyoto University Research Reactor Institute, the first injection of the spallation neutrons generated by the high-energy proton beams into a reactor core was accomplished on 4 March 2009. Using three detectors which located at near active core regions, the prompt and delayed neutron behaviors by proton injection were experimentally observed and the neutron beam characteristics at the beam duct were also watched by Gafchromic films. Under the subcritical condition with $0.76\%\Delta k/k$, an In wire irradiation experiment was accomplished horizontally. The $^{115}In(n,\gamma)^{116m}In$ reaction rate comparison was also performed by MCNPX simulation and its errors showed within the allowance of the experimental statistical errors.

Keywords Kyoto university critical assembly • FFAG accelerator • Accelerator-driven system • High-energy proton • MCNPX

1 Introduction

The Kyoto University Research Reactor Institute is going ahead with an innovative research project on Accelerator-Driven System (ADS) using a Fixed Field Alternating Gradient (FFAG) accelerator [2, 7]. The goal of the research project was to demonstrate the basic feasibility of ADS as a next-generation neutron source using the Kyoto University Critical Assembly (KUCA) coupled with a newly developed variable energy FFAG accelerator. At a new ADS with the FFAG accelerator, on 4 March 2009, the high-energy neutrons generated by spallation reactions with 100 MeV proton beams, which had a few pA intensity at a tungsten target, were successfully injected into a solid-moderated and -reflected core (A-core) in thermal neutron field of KUCA.

J.-Y. Lim (✉), C.H. Pyeon, T. Misawa, and S. Shiroya
Research Reactor Institute, Kyoto University, Osaka, Japan
e-mail: limjy@rri.kyoto-u.ac.jp; pyeon@rri.kyoto-u.ac.jp;
misa@rri.kyoto-u.ac.jp; shiroya@rri.kyoto-u.ac.jp

T. Yao (ed.), *Zero-Carbon Energy Kyoto 2009*, Green Energy and Technology,
DOI 10.1007/978-4-431-99779-5_9, © Springer 2010

In the previous studies [4–6], numerical results of combined MCNP-4C3 and nuclear data libraries (JENDL-3.3 and ENDF/B-VI.2) showed good agreement with those of ADS experiments with 14 MeV neutrons, in terms of reactivity and reaction rate analyses. And the foil activation method was found to be a useful measuring technique for examining the neutronic properties of ADS with 14 MeV neutrons at KUCA.

The static and kinetic experiments [4] on ADS with 14 MeV neutrons were conducted at KUCA and revealed the following: measurement and calculation methods were proved reliable for the evaluation of subcriticality effects down to $6\%\Delta k/k$. Anticipating a conventional ADS subcriticality level around $3\%\Delta k/k$ ($k_{eff}=0.97$), the measurement methodology and the calculation precision were considered convenient for the study of ADS at KUCA. And the calculations were reliable in describing reaction rate distributions: JENDL-3.3 nuclear library is convenient for the estimation of relative distribution within the experimental error.

For the neutron spectrum experiments [6] using the foil activation method in the subcritical systems, C/E values in reaction rates of the experiments and the calculations with JENDL/D-99 were found to be about a difference of 10%, although a large discrepancy was observed with some foils, especially in the center of the core. However, a special mention should be made of the fact that these experiments clearly revealed subcriticality dependence in reaction rate analyses.

Based on previous studies using a 14 MeV Cockcroft–Walton-type accelerator, we tried to observe the characteristics of ADS with FFAG proton accelerator. First of all, the feasibility of neutron multiplication by spallation neutrons from outside of core was investigated in this study. The verification of In reaction rate distributions was also performed by comparing with experimental data and Monte Carlo simulation.

2 Accelerator-Driven System with 100 Mev Protons

2.1 First Proton Injection Experiment

Using successful proton beams, ADS experiments were carried out at the KUCA A-core shown in Fig. 1. The spallation neutrons generated at a tungsten target, whose size was 80 mm diameter and 10 mm thickness, were successfully injected into the A-core for the first time in the world. In the ADS experiments, the main characteristics of the proton beams in the FFAG accelerator were 100 MeV energy; 30 Hz repetition rate; a few pA intensity. Level of the neutron flux yield obtained at the tungsten target was about 1×10^{6} n/s. The purpose of these ADS experiments was to establish a new neutron source using the ADS in combination with KUCA and a new FFAG accelerator [2, 7].

KUCA comprises two solid polyethylene-moderated and -reflected thermal cores designated A and B, and one water-moderated thermal core designated C. The A core is mainly composed of the normal, partial and special fuel assemblies; the polyethylene rods. From the viewpoints of the safety aspects in the core, the

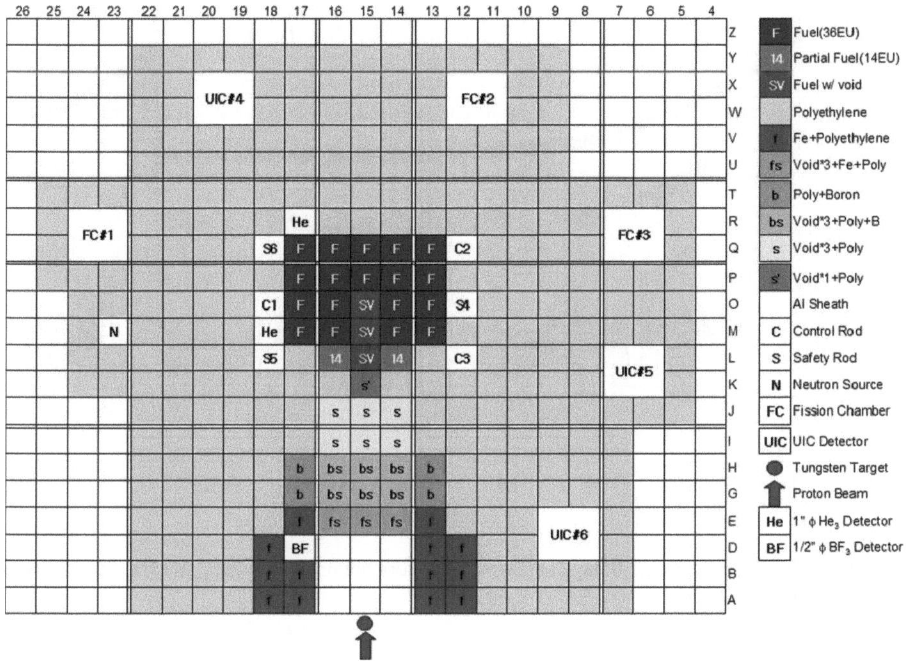

Fig. 1 Top view of the configuration of A-core in the ADS experiments with 100 MeV protons

tungsten target is located not at the center of the core but outside the position in the critical assembly. To provide the information on the detector position dependence of the prompt neutron decay measurement, the neutron detectors were set at three positions shown in Fig. 1: near the tungsten target [(17, D); 1/2" ϕ BF_3 detector]; around the core [(18, M); 1" ϕ He_3 detector and (17, R); 1" ϕ He_3 detector].

The prompt and delayed neutron behaviors were experimentally observed through a series of the time distribution of the neutron density: an exponential function and a decreasing tendency respectively as shown in Fig. 2. Namely, these behaviors demonstrated the fact that the neutron multiplication by an outer-source sustainable nuclear chain reaction using the spallation neutrons was surely occurred in the core. In these kinetic experiments, the subcriticality was deduced from the prompt neutron decay constant by the extrapolated area ratio method, and the relative difference between the results of this (0.76%$\Delta k/k$; (17, R) and 0.87%$\Delta k/k$; (18, M) in Fig. 1) and another experimental evaluation (0.76%$\Delta k/k$), which was obtained from the combination of both the control rod worth by the rod drop method and its calibration curve by the positive period method, was in the precision of within about 10%. The subcritical state was made by a full insertion of C1, C2 and C3 control rods into the core.

As an additional verification experiment, the production of spallation neutrons was confirmed using Gafchromic films which could be developed by neutrons. These films were located at four positions: at the target surface and 7/14/21 cm distance from target surface and the aluminum sheaths at 14, 16-A, D in Fig. 1 were

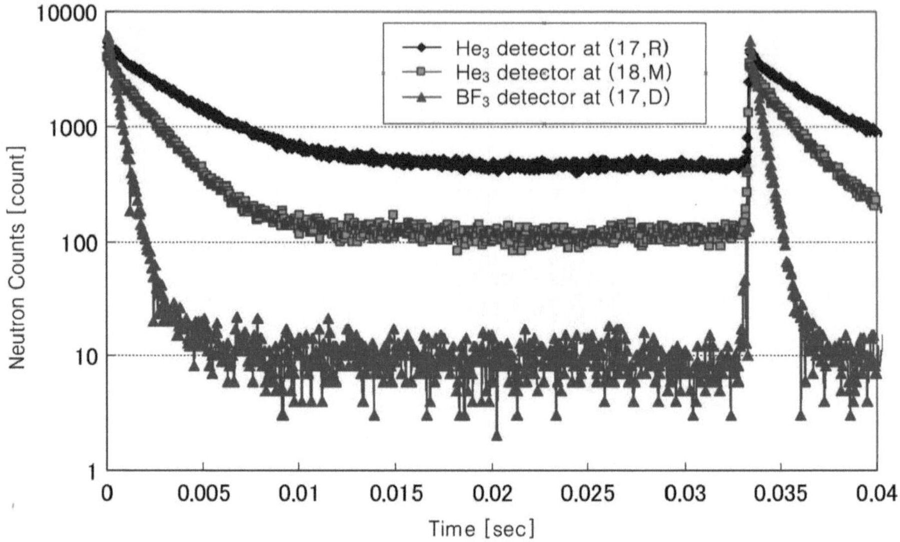

Fig. 2 Measured prompt and delayed neutron behaviors obtained from BF_3 and He_3 dectrctors in the A-core in Fig. 1

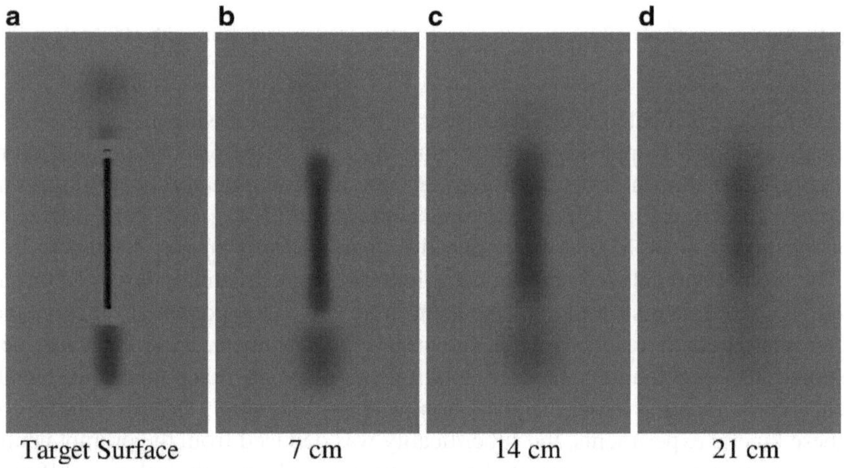

Fig. 3 Results of Gafchromic films varying the distance from target

withdrawn for the attachment of films. As shown in Fig. 3, the proton beams were injected into core with very narrow shape in vertical and it made the spallation neutrons at target surface be also narrow beam shape. However, after tungsten target surface, the spallation neutron beams were spread easily depend on distance from the surface. The peak value of these developed data at each Gafchromic films was reduced rapidly up to 9% at 21 cm distance.

2.2 Reaction Rate Distribution Comparison

Thermal neutron flux distribution was estimated through the horizontal measurement of $^{115}In(n,\gamma)^{116m}In$ reaction rates by the activation analysis of an indium wire of 1.0 mm diameter. The wire was set in an aluminum guide tube, from the tungsten target to the center of the fuel region (13, 14-A, P in Fig. 1), at the height of the center of the fuel assembly. In these static experiments, the subcritical system was made by the full insertion of C1, C2 and C3 rods as the same as the kinetic experiments, and the subcriticality was experimentally deduced to be 0.76%$\Delta k/k$. The numerical calculations were executed with the Monte Carlo multi-particle transport MCNPX [3] based on a nuclear data library ENDF/B-VII [1]. The source was represented by a 100 MeV proton isotopic source. Since the effect of the reactivity is not negligible, the In wire was included from tallies taken in the indium wire setting region.

The result of the source calculation was obtained after 2,000 cycles of 100,000 histories and the statistical error of the reaction rate was less than 1%. The measured and the calculated reaction rate distributions were compared to validate the calculation method for the ADS with 100 MeV protons at KUCA as shown in Fig. 4.

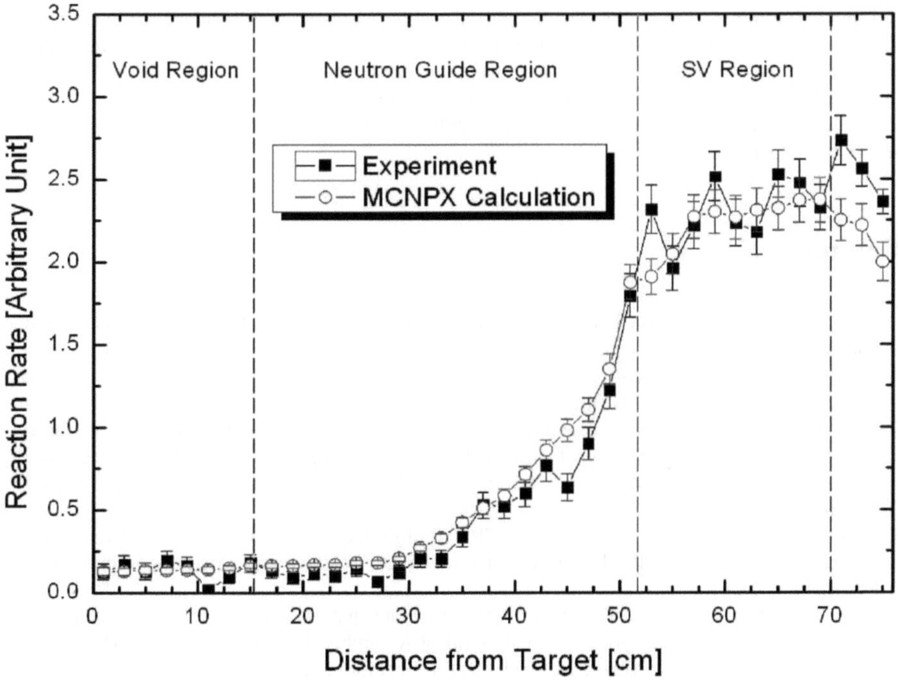

Fig. 4 Comparison of measured and calculated reaction rate distributions along the vertical of (13, 14-A, P) in Fig. 1

The calculation reaction rate distribution revealed approximately the reproduction of the experiment within the allowance of the experimental statistical errors, although these experimental errors larger relatively than those of the calculations. These larger errors of the experiments were attributed to the current status of the proton beams, including the beam intensity and the beam shaping at the target.

3 Concluding Remarks

At the KUCA A-core, the new ADS experiments with 100 MeV protons were conducted in the combination of the KUCA polyethylene-moderated and -reflected core and the FFAG accelerator. The static and kinetic experiments revealed the neutron multiplication by the outer-source sustainable nuclear chain reaction using the spallation neutrons generated by the interaction of the proton beams from the FFAG accelerator and the tungsten target. In the FFAG accelerator, a stable beam commissioning is still being under way, including the beam shaping and the intensity. As the final objective is to carry out experiments of the ADS with 150 MeV protons generated from the FFAG accelerator, the present experiments could be expected to contribute to further researches and development of both the experiments and the nuclear design in ADS at KUCA.

Acknowledgements This work was supported by the Ministry of Education, Culture, Sports, Science and Technology (MEXT) of Japan within the task of "Research and Development for an Accelerator-Driven Subcritical System Using an FFAG accelerator". A part of this study was supported by the Grant-in-Aid for Scientific Research form MEXT from Japan.

The authors are grateful to all the technical staff and the students of KUCA for their assistance during the experiments.

References

1. Chadwick MB, Oblozinsky P, Herman M et al (2006) ENDF/B-VII.0 next generation evaluated nuclear data library for nuclear science and technology. Nucl Data Sheets 107:2931
2. Mori Y (2006) Development of FFAG accelerators and their applications for intense secondary particle production. Nucl Instrum Meth A 562:591
3. Pelowitz DB (2005) MCNPX user's manual, version 2.5.0. Los Alamos National Laboratory
4. Pyeon CH, Hervault M, Misawa T et al (2008) Static and kinetic experiments on accelerator-driven system with 14 MeV neutrons in Kyoto University Critical Assembly. J Nucl Sci Technol 45:1171
5. Pyeon CH, Hirano Y, Misawa T et al (2007) Preliminary experiments for accelerator driven subcritical reactor with pulsed neutron generator in Kyoto University Critical Assembly. J Nucl Sci Technol 44:1368
6. Pyeon CH, Shiga H, Misawa T et al (2009) Reaction rate analyses for on accelerator-driven system with 14 MeV neutrons in Kyoto University Critical Assembly. J Nucl Sci Technol 46:965
7. Yonemura Y, Takagi A, Yoshii M et al (2007) Development of RF acceleration system for 150 MeV FFAG accelerator. Nucl Instrum Meth A 576:294

Part III
International Summer School on Energy Science for Young Generations (ISSES-YGN)

(i) Scenario Planning and Socio-economic Energy Research

Toward Education for Collaboration Between Different Fields: An Experiment of Facilitation Viewpoints Utilization for Reflecting Group Discussion

Kyoko Ito, Eriko Mizuno, and Shogo Nishida

Abstract Energy and environmental problems are much complicated and one of the measures against these problems is an appropriate education. Toward education of collaboration between different fields, group discussion is an appropriate place for learning awareness to collaboration. In this study, utilization of facilitation viewpoints in group discussion is focused on and an experiment was conducted to introduce facilitation viewpoints in group discussion. Based on the results, the possibility and proposal of the utilization were considered.

Keywords Education • Collaboration • Group discussion • Facilitation • Computer supported system • Reflection

1 Introduction

Energy and environmental problems are much complicated because of involving such points in controversy as the society, economy, safety, prospect, policy, and so on [1]. One of the important measures against these problems is an appropriate education for human resources expected to forge the future of the fields. The university and the graduate school as the higher education organizations are the educational areas of the students with advanced specialties, and it is expected to appear one after another of diverse human resources from the areas. There is, however, a limit in the alone idea of measures against the complex problems. The education of collaboration between different fields is expected. Group discussion is one of the education method, and functions as a simulation place of collaboration.

A utilization of "facilitation viewpoints" is focused on in the learning place of collaboration in this study. Facilitation is an approach of promoting intelligent

K. Ito (✉)
Center for the Study of Communication-Design, Osaka University, 1-3 Machikaneyama-cho, Toyonaka, Osaka, 560-0043, Japan
e-mail: ito@cscd.osaka-u.ac.jp

K. Ito, E. Mizuno, and S. Nishida
Graduate School of Engineering Science, Osaka University, Osaka, Japan

T. Yao (ed.), *Zero-Carbon Energy Kyoto 2009*, Green Energy and Technology,
DOI 10.1007/978-4-431-99779-5_10, © Springer 2010

interaction in group, and a support from neutral position [2]. The utilization of facilitation viewpoints is expected to act as overview of discussion and give opportunities to reflect the discussion to the participants.

In this study, a utilization of facilitation viewpoints is considered to reflect group discussion by participants toward education of collaboration between different field. And, an experiment is conducted to introduce facilitation viewpoints in group discussion. Based on the results, the possibility and a proposal of the utilization are considered.

2 Method

For giving the discussion participants opportunities to reflect their behavior and be aware of something for collaboration, the group discussion is set as follows:

Participants – students who major in different fields
Number of group members – for easiness to discuss (about 5–8)
Discussion theme – problems in the science and technology, and society [3]
Purpose of discussion – consensus building

The facilitation viewpoints are selected in order to promote the participants' reflection and awareness toward collaboration as follows:

(a) Time management as a constraint condition
(b) Balance of utterance rate as showing participants' strength
(c) Transition of argument as a process

The utilization timing of three facilitation viewpoints are during and after discussion. The screen for the presentation of the three viewpoints is designed. The ratio of elapsed time and remaining time, the ratio of each participant's all utterance time and structure of argument [4,5] are selected as time management, balance of utterance rate and transition of argument, respectively. The design of the screen is shown in Fig. 1.

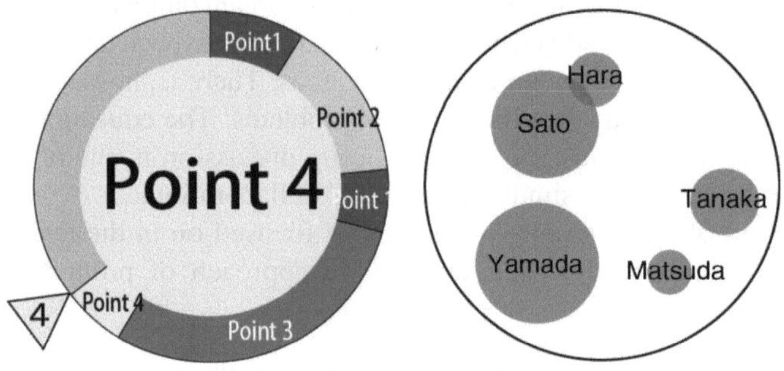

Fig. 1 Design of introducing facilitation viewpoints. The *left circle* shows transition of argument and a triangle of time management. The start point is *top of the circle*. The *right circle* shows the ratio of participants' utterance rate. The areas of the *inner circles* show the rate. Each *inner circles* includes each participant's name

In order to present the designed screen including the information, the category of argument, the uttering person and the ending time of utterance as the input, and three information based on the input as the output are necessary. A person for the input is arranged, and the person listens to discussion and inputs the necessary information.

3 Experiment

The purpose of the experiment is to consider the following points, based on the results of the experiment introducing facilitation viewpoints to group discussion:

- Problems and issues in the discussion by participants with different fields
- Possibility of facilitation viewpoints utilization in group discussion for high education

The method of the experiment is as follows:

Theme – right or wrong on the current policy of Japan on high level radioactive waste disposal.
Task – consensus building on the discussion theme (right or wrong, and the substantial reasons).
Group – two groups (Group 1 and Group 2), each for five participants.
Flow – leaning about basic knowledge on the discussion theme as advanced preparation.

On the day:

1. Group discussion (2 h)
2. Reflection about the group discussion
3. Questionnaire and interview about the discussion and the information presentation

Analyzed data – log data of balance of utterance volume and transition of argument, recoded video footage and results of questionnaire and interview.

The appearance of the discussion is shown in Fig. 2. As a result of the questionnaire and interview, it was shown that the participants were aware of the transition of argument and balance of utterance, and learned new viewpoint about group discussion. As the learned new viewpoint about group discussion, the examples are as follows:

- It is important to decide which argument is particularly emphasized.
- It is important to include of reconfirmation of the opinions.

As the utilization in the future, the utilization of utterance balance is considered. And, discussion after discussion for group reflection using transition of argument is considered.

Fig. 2 Appearance of the experiment. The five students in Group 2 discussed. The nearest person is arranged for input of the necessary information. The *left back* is the projected screen of the facilitation viewpoints

4 Conclusion

In this study, the introduction of facilitation viewpoints to group discussion was considered toward education of collaboration between different fields. An experiment using the proposed method was conducted. From the results of the experiment, it was found that the awareness of how to discuss was important, and a reflection method of group discussion using facilitation viewpoints was proposed. In the future, a proposal of extracting and sharing participants' awareness and development of educational usefulness will be considered.

Acknowledgements This research was partially supported by the Ministry of Education, Science, Sports and Culture, Grant-in-Aid for Scientific Research (C), 21500874, 2009.

References

1. The Ministry of Education, Culture, Sports, Science and Technology (eds) (2004) White paper on science and technology in 2004 – science and technology and society of the future
2. Kaner S, Toldi LC, Fisk S, Berger D (1996) Facilitator's guide to participatory decision-making. New Society, Philadelphia
3. Kobayashi T (2004) Who considers about science and technology? A trial of "Consensus conference". The University of Nagoya Press, Nagoya
4. Conklin J, Begeman ML (1988) gIBIS: a hypertext tool for exploratory policy discussion. In: Proc. ACM Conference on CSCW, pp 140–152
5. Horita M, Enoto T, Iwahashi N (2003) A pluralist approach to visualization of policy discourse: argumentation support as socio-technical systems. In: Proc. Research of Science and Technology for Society, pp 25–37

The Impact of Wind Power Generation on Wholesale Electricity Price at Peak Time Demand in Korea

Seunghyun Ryu, Shinyoung Um, and Suduk Kim

Abstract In this paper, wind power is analyzed to see its impact on wholesale electricity price or SMP (System Marginal Price) at peak time under the consideration of the stochastic characteristics of its power generation. For this purpose, future power supply curve is estimated considering future fuel price for power and the construction and decommission of power plant's plan. Future oil price is assumed to follow EIA's future oil price scenario and the relation between oil price and LNG price for power generation are examined to forecast LNG price for power generation. Information on future power plant's construction and decommission plan and power demand at peak time is used based on the 4th National Power Market Plan. Once future SMP is estimated using future power supply curve with the peak time demand, the stochastic characteristics of wind power generations is considered to see its impact on the changes in SMP. The result shows that SMP without wind power is estimated to be \$ 0.1568 per KWh, while it is shown to drop down to \$ 0.1501 per KWh, a 4.27% decrease with wind power on 2030.

Keywords SMP (System Marginal Price) • CBP (Cost Based Pool) • Wind • Merit-order effect • Renewable energy

1 Introduction

A CBP (Cost Based Pool) market was originally introduced in Korea in 2001 to be utilized temporarily until the full-scale industrial restructuring of electricity market is accomplished. Supply curve of CBP market is organized according to the order of the least variable cost of power plants. A day-ahead demand which is shown to be a perpendicular line with no price elasticity of power demand is estimated by KPX (Korea Power Exchange). Then SMP is defined as an equilibrium price where

S. Ryu, S. Um, and S. Kim (✉)
Department of Energy Studies, Division of Energy System Research, Ajou University, San 5, Woncheon Dong, Yeong Tong Gu, Suwon, Korea
e-mail: mirang98@naver.com; tls4768@naver.com; suduk@ajou.ac.kr

T. Yao (ed.), *Zero-Carbon Energy Kyoto 2009*, Green Energy and Technology, DOI 10.1007/978-4-431-99779-5_11,

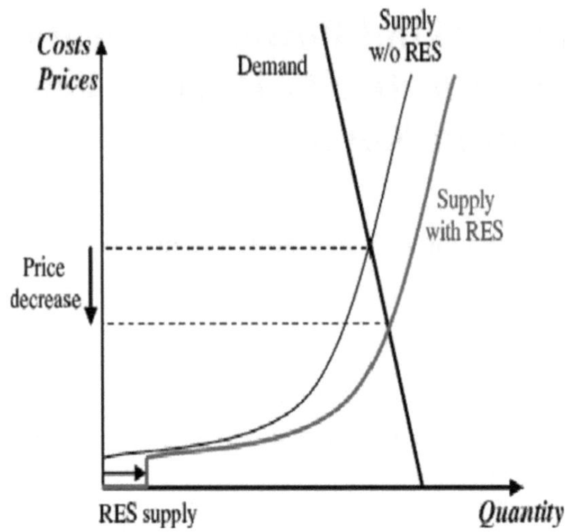

Fig. 1 Merit-order effect of renewable energy sources

this demand meets the supply curve of CBP market. All generators with lower variable cost than SMP will be dispatched.

Since the supply cost of electricity market is mainly affected by the cost of fuel for power generation, generators using the same type of fuel have similar variable costs. But there is a considerable price gap among generators using different fuel types and the supply curve looks like an upward sloped step function. Power generation by renewable sources has such characteristics as incurring no fuel cost and being purchased preferentially in market for its promotional purpose. Because variable costs of wind power generation is zero, existing supply curve of CBP market can be modified by moving it horizontality to the right as is depicted in Fig. 1.

2 A Methodology

For the purpose of this analysis, a base year supply curve (2008) is formulated based on the information of the heat rate (KWh/Mcal) and calories per unit cost (won/Gcal) of each generator on February 2008 (Fig. 2). Using this information, variable cost of each generator is calculated to form the supply curve according to its merit-order. After that, constructions and decommission of power plants' plan are reflected to the base year supply curve up to 2030 (Fig. 3). For the estimation of variable cost after 2009, future oil price is assumed to follow EIA's future oil price scenario and the relation between oil price and LNG price for power generation is estimated by simple regression. To see the impact of stochastic characteristics of wind power on the peak time demand, information on the estimated peak load from the 4th National Power Market Plan up to year 2022 is referred and is

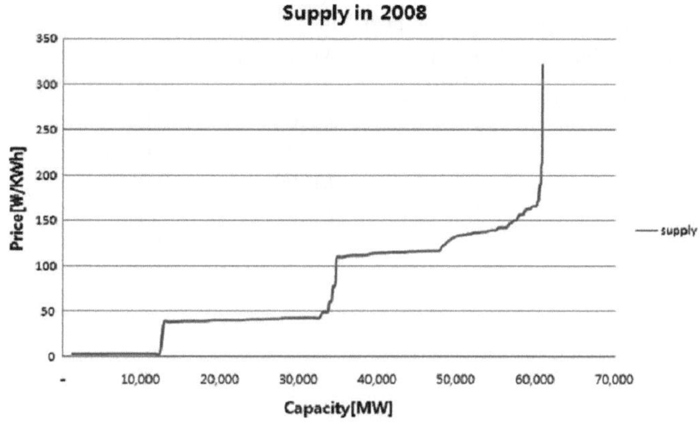

Fig. 2 Supply curve of Korea in 2008

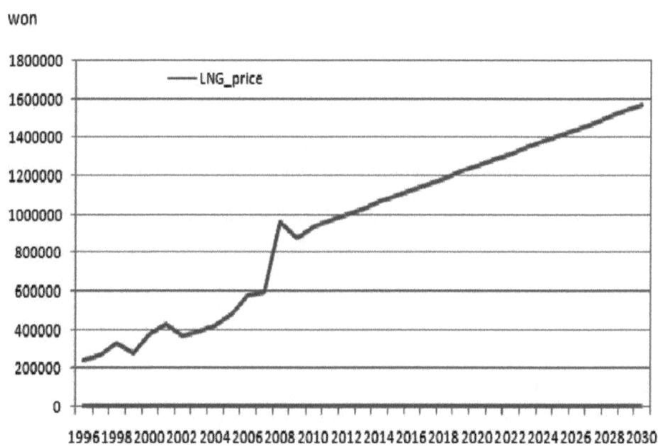

Fig. 3 Estimated LNG price of future

extended to year 2030 with the assumption that the load factor of year 2022 is continued to persist. For this purpose, the promotional target of wind power capacity reported on The 3rd Master Plan for The Promotion of New and Renewable Energy (2008) and the result of peak time impact of wind power from Korea East-West Power Co. (2009) is referenced.

3 Result

Table 1 shows the resulting impact of wind power generation on SMP at peak time for the whole period of analysis (Figs. 4–6). The won–dollar exchange rate assumed is 1,102.59 which is the annual average of year 2008.

Table 1 Impact of wind power on SMP at peak time

	Without_wind ($/KWh)	Wind_max ($/KWh)	Wind_min ($/KWh)	DIF ($/KWh)	Percent
2010	0.1269	0.1266	0.1269	0.0003	0.24
2015	0.1022	0.1018	0.1022	0.0004	0.39
2020	0.1090	0.1089	0.1090	0.0001	0.09
2025	0.1197	0.1190	0.1197	0.0007	0.58
2030	0.1568	0.1501	0.1568	0.0067	4.27

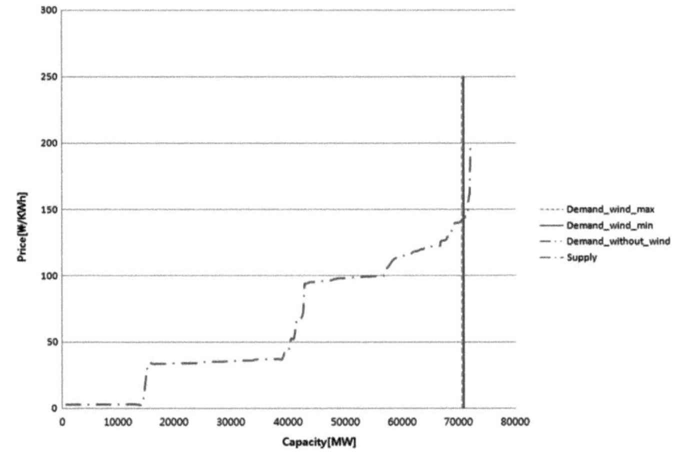

Fig. 4 Expected SMP in 2010

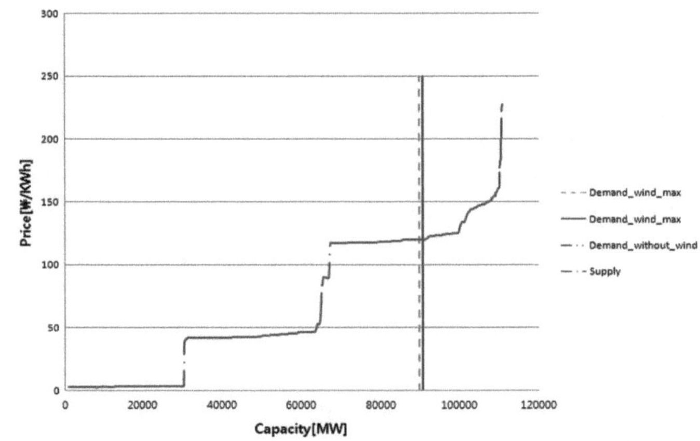

Fig. 5 Expected SMP in 2020

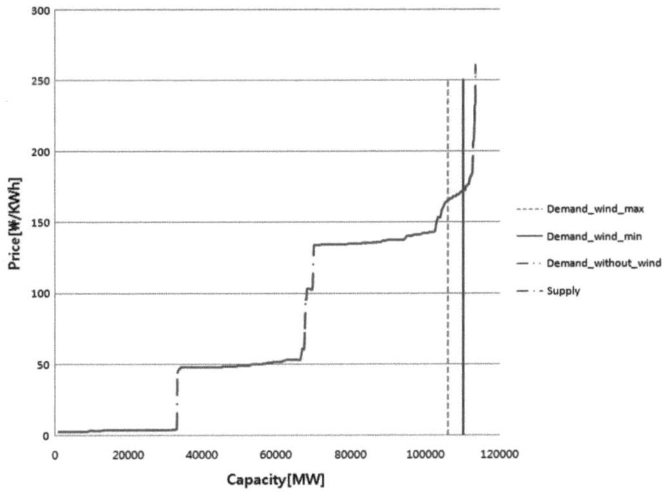

Fig. 6 Expected SMP in 2030

4 Conclusion

The purpose of this paper is to analyze wind power impact on wholesale electricity price or SMP at peak time under the consideration of its stochastic characteristics of power generation. The supply curves from 2009 to 2030 are constructed based on the power plants' construction and decommission plan with the forecast of LNG fuel cost. Peak time demand with explicit consideration of stochastic characteristics of wind power, its impact on SMP is analyzed. The result shows that SMP without wind power is estimated to be $ 0.1568 per KWh, while it is shown to drop down to $ 0.1501 per KWh, a 4.27% decrease with wind power on 2030. This is an interesting result considering the fact that the uncertainties of wind power such as undispatchability and intermittency could be a serious problem to overall power system in the future while wind power is expected to lower the wholesale price for the benefit of consumers.

References

1. Sensfuß F, Mario R, Massimo G (2008) The merit-order effect: a detailed analysis of the price effect of renewable electricity generation on spot market prices in Germany. Energy Policy 36:3086–3094
2. Weigt H (2009) Germany's wind energy: the potential for fossil capacity replacement and cost saving. Applied Energy 86:1857–1863
3. Ministry of Knowledge and Economy (2008) The 3rd plan for new and renewable energy technology development and promotion, Dec. 2008

4. Ministry of Knowledge and Economy (2007) The 4th national power market plan (2008–2022), Dec. 2007
5. Korea East-West Power Co. (2009) An analysis of renewable power facility and its peak time impact on overall power mix, Jan. 2009
6. Korea Energy Economics Institute (2008) The 3rd master plan for the promotion of new and renewable energy, Dec. 2008

An Analysis of Eco-Efficiency in Korean Fossil-Fueled Power Plants Using DEA

Hong Souk Shim and Sung Yun Eo

Abstract The joint production of goods and undesirable outputs such as pollutants which may not be disposable without cost makes it difficult to evaluate the environmental management of firms. In this paper, eco-efficiency is analyzed in the Korean power industry by focusing on information pertaining to power plants for the year 2007. This paper presents Data Envelopment Analysis (DEA) as a valuation model, and evaluates the relative eco-efficiency of fossil-fueled power plants. The dataset consists of total 26 fossil-fueled power plants operated by five different subsidiary power companies of KEPCO (Korean Electric Power Corporation) which are observed as Decision Making Units (DMU). Labor, capacity, and the amount of greenhouse gas emissions are used as inputs while power generation and sales are considered as outputs. In our analysis, six DMUs are found to be on the frontier with associated efficiencies designated as one. On the other hand, one DMU (#19) is found to be the least efficient. Results indicate that DMU 19 has the potential to reduce 74.7% of input and increase 76.3% of output. Efficient power plants can be used as a benchmark for inefficient power plants in efforts to confront climate change.

Keywords Climate change • Data envelopment analysis • Eco-efficiency

1 Introduction

Climate change is one of the main concerns of the contemporary world community. With the increasing consciousness about environmental problems and the burden placed by industrial activities on environmental quality, the environmental performance of companies has become increasingly important. Although power companies account for 1 out of every 4 tons of CO_2 emitted in Korea, proper

H.S. Shim (✉) and S.Y. Eo
Department of Energy Studies, Division of Energy Systems Research, Ajou University, San 5 Woncheon-Dong, Youngtong-Gu, Suwon, Korea
e-mail: symon99@ajou.ac.kr; Fishno001@naver.com

T. Yao (ed.), *Zero-Carbon Energy Kyoto 2009*, Green Energy and Technology,
DOI 10.1007/978-4-431-99779-5_12,

environmental evaluations of the power sector are lacking. Arguably the most important problem in such evaluations is overcoming difficulties in properly calculating environmental cost. With this in mind, DEA (Data Envelopment Analysis) could be a useful means of arriving at relative estimates of eco-efficiency by comparing similar DMUs (Decision Making Units). In this paper we evaluate the eco-efficiency of Korean fossil-fueled power plants for 2007.

The related research using DEA demonstrates potential for evaluating environmental problems in the Korean power industry. Golany et al. [5] evaluated the operational efficiency of power plants in the Israeli Electric Corporation. Pekka and Lubtacik [7] proposed the use of measuring technical efficiency and ecological efficiency separately. Those two efficiency indicators are then combined. The approaches are applied to measure the efficiencies of 24 power plants in a European country. Recently, Feroz et al. [4] analyzed the environmental production efficiency rankings of Kyoto Protocol nations and the relationship between a nation's ratification status and its environmental production efficiency rankings.

2 Eco-Efficiency

In environmental economics, the term eco-efficiency was coined by the World Business Council for Sustainable Development (WBCSD) in its 1992 publication "Changing Course." It is based on the concept of producing more goods and services using fewer resources and creating less waste and pollution. According to the WBCSD definition, eco-efficiency is achieved through the delivery of "competitively priced goods and services that satisfy human needs and bring quality of life, while progressively reducing environmental impacts of goods and resource intensity throughout the entire life-cycle to a level at least in line with the earth's estimated carrying capacity."

3 A CCR Model of DEA

A CCR model[1] is used as the analytic method for eco-efficiency evaluation. Suppose there are n DMUs: $DMU_1, DMU_2, ..., DMU_n$.

Some common input and output items for each of these are $j = 1, ..., n$. Measuring the efficiency of each DMU needs n optimizations, one for each DMU_j. We let the DMU_j to be evaluated on any trial be designated as DMU_o where o ranges over $1, 2, ..., n$. We solve the following fractional programming problem to obtain values for the input "weights" (v_i) ($i = 1, ..., m$) and the output "weights" (u_r) ($r = 1, ..., s$) as variables.

[1]Charnes et al. [1] initially proposed their model, which has become an important example of a DEA model.

$$Max\, h_0 = \frac{\sum_{r=1}^{s} u_r y_{ro}}{\sum_{i=1}^{m} v_i x_{io}}$$

$$s.t. \frac{\sum_{r=1}^{s} u_r y_{rj}}{\sum_{i=1}^{m} v_i x_{ij}} \le 1, \quad j = 1,\ldots, n$$

$$u_r \ge \varepsilon > 0, \quad r = 1,\ldots, s$$

$$v_i \ge \varepsilon > 0, \quad i = 1,\ldots, m$$

The constraints mean that the ratio of "virtual output" vs. "virtual input" should not exceed 1 for every DMU. The objective is to obtain weights (v_i) and (u_r) that maximize the ratio of DMU_o, the DMU being evaluated. The optimal object value θ is at most 1 by virtue of the constraints. Because fractional programming is non-linear and non-convex, we replace the above fractional programming by linear programming.

4 Methodology and Data

Suppose we have *n* DMUs each consuming m inputs and producing p outputs. The outputs corresponding to indices 1, 2, …, k are desirable and the outputs corresponding to indices k+1, k+2, …, p are undesirable outputs. Desirable outputs need to be produced as much as possible with the least production of undesirable outputs. Therefore, eco-efficiency will focus on minimizing undesirable outputs. According to the definition of eco-efficiency in Sect. 3, undesirable output might be used as an input.

The dataset of total 26 fossil-fueled power plants operated by five different subsidiary power companies of KEPCO (Korean Electric Power Corporation) are observed as Decision Making Units (DMU). Tables 1 and 2 summarize the data.

Table 1 Data for input/output

Inputs	*Mean*	*Standard dev.*	*Max*	*Min*
Capacity(MW)	1,364	1,192.1	4,000	105
Labor(man)	371	254.7	371	65
GHG(CO_2ton)	1,249,554	1,612,879.3	5,350,879	2,931
Outputs	*Mean*	*Standard dev.*	*Max*	*Min*
Sales(0.1bil won)	5,174	4,205.9	12,848	312
Generation(GWh)	8,328	9,170.9	29,354	63

GHG (Greenhouse Gas): Units of CO_2, CH_4 and N_2O were expressed in terms of equivalent tons of CO_2 based on the GWPs (Global Warming Potentials) of the 2nd IPCC (1995)

Table 2 Correlations of input/output

Correlation	Capacity	GHG	Labor	Sales	Generation
Capacity	1	0.05	0.78	0.89	0.96
GHG	0.05	1	−0.03	0.41	−0.07
Labor	0.79	−0.03	1	0.64	0.79
Sales	0.89	0.41	0.64	1	0.85
Generation	0.96	−0.07	0.79	0.85	1

Table 3 Eco-efficiency and rank

DMU	Eco-efficiency	Rank	DMU	Eco-efficiency	Rank
DMU21	1	1	DMU25	0.8026	14
DMU17	1	1	DMU2	0.6578	15
DMU10	1	1	DMU20	0.6417	16
DMU24	1	1	DMU11	0.6117	17
DMU6	1	1	DMU4	0.6023	18
DMU22	1	1	DMU13	0.5594	19
DMU12	0.9960	7	DMU18	0.5280	20
DMU16	0.9877	8	DMU26	0.4916	21
DMU1	0.9647	9	DMU3	0.4617	22
DMU8	0.9591	10	DMU9	0.4608	23
DMU5	0.9566	11	DMU23	0.4541	24
DMU15	0.9168	12	DMU14	0.4305	25
DMU7	0.8710	13	DMU19	0.2533	26

Table 4 The comparison between actual and target values

	Actual	Target	Potential Improvement (%)
Capacity	1,150	291.26	−74.67
GHG emission	372,555	94,356.47	−74.67
Labor	331	83.83	−74.67
Sales	1,559	1,559	
Generation	637	1,123.19	76.33
Pair group		DMU 10, 21, 24	

5 Result of Evaluation

In the analysis, the efficiency result is described above according to the CCR input-oriented DEA model[2] of the plants. DMU nos. 21, 17, 10, 24, 6 and 22 are on the frontier with associated efficiency vales of 1. On the other hand, DMU 19 has the worst efficiency. This suggests that DMU 19 needs to alter its operation to achieve less environmentally damaging impacts.

[2] For more explanation, see Data Envelopment Analysis, Chap. 3 written by Cooper [2].

Table 3 shows us how DMU 19 might engage in such alterations. The pair-wise grouping consisting of DMUs 10, 21, 24, on the one hand, and DMU 19 on the other, offers benchmarking values for DMU 19. Comparable values from the benchmarked group suggest that DMU 19 may be able to reduce 74.67% of capacity, GHG emission and labor, and increase 76.33% of power generation to be eco-efficient (Table 4).

6 Conclusion

In this paper we presented an evaluation of eco-efficiency and target values for potential improvements. While the traditional analyses using DEA do not consider undesirable outputs, we explicitly considered undesirable outputs. This offers one means of evaluating the eco-efficiency of power plants with a view towards identifying eco-inefficient plants and suggesting measures to close the gap with their more efficient peers. Thus, analyses of this type can be a useful addition to the formulation of policies to combat climate change. As a topic for further research, it would be useful to broaden the analysis to include foreign fossil-fueled power plants along with Korean plants.

Acknowledgments We are grateful to Professor Ho Jung Park at Korea University for sharing data and to Professor Su Duk Kim, who served as our advisor and helped to edit this paper.

References

1. Charnes A, Cooper WW, Rhodes E (1978) Measuring the efficiency of decision making units. Eur J Oper Res 2:429–444
2. Cooper WW (1999) Data envelopment analysis: a comprehensive text with models, applications, references and DEA-solver software. Kluwer, Dordrecht, pp 41–71
3. Fare R, Grosskopf S, Tyteca D (1996) An activity analysis model of the environmental performance of firms: application to fossil fuel fired electric utilities. Ecol Econ 18:161–175
4. Feroz E, Raab LR, Ulleberg GT, Alshrif K (2009) Global warming and environmental production efficiency ranking of the Kyoto Protocol nations. J Environ Manage 90:1178–1183
5. Golany B, Roll Y, Rybak D (1994) Measuring efficiency of power plants in Israel by Data Envelopment Analysis. IEEE Trans Eng Manage 41(3):291–301
6. Park SU, Lesourd JB (2000) The efficiency of conventional fuel power plants in South Korea: a comparison of parametric and non-parametric approaches. Int J Prod Econ 63:59–67
7. Pekka JK, Lubtacik M (2004) Eco-efficiency analysis of power plants: an extension of data envelopment analysis. Eur J Oper Res 154:437–446

An Analysis of Energy Efficiency Using DEA: A Comparison of Korean and Japanese Economic Regions

Jayeol Ku

Keywords DEA • Energy efficiency • Economic region

1 Introduction

Reflecting growing concerns regarding global warming and climate change, the Japanese Prime Minister Fukuda announced a climate change policy named the "Fukuda Vision" in 2008. Japan set out a long term plan to reduce its carbon emissions by 60–80% by 2050, and proposed a sectoral approach for improving energy efficiency. The Korean government also announced a new paradigm for "Low-Carbon, Green Growth," emphasizing sustainable economic growth by the improvement of energy efficiency to reduce carbon emissions. It is clear that energy efficiency improvement has become a main key to environmental management policy for reducing greenhouse gas (GHG) emissions.

For the purposes of conducting a comparative economic study involving both countries, reliance on administrative districts vis-à-vis economic districts is less than adequate. Recently, Korean regions have been regrouped into seven economic regions which aim to enhance the competitiveness of regions' efficiencies through inter-regional networking and cooperation.[1] Also, Japan has already finished discussion of its National Land Sustainability Plan. The "Wide-area Regional Plan," in particular, provides a regrouping of the 47 administrative districts into 10 economic regions.[2] These economic regions are set against similar political-administrative backgrounds in Korea and Japan, thus facilitating the use of these units for a comparative study.

J. Ku (✉)
Department of Energy Studies, Division of Energy Systems Research, Ajou University, San 5 Woncheon-Dong, Youngtong-Gu, Suwon, Korea
e-mail: ken9410@gmail.com

[1] 5+2 Economic Region: Seoul Metropolitan, Kangwon, Chungcheong, Honam, Dongnam, Daegyeong, Jeju Area.

[2] Tokyo Metropolitan Area, Kinki, Chubu, Tohoku, Hokuriku, Chugoku, Shikoku, Kyushu Area. Prefectures comprised in each region are different from traditional delineations. Details are available at http://www.mlit.go.jp/kokudokeikaku/zs5-e/part3.html.

T. Yao (ed.), *Zero-Carbon Energy Kyoto 2009*, Green Energy and Technology,
DOI 10.1007/978-4-431-99779-5_13,

Expressing energy intensity as a direct ratio of the energy input to GDP is widely used to indicate energy efficiency. However, the energy efficiency indicator in conjunction with labor and capital can provide useful insights [11]. Hu and Wang [6] also proved that a TFEE (total-factor energy efficiency) index computed with production factors is more reasonable than a PFEE (partial-factor energy efficiency) index. Hu and Honma [5] analyzed the energy efficiency of the Japanese administrative regions by employing DEA (Data Envelopment Analysis). However, these results are not necessarily informative for newly designed regional policies.

This paper analyzes the energy efficiency of economic regions in Korea and Japan using DEA. This is in contrast to much of the extant research which makes use of political-administrative units. Results provide an energy saving target and TFEE index for indicating the efficiency level of regional energy use.

2 Methodology and Data Description

Data envelopment analysis (DEA) involves the use of linear programming methods to construct a non-parametric piece-wise surface (or frontier) over the data. This paper uses input-orientated measures following Farrell [4] and the constant returns to scale (CRS) DEA model [3]. First, let us define the notation. There are data on K inputs and M outputs for each of N objects. The i-th object is represented by the column vectors x_i and y_i, respectively. The $K \times N$ input matrix, X, and the $M \times N$ output matrix, Y, represent the data for all N forms.

$$\mathrm{Min}_{\theta,\lambda}\theta \text{ such that } -y_i + Y\lambda \geq 0,\ \theta_{x_i} - X\lambda \geq 0,\ \lambda \geq 0,$$

where θ is a scalar and λ is a $N \times 1$ vector of constants. The value of λ obtained will be the efficiency score for the i-th object.

Efficiency is defined by the distance from the "best practice" production frontiers. By comparing the relative practice of various inputs and outputs in different objects, we can identify the energy saving target for those not on the frontier. The Energy saving target is calculated by summation of slack and radial adjustments of energy input in the DEA model. Then a total-factor efficiency indicator can be provided as follows:

$$\mathrm{TFWW}_{(i,t)} = 1-(\text{Energy Saving Target}_{(i,t)}/\text{Actual Energy Input}_{(i,t)}).$$

A dataset of five economic regions in Korea and eight regions in Japan for the period 1997–2006 is constructed [1,2,7–10].[3] There are two factors of production (labor and capital) and six energy factors (electric power, gasoline, kerosene, heavy oil, diesel and LPG) as inputs. The Gross Regional Domestic Product (GRDP) is the only output. The correlation coefficients of the input and output are all positive. The sources of data are disclosed in the references.

[3] Small-sized economic regions are excluded: Kangwon and Jeju areas in Korea; Hokkaido and Okinawa areas in Japan.

3 Empirical Results

Table 1 shows the overall technical efficiency scores for the economic regions in Korea and Japan from 1997 to 2006. Most of the Japanese economic regions have high efficiency scores compared with Korean ones for the entire period. The Tokyo Metropolitan and Kinki areas are the most efficient regions. These regions are regarded as the benchmark for the energy saving target for other regions and their efficiency is scored as one.

No Korean economic regions ranked among the highest 20% in terms of efficiency while Japan had no regions which ranked in the lowest 40%. Although all regions in Korea scored poorly against Japanese regions, their efficiency has been increasing since 1998 partially due to the increasing rate of energy consumption being lower than the rate of economic growth.

The figures for total-factor energy efficiency in electric power of Korean and Japanese regions are shown in Table 2. According to TFEE index, all economic regions in Korea should reduce their electric power usage by more than half of current usage to reach efficient levels in energy use. Both the Tokyo Metropolitan and Kinki areas use great amounts of energy because they are the main social and industrial areas. Nevertheless, they are still on the frontier in terms of total-factor energy efficiency in electric power, in keeping with the results seen for overall technical efficiency. The TFEE index figures for other energy factors, such as gasoline, kerosene, heavy oil, diesel and LPG, demonstrate similar results.

4 Conclusion

This paper analyzes energy efficiency using DEA and compares the TFEE of economic regions as defined by new national plans of Korea and Japan. Data on energy sources and other production factors are analyzed in a DEA model for the period of 1997–2006. GRDP is considered as the only output.

Even if Japan, in general, is regarded to be one of the most energy efficient countries, this paper shows that Japanese regions, in comparison with the Tokyo Metropolitan Area, have additional energy saving potential.

While it is generally said that Korea has up-to-date technology in the energy-intensive industries, most Korea economic regions show low energy efficiency relative to their Japanese counterparts. It may be that low efficiency results from a distorted price structure and the lack of additional efforts for the improvement of energy efficiency. The Japanese economic regions, particularly the Tokyo Metropolitan Area, could be a good benchmark for Korean regions.

Further plans for improvement of energy efficiency can be taken at the regional level. Toward this end, industrial sector analysis at the regional level would provide better insights for improving energy efficiency. Such an analysis would require a more detailed dataset, and thus remains a task for further research.

Table 1 Overall technical efficiency scores of economic regions in Korea and Japan

	1997	1998	1999	2000	2001	2002	2003	2004	2005	2006
Seoul Metropolitan	0.870	0.737	0.767	0.774	0.765	0.783	0.752	0.760	0.807	0.878
Chungcheong	0.781	0.661	0.698	0.694	0.674	0.684	0.687	0.725	0.718	0.754
Honam	0.837	0.694	0.692	0.669	0.652	0.643	0.697	0.703	0.667	0.715
Dongnam	0.832	0.773	0.795	0.776	0.780	0.781	0.782	0.840	0.842	0.853
Daegyeong	0.718	0.606	0.633	0.636	0.634	0.636	0.666	0.754	0.764	0.776
Tokyo Metropolitan	1.000	1.000	1.000	1.000	1.000	1.000	1.000	1.000	1.000	1.000
Kinki	1.000	1.000	1.000	1.000	1.000	1.000	1.000	1.000	1.000	1.000
Chubu	0.864	0.875	0.897	0.903	0.911	0.920	0.923	0.931	0.952	0.962
Tohoku	0.961	0.951	0.944	0.932	0.937	0.926	0.908	0.904	0.900	0.949
Hokuriku	0.854	0.854	0.855	0.851	0.842	0.854	0.853	0.846	0.854	0.851
Chugoku	0.844	0.843	0.847	0.838	0.839	0.852	0.855	0.847	0.855	0.868
Shikoku	0.835	0.856	0.832	0.841	0.870	0.872	0.879	0.846	0.817	0.819
Kyushu	0.832	0.842	0.834	0.849	0.863	0.881	0.901	0.866	0.869	0.874

Table 2 Total-factor energy efficiency in electric power of Korean and Japanese regions

	1997	1998	1999	2000	2001	2002	2003	2004	2005	2006
Seoul Metropolitan	0.521	0.491	0.504	0.489	0.469	0.484	0.457	0.456	0.485	0.525
Chungcheong	0.362	0.357	0.368	0.348	0.320	0.322	0.311	0.324	0.310	0.311
Honam	0.382	0.340	0.336	0.316	0.300	0.305	0.341	0.339	0.317	0.340
Dongnam	0.332	0.344	0.355	0.339	0.337	0.341	0.341	0.375	0.379	0.378
Daegyeong	0.293	0.262	0.273	0.268	0.267	0.269	0.286	0.332	0.338	0.343
Tokyo Metropolitan	1.000	1.000	1.000	1.000	1.000	1.000	1.000	1.000	1.000	1.000
Kinki	1.000	1.000	1.000	1.000	1.000	1.000	1.000	1.000	1.000	1.000
Chubu	0.724	0.731	0.769	0.769	0.775	0.777	0.770	0.772	0.779	0.771
Tohoku	0.961	0.951	0.923	0.888	0.898	0.877	0.839	0.833	0.823	0.874
Hokuriku	0.743	0.740	0.731	0.706	0.701	0.700	0.688	0.673	0.672	0.659
Chugoku	0.627	0.622	0.619	0.602	0.603	0.617	0.623	0.601	0.591	0.597
Shikoku	0.636	0.650	0.639	0.650	0.670	0.676	0.681	0.654	0.630	0.629
Kyushu	0.793	0.803	0.796	0.818	0.819	0.843	0.862	0.822	0.827	0.833

References

1. Agency for Natural Resources and Energy, Regional Energy Consumption Statistics (Japanese only). http://www.enecho.meti.go.jp/
2. Cabinet Office, Government of Japan, Annual Report on Prefectural Accounts. http://www.esri.cao.go.jp/
3. Charnes A, Cooper WW, Rhodes E (1978) Measuring the efficiency of decision making units. Eur J Oper Res 2:429–444
4. Farrell MJ (1957) The measurement of productive efficiency. J R Stat Soc Series A 120(Part 3):253–290
5. Hu JL, Honma S (2008) Total-factor energy efficiency of regions in Japan. Energy Policy 36:821–833
6. Hu JL, Wang SC (2006) Total-factor energy efficiency of regions in China. Energy Policy 34:3206–3217
7. Japan LP Gas Association. http://www.j-lpgas.gr.jp/
8. Korea National Oil Corporation, Petroleum Demand & Supply Information System (Pedsis). http://www.pedsis.co.kr/
9. Korea National Statistical Office, Korean Statistical Information Service (KOSIS). http://www.kosis.kr/
10. Ministry of Economy, Trade and Industry, Yearbook of Mineral Resources and Petroleum Product Statistics. http://www.meti.go.jp/
11. Patterson MG (1996) What is energy efficiency? Concepts, indicators and methodological issues. Energy Policy 24:377–390

The Role of Nuclear Power in Energy Security and Climate Change in Vietnam

Dinhlong Do, Il Hwan Ahn, and Suduk Kim

Abstract The purpose of this paper is to discuss the role of nuclear power in terms of energy security and climate change in Vietnam. To facilitate comparison with Vietnam's energy development plan, the paper also discusses the Korean experience from various perspectives: past history of nuclear energy supply among other energy sources, energy consumption, energy policy and nuclear power development, consumer reactions and corporate social responsibility. The study shows that nuclear power will play an important role in Vietnam's energy security and greenhouse gas mitigation (GHG) in the future. However, the paper also argues that such an ambitious nuclear development plan may also contain risks due to poor infrastructure and the lack of human resources. As a conclusion, a more cautious nuclear power development plan can be regarded as suitable for Vietnam.

Keywords Climate change • Energy security • Nuclear power

1 Introduction

Energy security can be defined as the adequacy, reliability and affordability of energy supply to support the functioning of the economy and social development. In other worlds, energy security is the provision of sufficient energy to support economic and social activity with minimal disruption and at reasonable prices. Vietnam, a country that has emerged as one of the most active economies in the world in recent economic history, is also facing a visible set of challenges relating to energy security and climate change. An oil and gas production slowdown, electricity shortages, coal exploitation difficulties, and rapid energy demand will lead

D. Do, Il.H. Ahn, and S. Kim (✉)
Department of Energy Studies, Division of Energy System Research, Ajou University, San 5, Woncheon Dong, Yeong Tong Gu, Suwon, Korea
e-mail: tuebadolong@yahoo.com; ahnih4336@empal.com; suduk@ajou.ac.kr

T. Yao (ed.), *Zero-Carbon Energy Kyoto 2009*, Green Energy and Technology, DOI 10.1007/978-4-431-99779-5_14, © Springer 2010

Vietnam to become a net importer of energy after the year 2010 [3]. In this context, energy security has risen to be one of the top priorities of Vietnam's national energy policy. Moreover, although Vietnam's energy consumption is still small with 25.9 million tons of oil equivalent (MTOE) in 2007, the country will be significantly affected by climate change. The objectives of this paper are to discuss the role of nuclear power in energy security and climate change in Vietnam, identify barriers to nuclear power development and finally, by referring to Korean experiences, suggest a more cautious nuclear power development plan for Vietnam.

2 The Challenge of Sustainable Energy Supply

Energy demand and energy supply gap: Vietnam is a net exporter of energy with 15 million tons of crude oil and 31 million tons of coal exported in 2007. In fact, the country has imported oil products because the first refinery plant has just begun operations this year. Recently, due to a shortage of electricity, Vietnam had to import power from China amounting to 550 MW in 2007. Moreover, Vietnam has also imported about 200,000 tons of fat coal per year for industry production for the last three years. By the year 2025, it is forecasted that Vietnam will need to increase its primary energy supply by at least three to four times and its electricity generation by six or seven times the levels seen in the year 2007. A comparison of primary energy demand and supply shows that after the year 2010, in the base scenario (BS) and high scenario (HS), the capacity of energy exploitation will be lower than primary energy demand. The amount of energy shortage will come to 12 MTOE in 2015 and 41.3 MTOE in 2020 [4]. Given the current statistics of energy access and shortages and the likely needs for energy in the future, Vietnam faces a formidable challenge in meeting its energy needs and providing adequate and affordable energy to all sections of society in a sustainable manner.

Electricity balance: Vietnam's electricity demand in the period 2006–2015 will increase 17% per year in the BS and 20% per year in the HS. In 2020, Vietnam will need more than 400 billion kWh, about six times of electricity generation in 2007 even after taking into account promotion of energy conservation. The future electricity generation mix of Vietnam will be governed mainly by the supply potential of the indigenous energy resources. Table 1 shows that indigenous power generation will be lower than electricity demand in the future.

In view of the limitations of indigenous energy resources, import of electricity from neighboring countries should be planned. Although there are some advantages in importing electricity from neighboring countries, such as the possibility of exploiting hydropower sources in these countries without the need for investment capital and no environmental pollution, there are some disadvantages, including possible limitations in electricity import and the increase of trade deficit. In addition, import of gas through the regional pipeline and liquefied natural gas (LNG) should also be considered. Nevertheless the limitation of gas import and the dependence on energy prices become noticeable hindrances. New and renewable energy (RE) is considered as an energy resource of the future. However, the high cost

Table 1 Electricity balance

Year	2015 Domestic supply	Electricity (10^9 kWh)	2020 Domestic supply	Electricity (10^9 kWh)
Coal (Mt)	42.1	105.3	56.8	142.1
Gas (10^9 m^3)	13.1	65.5	14.2	75
Hydro (MW)	16,300	58.7	19,500	60
RE (MW)	1,420	4.7	2,800	10.7
Total supply		234.2		287.8
Total demand		241/297		403/514
+/– (10^9 kWh)		–6.8/–62.8[a]		–115.2/–226.2

[a] Base scenario / High scenario Source: Ministry of Trade and Industry (2008) [4]

of RE is still a big obstacle and the fact that some RE sources depend on weather conditions makes it difficult for large scale electricity production. Furthermore, Vietnam may need to import up to 26.6 million tons (Mt) of coal in 2015 and 111 Mt in 2020 [6]; import of coal from Australia and Indonesia can be an appropriate solution. On the other hand, too much coal import will also damage the environment at ports, emit more polluted substances and make a larger trade deficit. In this context, nuclear power development can significantly help in securing energy supply, diversifying energy supply resources, reducing the energy import dependence and in creating energy security in the long run.

3 Plan for Nuclear Power Development and Identification of Barriers to Vietnam's Nuclear Power Development

Plan for nuclear power development: Vietnam will have four nuclear power plants (NPPs) with a total of 11,000 MW within five years from 2020 to 2025 [3]. In 2008, faced with the risk of electricity shortage and the price increase of fossil fuels, the Vietnamese government decided to accelerate and double the scale of the first nuclear power project that includes four 1,000 MW-class NPPs with total capacity of 4,000 MW. The four units are expected to be commissioned from 2019 to 2021.

Reducing Greenhouse Gas (GHG) Emissions: When the first two NPPs come into operation, they can produce about 30 billion kWh per year. Annually, NPPs will lessen CO_2 emission by about 29.5 million tons compared to coal-fired power generation. From 2025, with over 90 billion kWh electricity production per year, nuclear power can annually reduce CO_2 up to 80 million tons.

Barriers to nuclear power development:
High investment capital: Building NPPs requires huge investment cost and long construction time with complicated technology that surpasses Vietnam's current technology level. With the capital over 2 billion USD for one unit of 1,000 MW, financing nuclear power development will be a challenge for Vietnam [5].

Infrastructure: Vietnam has one nuclear reactor with a capacity of 500 kW which has been operated since 1984 [7]. However, infrastructure for nuclear energy development is still at the primary stage with equipment deficiencies and backward technology. Although Vietnam has already approved nuclear laws, the detailed regulations for nuclear power development in Vietnam are insufficient and they lack synchronization.

Lack of human resources: In 2007, Electricity of Vietnam (EVN) had a total of 13 people who majored in nuclear power. However, they were working in different positions and mostly not directly involved in nuclear power-related activities. Totally, Vietnam has over 500 engineers and scientists who are working in the nuclear sector. However, we can say that Vietnam still lacks the necessary manpower, especially that of highly specialized nuclear experts, both in term of quantity and quality [5].

Social problems and public acceptance: Social awareness of the role of nuclear energy is insufficient. Social problems such as awareness of law execution and an inadequate understanding of nuclear safety are obstacles for nuclear power development. Recently, public acceptance problems have arisen regarding utilization of NPPs, especially near areas intended for NPPs.

4 Korean Experiences

In 2007, Korea's final energy consumption was about sevenfold that of Vietnam's with 180.54 MTOE [1]. Korea is highly dependent on foreign sources for its energy needs. In particular, its high reliance on the Middle East for oil and natural gas makes Korea vulnerable to international energy crises. In response, the Korean government has promoted a policy of decreasing the dependence on petroleum and increasing the use of LNG, nuclear energy and bituminous coal in an effort to stabilize energy supply [2]. Furthermore, stability in supply is continuously promoted by lowering the risks attendant with fluctuations in supply volume and price through diversification of the supply channels for oil and natural gas. In 1967, the Korean government decided to build its first two NPPs of 500 MW each and put them into operation in 1976. However, an increase of material prices and a shortage of construction materials delayed the project two more years. The first NPP named KORI I began commercial operation in 1978 with capacity of 587 MW and its invested capital was 4.5 times more than the initial estimate. In 1978, the share of nuclear energy was 1.5% of total primary energy supply. In spite of the efforts made by the Korean government, the second NPP was not commissioned until five years later in 1983 with capacity of 650 MW. The construction of the first NPP in Korea had to overcome strong opposition of local people by active propaganda and satisfactory compensation. At the time of starting nuclear power programs, Korea did not have sufficient domestic laws which can regulate nuclear activities. Therefore, the Korean government had applied most of the technical standards of the United States. Following the introduction of NPPs in the 1970s, Korea accumulated its nuclear technology in the 1980s, achieved independence from the import of foreign technology in the 1990s and demonstrated the advanced level of its nuclear technology capabilities in the 2000s. Currently, nuclear power has become the most important energy source in Korea. It accounts for 14.7% of total primary energy consumption, 27% of the total installed capacity with 17,716 MW and 36% of total electricity generation with 142.94 billion kWh. Four 1,000 MW-class and two 1,400 MW-class NPPs are under construction. By 2014, when six new NPPs will be completed, nuclear power capacity will grow to 24,516 MW, making it Korea's largest power supplier. Further, Korea will have an additional 8,400 MW by the year 2022 and total nuclear power capacity will be 34% of total generating facilities with approximately 33,000 MW. Therefore, as a major source of electricity generation in Korea [8], nuclear power contributes greatly to the stability of national electricity supply and energy security.

5 Implications and Conclusions

Korean experiences show that it takes several years to train an operator and Vietnam may need over ten years to train nuclear experts in order to master the nuclear technology. Moreover, it is uncertain whether foreign countries can help Vietnam train real nuclear experts, especially whether they can transfer nuclear know-how to Vietnam. As a result, it is likely that Vietnam will be dependent on foreign technology and that may lead to economic dependence and even to safety problems. In comparison with Korea, Vietnam's initial nuclear development plan is more ambitious. Building two sites of two nuclear power units of 1,000 MW each at the same time will be a big challenge for Vietnam because of the low capability of management for such a complex project, poor infrastructure including transmission system and the lack of manpower. Thus, this plan poses great safety concerns and its success is precarious. In conclusion, even though it is certain that nuclear power will play a major role in ensuring Vietnam's energy security and mitigating GHG, we would like to recommend a cautious approach to nuclear power development that progresses in the following fashion: first, introduction of the first NPP with total of 1,000 or 2,000 MW of capacity, second, cautious expansion of NPPs, striving to accumulate Vietnam's nuclear technology within 5 or 6 years, and finally, achieving independence from the import of foreign technology within 10–15 years and developing domestic nuclear technology.

References

1. KEEI (2008) Yearbook of energy statistics, 2008
2. Ministry of Commerce, Industry and Energy of Korea (2004) Energy policies of Korea
3. Ministry of Trade and Industry of Vietnam (2005) Master plan for power development in the period 2006–2015 with prospect to 2025. Institute of Energy, 2005

4. Ministry of Trade and Industry of Vietnam (2008) Vietnam's energy sector after two years joining the WTO and demand for sustainable development
5. Ministry of Trade and Industry of Vietnam (2007) Action plan for the implementation of the strategy for peaceful utilization of atomic energy up to 2020
6. Ministry of Trade and Industry of Vietnam (2008) The modified master plan of coal sector development 2006–2015 periods, vision up to 2025. VINACOMIN 2008
7. Ministry of Trade and Industry of Vietnam (2006) The national strategy for peaceful utilizations of atomic energy up to 2020
8. Ministry of Knowledge Economy of Korea (2008) The 4th basic plan of long-term electricity supply and demand (2008–2022). December 2008

Opportunities and Challenges of Renewable Energy and Distributed Generation Promotion for Rural Electrification in Indonesia

Zulfikar Yurnaidi

Abstract This paper reviews the opportunities Distributed Generation (DG) has to solve electrification problems in Indonesia, especially for rural areas. The main approach used in this paper is comparison, with a focus on other countries who share similar conditions. Renewable energy sources are examined from the perspective of sustainability and environmental friendliness. For rural areas, geographical disadvantages serve to increase the competitiveness of renewable energy and DG. An abundant resources also becomes another strong point for renewable energy utilization. Nevertheless, there are some challenges – technological, financial and social – that DG must face. Lessons from other countries provide insights and suggestions to overcome those challenges.

Keywords Distributed generation • Renewable energy • Remote area • Electrification

1 Introduction

Distributed Generation (DG) – also known as dispersed generation or decentralized generation – refers to the concept of generating electricity near the user. Recently, the re-emergence of DG technology has become an issue in developed countries, mainly because of the liberalization of electricity markets and environmental concerns [5]. As for developing countries, DG is perceived as an alternative solution for electrification of geographically disadvantaged rural and remote areas [4].

Indonesia by its nature has problems of electricity penetration. As the largest among archipelagic countries with around 17,000 islands, Indonesia faces a severe electrification problem due to scattered populations across varying geographic conditions. Here DG could offer a solution for large countries with scattered rural

Z. Yurnaidi (✉)
Department of Energy Studies, Division of Energy Systems Research, Ajou University, San 5, Woncheon-Dong, Youngtong-Gu, Suwon, Korea
e-mail: viczhoel@ajou.ac.kr

T. Yao (ed.), *Zero-Carbon Energy Kyoto 2009*, Green Energy and Technology, DOI 10.1007/978-4-431-99779-5_15, © Springer 2010

populations such as Indonesia. Because electricity is generated at the consumer end, thereby avoiding transmission and distribution costs, DG offers a better solution than directly connecting the consumer to a national grid [9].

The World Alliance for Decentralized Energy [1] defines "decentralized energy" – yet another term of DG – as the production of electricity at or near the point of use, irrespective of size, fuel or technology. DG can be grid-connected (on-grid) or stand alone (off-grid) and can be powered by wide variety of fossil fuels. In general, DG technologies can be broken into two divisions: first, cogeneration or combined heat and power (CHP); second, renewable energy systems (RES) and energy recycling technologies. This paper will focus on the latter.

2 Opportunities of Renewable Energy and Distributed Generation

In February 2006, the government of Indonesia enacted Presidential Decree No. 5/2006 for National Energy Policy. In this policy, the share of new and renewable energies in energy mix will be increased from around 5% in 2007 to more than 17% in 2025, which are composed of bio-fuel (5%), geothermal (5%), coal liquefaction (2%), and others, including nuclear (5%). Supporting the policy, abundant resources are already available in Indonesia. Table 1 summarizes the national energy potency of non-fossil energy. The utilization of renewable sources can be increased to diversify the energy supply mix, ensuring both energy security and sustainability. The global concern of climate change also gives strong rationale for renewable energy utilization since it produces little or no emissions.

For large, dispersed countries with vast rural areas, renewable sources can be utilized in a DG system to generate electricity. One developing country with significant experience in this regard is India. Currently, India has nearly 600,000 villages, yet only 44% of India's 138 million rural households use electricity for lighting while the remaining 55% predominantly use kerosene, which has lower efficiency. The government of India is keen on increasing the share of RES in power generation capacity by an additional 12,000 MW by 2012.

Case studies of different Renewable Energy Systems (RES) involving decentralized power generation systems – DGs – can be differentiated into bio-energy, solar

Table 1 National non-fossil energy potency [2]

Non-fossil energy	Resources	Power equivalent (GW)	Installed capacity (GW)
Hydro	845	75.67	4.2
Geothermal	219	27	0.8
Mini/micro hydro	0.45	0.45	206
Biomass	49.81	49.81	0.3
Solar	4.8 kWh m^{-2} per day	–	0.01
Wind	9.29	9.29	0.0006
Uranium (nuclear)	24.112 ton	3 (11 years)	–

power and other renewable energies. Based on surveys in several solar villages in India, these are some of the benefits people received through solar PV: power available for longer periods of study, trade and business, increased income from extended hours of work, and time saved while cooking.

In addition to government plans, voluntary agents and groups also contribute to development and implementation of various DG projects. Some successful initiatives, such as Decentralized Energy Systems India Power, Barefoot College in Tilonia, and Nimbkar Agriculture Research Institute in the Phaltan and Sundarbans Islands in West Bengal demonstrate the potential of DG projects [3]. The basic idea is to develop "independent rural power producers". The goal is not simply the production of power, but also to help in economic development of villages.

Rural electrification can be started by implementing small-scale Renewable Energy Technologies (RETs) to ensure minimum electricity needs such as lighting and water pumping. But afterwards, it can be expanded to a mini-grid with higher capacity. The greater supply of electricity can be used to support small industry activity or enhancement of basic agriculture activity, which could give a better chance for the villagers to boost their economic activity. An example is the usage of electronic milk testing which has the ability to correctly estimate the fat content of milk. The milk produced by farmers assisted by this technology will have a higher selling price in the market. Furthermore, this DG system can also be connected to the main grid which could provide better stability and quality of electricity supply. This concept can be summarized by the framework provided in Fig. 1 to depict DG schemes for electricity provision in remote rural areas.

As with other remote rural areas, the concept of renewable based DG can be applied as well to electrification of isolated islands. Several projects have illustrated

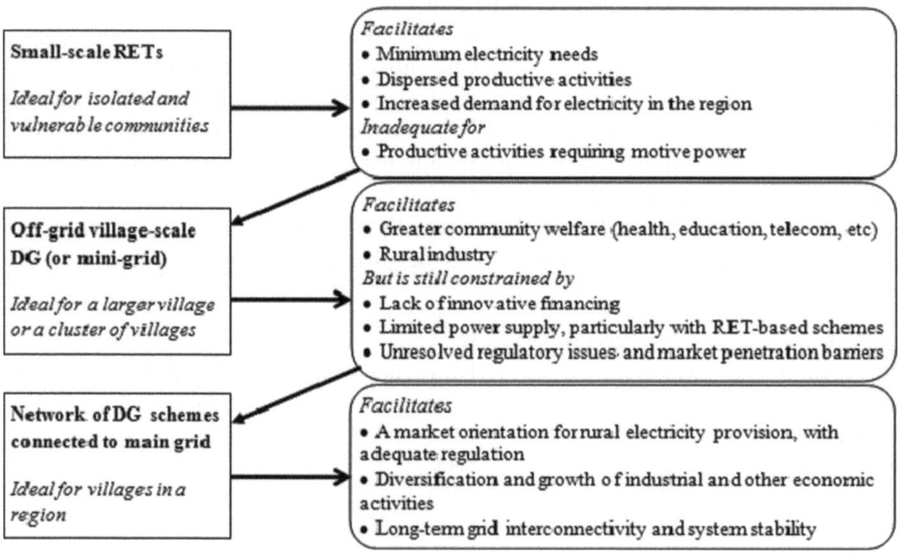

Fig. 1 Framework of DG schemes for electricity provision in remote areas

this. Renewislands [6] is a concept of integrating intermittent renewable energy supply (RES), fuel cell and hydrogen infrastructure to promote greater innovative decentralized power systems penetration, especially in islands. Several projects in the Canary Islands (Spain), Aero Islands (Denmark), Greek islands, Madeira Islands, Azores Archipelago and Cape Verde Islands offer good examples of the concept in practice. It shows that the deployment of renewable energy in islands is a great opportunity to test new technologies, where more conventional technologies are costly. As Indonesia consists of so many islands, this study provides valuable insight into dealing with issues of island electrification.

Another technology that can be utilized for rural electrification is a hybrid system utilizing both fossil and renewable sources. A mini-grid system in the Kythnos Island utilizes fossil (diesel) and renewable (photovoltaic) energies as well as a storage system (battery). Mitra et al. [8] shows the successful operation of the system using primary plant data analysis. This system is also an example of the second stage of the framework described in Fig. 1. There is a possibility in the future that this system will be included in the main grid of Greece.

The analysis above which is based on various researches gives meaningful insights to help Indonesia with its problem of electrification rate. India's experience gives lessons that the DG system can provide a solution for penetrating electricity to rural areas. It also can enhance the standard of living of the society living there. As an archipelagic country, the experience of several other islands in the world shows that the renewable based DG technologies can be regarded as solution of electricity penetration to the 17,000 islands Indonesia possesses. Therefore, Indonesia must start to utilize its potential in power generation using renewable energies, especially to fulfill the target of electrification rate as high as 90% in 2025.

3 Challenges of Renewable Energy and Distributed Generation

The previous section elaborated the "benefits" resulting from DG. However, DG systems certainly entail "costs". Here are the major issues: high financial cost; economic efficiency; environmental protection; energy security; power quality, including system frequency and voltage level; and connectivity issues. Other than that, the lack of supporting institutions and infrastructures in Indonesia also creates difficulties in implementing DG.

As mentioned before, in the case of rural electrification the renewable technologies can have better competitiveness. Pre-study and simulation of a DG project should be done before implementing the project. For example, a study using the HOMER (Hybrid System Optimization Model for Electric Renewables) by Rehman et al. [7] shows the specific economic feasibility of a wind–diesel hybrid system. Conditions of more than 6.0 m s^{-1} wind speed and more than \$0.1 per liter of fuel price are the pre-requisite of the hybrid system. It means that with proper handling, the challenges above can be overcome for specific cases.

Based on implementation cases in India, there are some solutions that are proven useful to solve those problems. First is the involvement of local community. In the past, the PV project in Indonesia failed to survive since there was no proper anticipation of maintenance system. As done by several voluntary agents in India, the awareness of DG's importance and benefits must come first.

A second important factor involves non-government institutions. Public initiatives must be raised. Also, the role of entrepreneurs cannot be ignored. Some local entrepreneurs who could manage the generation and distribution of electricity may as well sell it to others in a profitable way. The main purpose of non-government institutions' involvement is to motivate, train and assist the villager in how to implement and manage a DG system in their area.

The third factor is government's role. The government must optimally use its authority to maintain the best climate for the development of renewable energy and DG. It can be done by creating incentives for the players in the field and creating policies in favor of renewable energy and DG. Since ensuring all citizens can enjoy electricity is government's responsibility, naturally the government should become the most active player in this field.

Lastly, we should consider the role of international cooperation. The role of it is to patch the holes left by other national institutions, particularly with technology and financing. The global awareness of climate change can be used in favor of this. As instituted under the Kyoto Protocol, the Clean Development Mechanism (CDM) provides a mean for cooperation between developed and developing countries. Renewable energies are included and it creates paths of technology and fund transfer to finance DG projects.

4 Concluding Remarks

Indonesia has potential to implement Distributed Generation (DG) systems utilizing renewable sources that are abundantly available. Lessons from other countries show that it can be an effective solution for remote rural areas' electrification. Yet the participation of community, institutions, government, and international bodies must be optimally maintained to achieve successful implementation.

Acknowledgment I would like to express my sincere acknowledgement to Prof. Suduk Kim, my advisor at Ajou University for his nice comments and great help in improving this paper.

References

1. WADE (2003) Guide to decentralized energy technologies. WADE, Edinburgh
2. Ministry of Energy and Mineral Resources Republic of Indonesia (2006) Blueprint of National Energy Management 2006–2025. Government of Indonesia, Jakarta
3. Sharma DC (2007) Transforming rural lives through decentralized green power. Futures 39:583–596

4. Chaurey A, Ranganathan M et al (2004) Electricity access for geographically disadvantaged rural communities: technology and policy insights. Energy Policy 32:1693–1705
5. Pepermans G, Driesen J et al (2005) Distributed generation: definition, benefits and issues. Energy Policy 33:787–798
6. Chen F, Duic N et al (2007) Renewislands: renewable energy solutions for islands. Renewable Sustain Energy Rev 11:1888–1902
7. Rehman S, El-Amin IM et al (2007) Feasibility study of hybrid retrofits to an isolated off-grid diesel power plant. Renewable Sustain Energy Rev 11:635–653
8. Mitra I, Degner T et al (2008) Distributed generation and micro-grids for small island electrification in developing countries: a review. SESI J 18:6–20
9. Hiremath RB, Kumar B et al (2009) Decentralized renewable energy: scope, relevance and applications in the Indian context. Energy Sustain Develop 13:4–10

Wind Power Generation's Impact on Peak Time Demand and on Future Power Mix

Jinho Lee and Suduk Kim

Abstract Although wind power is regarded as one of the ways to actively respond to climate change, the stability of the whole power system could be a serious problem in the future due to wind power's uncertainties. These uncertainties include intermittency and the fact that wind power cannot be relied upon to supply energy "on demand," including and especially during periods of peak demand. From this perspective, the peak-time impact of stochastic wind power generation is estimated using a simulation method that extends to 2030 based on the 3rd master plan for the promotion of new and renewable energy. Results show that the highest probability of wind power impact on peak time power supply could be up to 4.41% of total installed capacity in 2030. The impact of wind power generation on overall power mix is also analyzed up to 2030 using the screen curve method (SCM). The impact turns out to be relatively small, but the estimated investment cost to make up any lack of power generation through reliance on LNG power generation facilities is shown to be a significant burden on existing power companies.

Keywords SCM (Screening Curve Method) • Wind power generation • Peak time demand

1 Introduction

Increasing concerns about environmental problems due to climate change are driving many countries to supply electricity using new and renewable resources. The Korean government embraced the concept of "Green Growth" on August 15, 2008 to mitigate greenhouse gases and environmental pollution. Green growth is said to be accomplished by promoting green industries and it is also expected to create a number of jobs. The supply of new and renewable energy has an important role in green energy, but wind power in particular presents some inevitable problems.

J. Lee and S. Kim (✉)
Department of Energy Studies, Division of Energy Systems Research, Ajou University, San 5 Woncheon-Dong, Youngtong-Gu, Suwon, Korea
e-mail: easyknow23@naver.com; suduk@ajou.ac.kr

T. Yao (ed.), *Zero-Carbon Energy Kyoto 2009*, Green Energy and Technology, DOI 10.1007/978-4-431-99779-5_16,

These problems include intermittency in the supply of electric power, and intrinsic inability to respond to fluctuations in demand [1]. With increasing supply of new and renewable power, the uncertainties that affect the whole electrical supply system have to be explicitly considered.

In order to analyze the impact of wind power on the peak time power supply, a simulation method is used which extends to the year 2030. The long-term impact of wind power on the power mix is also analyzed by using SCM to calculate investment cost to make up the lack of power generation in terms of combined-cycle LNG power generation facilities.

In the next section, we outline our assumptions and discuss our two models. The first model is for the analysis of peak-time impact of wind power. The second model is used to analyze the impact of wind power on the future power mix. In both cases, simulation methods are applied. The results of the simulations are summarized for both cases. Conclusions follow in the last section.

2 Simulation for Analyzing the Impact of Wind Power Generation on Peak Time Power Supply and SCM

2.1 Assumptions for Simulation

Based on the 3rd Master Plan of Electric Power Demand and Supply (2006) announced by the Ministry of Knowledge and Economy, an estimation of hourly patterns of electric power demand is performed. Yearly and hourly data of electric power generation are estimated using hourly power demand data from 2008 with reference to the standard load factor reported in the National Energy Master Plan (2008). A load factor of 76.1% is applied to calculate peak load for the 10-year period of 2020–2030.

The following explains the detailed method used to estimate hourly power pattern in the future. Let L_{it}, L_{Mt}, $\bar{L}_t$, and $G_t = \sum_t L_{it}$ be hourly load, peak load, the average of hourly load, and annual power generation, respectively, at the ith hour in year t (i.e., $\bar{L}_t = G_t/8760$). We can normalize hourly load such that $I = \{I_1, I_2, ..., I_{8760}\}$ where $I_i = L_i/G_{2008}$ and $\sum L_i = 1$. First, $\hat{L}_t = G_t \times I$ is used to find preliminary future load pattern, and then we define $d\hat{L} = \hat{L} - \bar{\hat{L}}_{it}$, $d\hat{L}_{max} = \max\{d\hat{L}_{it}\}$ to make a new series of future hourly power demand pattern, $\tilde{L}_{it} = \dfrac{d\hat{L}}{\max\{d\hat{L}_{it}\}} \times (L_{Mt} - \bar{\hat{L}}_t) + \bar{\hat{L}}_t$. With this type of modification, the newly obtained hourly demand pattern satisfies both the future annual power generation and peak load announced in the 3rd Master Plan. The 3rd Master Plan for the Promotion of New and Renewable Energy contains a road map for wind power which involves expanding its capacity up to 37 times the current capacity (199 → 7,301 MW) by 2030. Based on this information, the total amount of future wind power generation is calculated.

2.2 Analyzing Peak Time Impact of Wind Power

For the convenience of our analysis, both the power curve of a wind turbine and wind speed at the national level is simulated with certain constraints. Wind speed data for the areas of Youngdeok, Jeju-hanrim, Saemangeum, Seosan, and Samcheok from Sep. 12, 2000 to Sep. 11, 2000 from the wind map of KIER (Korea Institute of Energy Research) is selected for such purposes. For practical simulation, 365 × 5 random numbers are created first to select 365 days' wind speed patterns for 24-h intervals. Selected wind speed is overlapped with the obtained power curve for numerical integration. The simulation is done 1,000 times [5].

The promotion target for new and renewable power by the government is adopted as a base scenario. A total of four scenarios for simulation purposes are used. These scenarios represent combinations of fixed or variable annual wind power generation, and original or new peak load after the impact of wind power generation. However, only the first scenario involving the combination of fixed wind power generation with its original peak case is presented here. The same is applied for the analysis of SCM. In this scenario, stochastic output of hourly wind power generation is subtracted from peak load, and then the result is summarized in terms of maximum, 97.5%, 50%, 2.5%, and minimum impact cases.

Simulation results show that even the largest estimated impact on 2010 is less than 0.33%. The comparable figures for 2020 and 2030 are 1.20% and 4.41%, respectively. Considering the reserve margin of power was only around the 7% level on 2007, this impact may be quite significant.

2.3 Analyzing the Impact of Wind Power on Power Mix Using SCM

An analysis of stochastic wind power generation and its impact on the power mix is performed using the typical method of SCM (Fig. 1). SCM is used to identify the

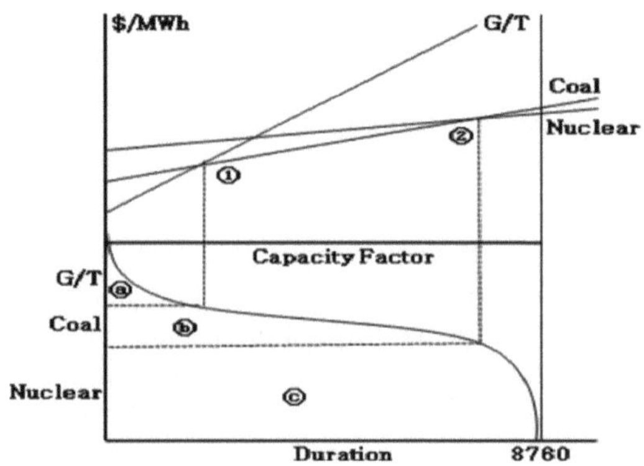

Fig. 1 Illustration of SCM

Table 1 Estimated investment cost required for each power company in 2030 (unit: million dollars/year, Exchange rate: 1,300 won/$)

KOSEP	KOMIPO	Western Power	KOSPO	EWP	Kwater	GS Power	GSEPS	Others
1,760	5,046	4,534	5,353	4,805	197	2,218	1,109	617

appropriate power mix by identifying the intersections among least-cost energy sources for power generation for a given load duration curve (LDC).

Results show that the operation of a 635 h LNG 700 MW facility, 7,227 h coal 1,000 MW facility, and a 8,760 h nuclear 1,400 MW facility are required in terms of given cost information and LDC. Considering the fact that wind power generation is treated as a must-run, its long term impact on power mix is analyzed in this model. Resulting investment cost to make up such lack of power generation in terms of combined-cycle LNG power generation facilities is also calculated. Impact of the lack of power generation due to wind power on peak load facilities is estimated to be 1.17%, intermediate load facilities, 1.52%, and base load facilities, 0.02%, respectively, in 2030. The impact is considerably small at present but it is expected that the scale of power generation required to satisfy the lack of power generation will become larger as the installed capacity becomes larger.

Additional investment cost required is estimated to prepare for the uncertainty of wind power generation. A minimum of 31.5 million dollars and a maximum of 28 billion dollars of additional investment are estimated to be required in terms of LNG 700 MW facility. Table 1 summarizes the result when the above required cost is allocated to each power company based on their current installed capacity.

3 Conclusion

In this study, the impact of wind power generation on peak-time power supply and on overall power mix is examined considering the intrinsic uncertainty of wind power. As a result, the impact of wind power on peak time supply on 2030 is turned out to be a non-negligible level of up to 4.41%. For additional investment cost, a minimum of 31.5 million dollars and a maximum of 28 billion dollars of additional investment are estimated to be required in terms of LNG 700 MW facilities. The result is also calculated for each power company to obtain a better understanding of its impact at the industry level.

Current operation of the electricity market, while treating renewable power as a must-run resource, may include additional costs for existing utilities to address the variability and uncertainty of wind power. Moreover, it is quite clear that these additional backup facilities should be prepared by all means. From this perspective, we may understand that new and renewable energy resources and existing resources are not substitutes but complements, thus differing from the conventional understanding of new and renewable energy.

References

1. Moon SI (2009) The vision of smart grid in Korea. In: The 1st conference on Electric Power Issues
2. Ministry of Knowledge and Economy (2006) The 3rd master plan of electric power demand and supply
3. Ministry of Knowledge and Economy (2008) The 1st national energy master plan 2008–2030
4. Ministry of Knowledge and Economy (2008) The 3rd master plan for the promotion of new and renewable energy
5. Kim SD, Ha JW, Park JB (2007) Probabilistic impact analysis of the power generation of wind farms on peak electricity demand. Appl Econ 9(3):39–59

Development of LiPb–SiC High Temperature Blanket

Dohyoung Kim, Kazuyuki Noborio, Takayasu Hasegawa, Yasushi Yamamoto, and Satoshi Konishi

Abstract This paper reports the development LiPb–SiC blanket concept aimed at the high performance liquid blanket to be feasible in the near future.

It is based on the current LiPb liquid tritium breeder concept with reduced activation ferritic/martensitic steel (RAFM), but with cooling panels made of SiC/SiC material that thermally and electrically insulate RAFM from LiPb. Extraction of thermal energy over 900°C is expected to be possible, that can be used for high efficiency electricity generation or thermochemical hydrogen production.

In the activities for the research of this concept, development of the SiC composite cooling panel, permeation behavior of hydrogen isotopes through SiC materials, LiPb–hydrogen chemistry, magneto-hydro-dynamic (MHD) pressure drop in the insulating SiC flow channels, neutronics analysis, etc., are studied. In addition, the experiments are carried out with LiPb at the temperature over 900°C.

Keywords LiPb–SiC blanket • Tritium breeding rate (TBR) • Magneto-hydro-dynamic (MHD) pressure drop

1 Introduction

The liquid metal breeder blanket using lithium lead (LiPb) has been explored extensively in the world and is known as an advanced blanket. This material can perform demanded abilities of blanket which can be the tritium breeding, neutron shielding and coolant by only one material due to their potential attractiveness of economy, safety and relatively mature technology base [1–5]. There are some advantages with use of LiPb as a breeder. For example, high temperature over 1000°C can be obtained and the radiation damage does not make a big difference because of liquid state. Although the blanket has an ability to supply high temperature as mentioned, high heat load should be considered especially at the surface of first

D. Kim (✉), K. Noborio, T. Hasegawa, Y. Yamamoto, and S. Konishi
Institute of Advanced Science, Kyoto University, Gokasho, Uji, Kyoto, 611-0011, Japan
e-mail: dohyoung@iae.kyoto-u.ac.jp

T. Yao (ed.), *Zero-Carbon Energy Kyoto 2009*, Green Energy and Technology,
DOI 10.1007/978-4-431-99779-5_17,

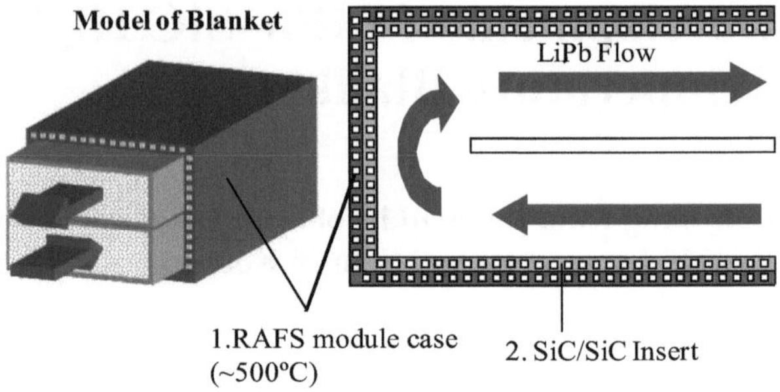

Fig. 1 The flow route of LiPb

wall which is exposed to plasma. We have adopted helium (He) as a coolant. Since previous work, we have designed and investigated a LiPb flow route as the Fig. 1 which has two opposing flow routes and is separated by SiC/SiC composite having He coolant channels from reduced activation ferritic steel (RAFS). We obtained parameters related on the velocity of He and LiPb to make a temperature at a part of border between RAFS of first wall and SiC/SiC composite less than 500°C which is regarded as a restricted temperature. On condition that thickness of LiPb is 50 cm, it estimated the local TBR of blanket at 1.25 by ANISN [6], value of which was considered not enough from the view point of blanket coverage. In our recent study, there are two ways to increase TBR, the one is increasing volume of TBR and the other is using multipliers. The former is not reasonable way from the view point of a cost of coil surrounding blankets. Then, we attempted to employ beryllium (Be) as the multiplier and investigate Be works with LiPb. And we proposed the model of blanket which has the multiplier only in front of breeder and installed at both front and back of breeder. The value of TBR was obtained more than 1.4 on a certain condition [7]. The magnetic field of fusion reactor influences on the LiPb flow. Therefore, we designed the shape of LiPb–SiC blanket and calculated the demanded flow velocity of He coolant and LiPb for the outlet temperature of LiPb over 900°C in this study. And the magnetic field strength and the pressure of LiPb were measured with SiC insert in the test section of LiPb loop.

2 Measurement Method

2.1 A Model to Design LiPb Route

The model is shown in Fig. 2. It defines the LiPb critical temperature on the side of blanket surface to the flow velocity of He, and is considered on the model with LiPb flowing from the back side to front side and to back side again of blanket. The calculated MHD pressure loss is shown as following;

$$P_{loss} = \sigma B^2 v d_1 \quad (1)$$

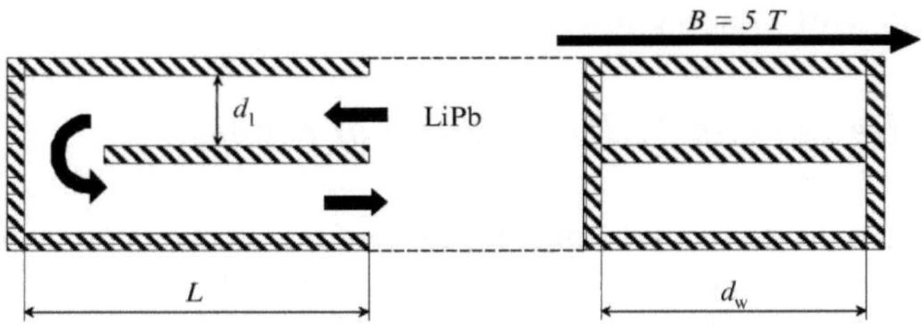

Fig. 2 The model of calculation for MHD pressure tests

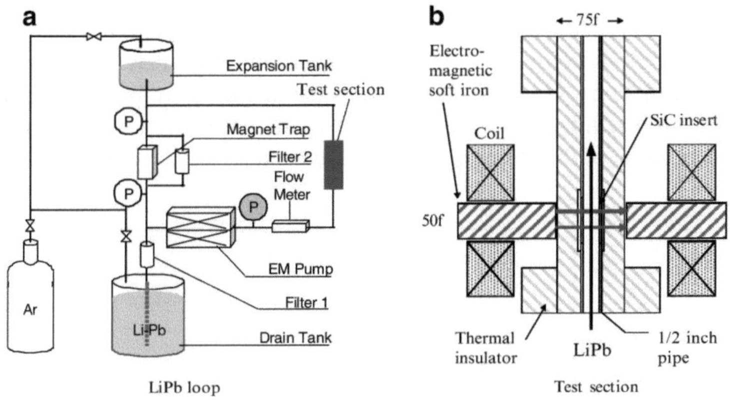

Fig. 3 Schematics of the LiPb loop for MHD

Table 1 Operating condition of LiPb loop

Temperature (°C)	350–450
Inner volume (l)	6
Flow rate (l min^{-1})	0–3
Fluid	LiPb
Structural material	SUS316

where σ is electrical conductivity, B is magnetic flux density, v is the velocity of LiPb and d1 is the height of LiPb route. The model of flow channel was investigated by the calculated MHD pressure loss and the demanded flux for getting heat.

2.2 *MHD Tests at LiPb Loop*

We installed LiPb loop to measure MHD pressure loss. Figure 3 is the schematic of the LiPb loop and we investigated the influence of SiC insert. Table 1 is operation conditions of LiPb loop for MHD.

3 Results and Discussion

3.1 The Design and MHD by Shape of Blanket

We investigated the design of blanket for LiPb of high temperature at the outlet of loop. Conditions of calculation in Fig. 2 were as the following; Total thickness of blanket was 54.5 cm, Thickness of LiPb was 50 cm, $_6$Li concentration rate was 90%, TBR was 1.2, Inlet temperature of LiPb was 300°C and Inlet temperature of He was 300°C [7]. Figure 4 is the outlet temperature of LiPb loop by each He velocity and d_1 in Fig. 2. It can obtain high outlet temperature of LiPb as decreasing d_1 and increasing He velocity which critical wall temperature is considered. Table 2 is the flow velocity of LiPb fallen below critical wall temperature on flow velocity of He by height d_1. If it lessens the height, there is a possibility which will take a high temperature. If the height is increased, fast velocity of LiPb is demanded because of the temperature increase by slow velocity on the wall surface.

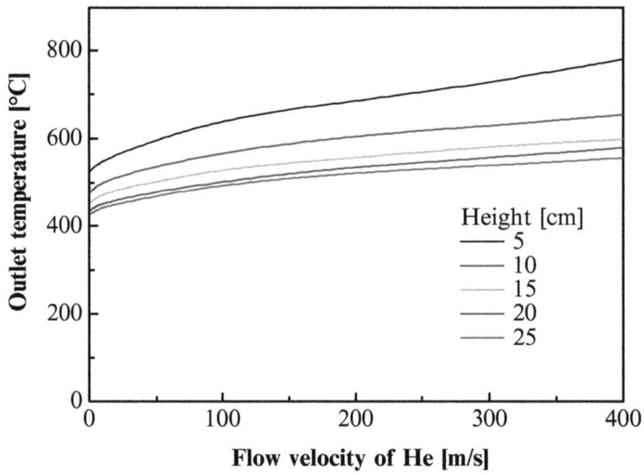

Fig. 4 Influence of He velocity on outlet temperature of Li–Pb considering critical wall temperature

Table 2 LiPb velocity by d_1 and He velocity

		Height (cm)				
		5	10	15	20	25
Flow velocity of He (cm s^{-1})	1	1.7	2.6	3.5	4.5	5.4
	10	1.5	2.2	3.0	3.9	4.7
	100	1.1	1.7	2.3	3.0	3.5
	200	1.0	1.5	2.1	2.6	3.1
	300	0.9	1.4	1.9	2.4	2.9
	400	0.8	1.3	1.8	2.2	2.7

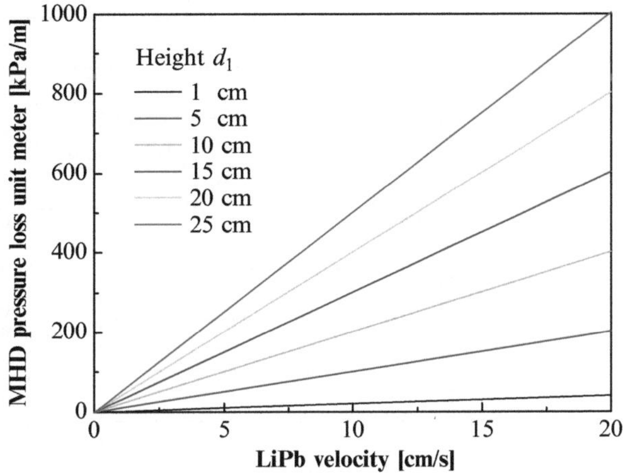

Fig. 5 MHD pressure loss by the shape and flow velocity of LiPb (B = 5 T)

MHD pressure loss increase as increasing d_i and LiPb velocity in Fig. 5. These results are assumed as a conductor. MHD pressure loss does not become a problem in this condition (1 cm s^{-1} or less). To decrease MHD pressure loss, the height of channel has to be low. However, TBR decreases due to the decrease of projective area by the departmentalized channels. However, the pressure loss in the breeding zone is very small compared to the one occurring in the liquid metal piping and manifolds [8]. Therefore the MHD pressure loss must be considered throughout liquid metal loop.

3.2 MHD at the Test Section

Figure 6 is the magnetic field strength on coil current. The condition of this test in Fig. 3 was as follows; the maximum magnetic field was 0–0.16 T and this field was almost constant within ±0.5 cm. Then, magnetic field shows high values at the center including z axis of SiC insert. Figure 7 shows the relations with the pressure and flow rate of LiPb by SiC insert in Fig. 3. With SiC insert, there is almost no pressure difference with change of magnetic field. However, the difference is so small, and there may be unknown pressure changes by other reasons. Magnetic field is not enough in these experiments, and stronger magnet is required to get meaning full data.

4 Conclusions

We designed the LiPb high temperature blanket included Heat Insulation by SiC panel with He coolant channel. The evaluation of MHD pressure loss has been carried out, and possible design windows were identified as follows;

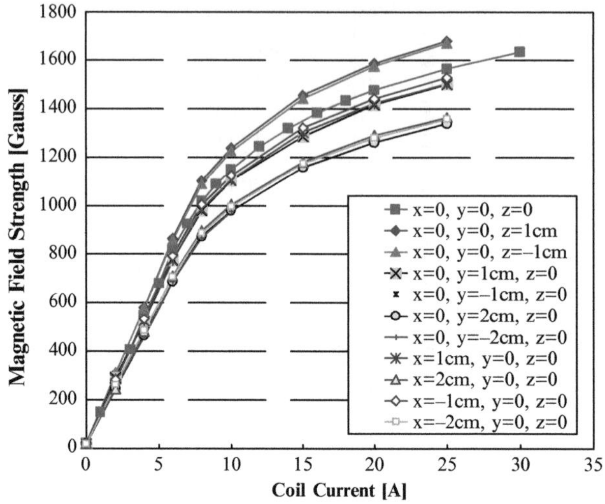

Fig. 6 Magnetic field strength as a function of coil current

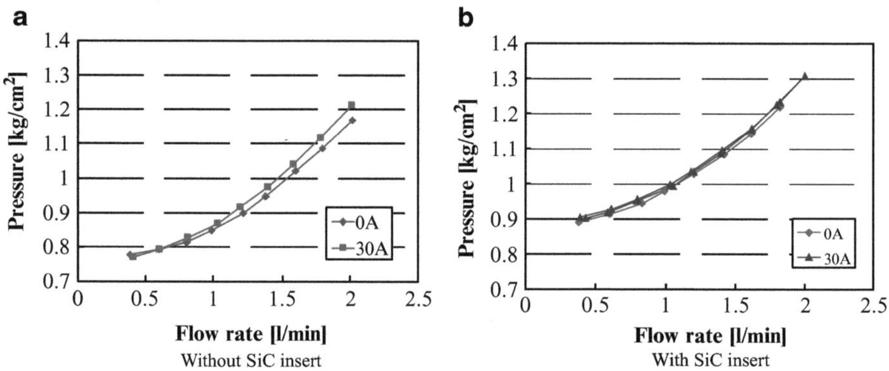

Fig. 7 Influence of SiC insert on pressure

To decrease MHD pressure loss, the height of channel has to be low. However, TBR decreases due to the decrease of projective area by the departmentalized channels. At SiC insert section, pressure did not change with change of magnetic field strength.

References

1. Maisonnier D, Cook I, Pierre S, Lorenzo B, Luigi DP, Luciano G et al. (2006) DEMO and fusion power plant conceptual studies in Europe. Fusion Eng Design 81:1123–1130
2. Norajitra P, Buhler L, Fischer U, Malang S, Reimann G, Schnauder H (2002) The EU advanced dual coolant blanket concept. Fusion Eng Design 61–62:449–453

3. Sze DK, Tillack M, El-Guebaly L (2000) Blanket system selection for ARIES ST. Fusion Eng Design 48:371–378
4. Tillack MS, Malang S (1997) High performance PbLi blanket. In: Proceedings of the 17th IEEE/NPSS Symposium on Fusion Energy, San Diego, CA, pp 1000–1004
5. Malang S, Bojarski E, Buffier L, Deckers H, Fischer U, Norajitra P et al (1992) Dual-coolant liquid metal breeding blanket. Fusion Technol 17:1424–1428
6. Engle WW Jr (1967) A user's manual for ANISN, "One dimensional discrete ordinates transport code with anisotropic scattering". Oak Ridge Report K-1693
7. Hasegawa T, Yamamoto Y, Konishi S (2007) Conceptual design of advanced blanket using liquid LiPb. In: Proceedings of 22nd IEEE/NPSS Symposium on Fusion Engineering (SOFE07), 17–21 June 2007, Albuquerque, NM
8. Li Puma A, Berton JL, Branas B, Buhler L, Doncel J, Fischer U, Farabolini W, Giancarli L, Maisonnier D, Pereslavtsev P, Raboin S, Salavy J-F, Sardain P, Szczepanski J, Ward D (2006) Breeding blanket design and systems integration for a helium-cooled lithium–lead fusion power plant. Fusion Eng Design 81:469–476

(ii) Renewable Energy Research and CO_2 Reduction Research

Lipid-Domain-Selective Assembly of Photosynthetic Membrane Proteins into Solid-Supported Membranes

Ayumi Sumino, Toshikazu Takeuchi, Masaharu Kondo, Takehisa Dewa, Hideki Hashimoto, Alastair T. Gardiner, Richard J. Cogdell, and Mamoru Nango

Abstract In a bacterial photosynthesis, light-harvesting complex 2 (LH2) and the light-harvesting-reaction center complex (LH1-RC) play the key roles of capturing and transferring light energy and subsequent charge separation. Here, we present a novel strategy to assemble both LH2 and LH1-RC into a solid-supported lipid bilayer. Spectroscopic and microscopic analyses revealed that both LH2 and LH1-RC were incorporated into the solid-supported lipid bilayer without denaturation. Through a stepwise-assembling procedure, LH2 and LH1-RC were separately organized in the lipid-domain-structured area.

Keywords Photosynthetic membrane protein • Light-harvesting complex • Protein assembly • Lipid domain • Supported lipid bilayer • AFM • TIRF microscopy

A. Sumino, T. Takeuchi, M. Kondo, T. Dewa (✉), and M. Nango (✉)
Graduate School of Engineering, Nagoya Institute of Technology, Nagoya 466-8555, Japan
e-mail: takedewa@nitech.ac.jp; nango@nitech.ac.jp

H. Hashimoto
Department of Physics, Graduate School of Science, Osaka City University, 3-3-138, Sugimoto, Sumiyoshi-ku, Osaka 558-8585, Japan

T. Dewa
PRESTO/JST, Saitama, Japan

H. Hashimoto and M. Nango
CREST/JST, Tokyo, Japan
e-mail: nango@nitech.ac.jp

A.T. Gardiner and R.J. Cogdell
University of Glasgow, Glasgow G12 8TA, Scotland, UK

T. Yao (ed.), *Zero-Carbon Energy Kyoto 2009*, Green Energy and Technology,
DOI 10.1007/978-4-431-99779-5_18, © Springer 2010

1 Introduction

In biological membranes, various membrane proteins and lipids are laterally organized so as to establish highly-efficient reaction systems. In purple bacterial photosynthetic membranes, two types of membrane protein-pigment complexes light-harvesting complex 2 (LH2) and light-harvesting-reaction center complex (LH1-RC) perform the transfer and capture of light-energy, followed by charge separation in the primary photochemical event. A recent atomic force microscopic (AFM) study has revealed the existence of supramolecular arrays consisting of LH2 and LH1-RC in native photosynthetic bacterial membranes [1]. It has been an open question as to how the arrangement in these arrays affect the function of the primary events. Addressing this issue should provide an insight into molecular-level strategies for the construction of artificial systems for light-energy conversion; however, so far few methods have been established for the controlled assembly of such membrane proteins. We have previously demonstrated the lateral organization of LH2 molecules on a domain-structured lipid bilayer supported on a coverslip [2]. In this method the LH2 molecules, which are incorporated into a cationic vesicle, are integrated into a negatively charged planar membrane via a vesicle fusion event. This strategy prompted us to develop a more advanced method to assemble both LH2 and LH1-RC into a "lipid-domain (area of planar membrane) in a selective" manner to study the relationship between protein arrangement and function, i.e., the stepwise formation of planar membranes incorporating LH2 and LH1-RC into different areas. In this report, we demonstrate the formation of a planar membrane incorporating LH2 and LH1-RC and the evaluation of it using AFM and total internal reflection (TIRF) microscopy.

2 Materials and Methods

2.1 *Materials*

Unless stated otherwise, all chemicals and reagents were obtained commercially and used without further purification. The phospholipids used were 1,2-dioleoyl-*sn*-glycero-3-[phospho-*rac*-(1'-glycerol)] (DOPG), 1,2-dimyristoyl-*sn*-glycero-3-phospho-(1'-*rac*-glycerol) (DMPG), 1,2-dioleoyl-*sn*-glycero-3-phosphocholine (DOPC), 1,2-dioleoyl-*sn*-glycero-3-ethylphosphocholine (EDOPC), and 1,2-dioleoyl-*sn*-glycero-3-phosphoethanolamine-*N*-lissamine-rhodamine B sulfonyl (*N*-Rh-DOPE). The surfactants used were *n*-octyl-β-D-glucoside (OG) and *N,N*-dimethyldodecylamine *N*-oxide (LDAO). LH2 and LH1-RC were isolated from purple photosynthetic bacteria *Rb. sphaeroides* 2.4.1. (LH2), *Rps. acidophila* 10050 (LH2), *Rb. sphaeroides* puc705BA (LH1-RC), and *Rps. palustris* 2.1.6. (LH1-RC). For the TIRF microscopic observation, LH1-RC was labeled with a fluorescent probe, OregonGreen 488 maleimide, on the cysteine residue at H156 of RC.

2.2 *Liposome Preparation*

DMPG giant vesicles were prepared by the electroformation method [3]. Proteoliposomes containing LH2 or LH1-RC were prepared by the dialysis method from a co-micellar solution of lipid, protein, and OG. The resulting liposome solution was extruded through a 0.1 μm poly carbonate membrane.

2.3 *Microscopic Analyses*

AFM images were taken with a modified JSPM-5200 (JEOL) equipped with a cantilever NCH (NanoWorld), in an aqueous condition, in a liquid cell at room temperature. TIRF microscopic observation was performed using an objective-type TIRF microscope, TE-2000U (Nikon), and a cooled CCD camera (ORCA-ER: HAMAMATSU).

2.4 *Formation of Planar Lipid Bilayer on a Coverslip*

A chemically washed coverglass (Matsunami NEO) was treated with 3-aminopropyl triethoxysilane (APS) in a dry benzene solution for 4 h at 80°C [4]. An anionic vesicle solution was applied onto the APS-modified coverslip and spontaneously formed a planar lipid membrane.

3 Results and Discussion

3.1 *Reconstitution of Photosynthetic Membrane Proteins into Liposomes*

Figure 1 shows the absorption spectra of LH2 (A) and LH1-RC (B) in a Tris buffer solution (20 mM *p*H 8.0, 0.1w% LDAO) (dashed line) and in a reconstituted liposomal membrane (dotted line). The characteristic absorption bands attributed to Qy of bacteriochlorophyll *a* in these proteins are essentially identical in both conditions. The LH2 complexes reconstituted into the lipid bilayer can be successfully visualized with high resolution AFM. When the liposomal solution containing reconstituted LH2 was applied onto a mica substrate, the vesicle ruptured on the surface to form a planar membrane patch. Figure 2 shows the AFM image of such a patch. The LH2 molecules whose structure is cylindrical ($d = 6$ nm) can be clearly seen. This exactly corresponds to their structure observed by X-ray crystallography [5].

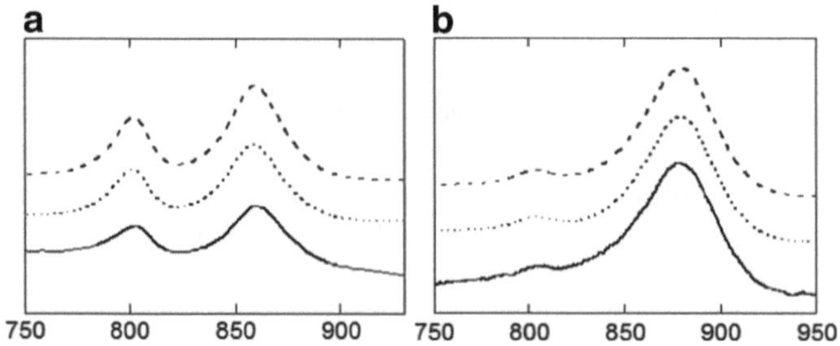

Fig. 1 Absorption spectra of LH2 from *Rps. acidophila* 10050 (**a**) and LH1-RC *Rps. palustris* 2.1.6. (**b**) in micellar solutions (*dashed line*), proteoliposome solutions (*dotted line*), and on supported membranes (*solid line*)

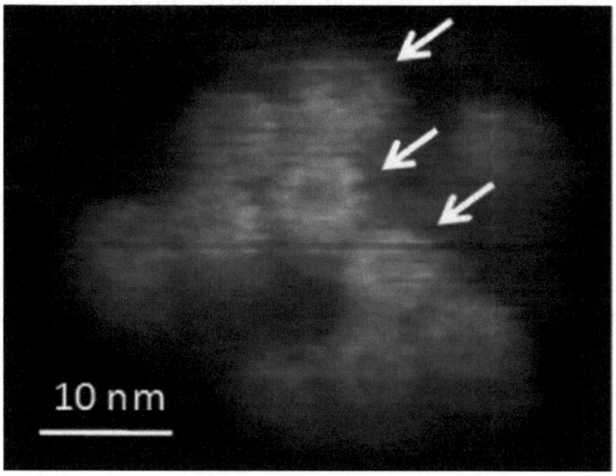

Fig. 2 AFM image of LH2 from *Rb. sphaeroides* 2.4.1. in the planar membrane patch (DOPC). The observed ring structures indicated by *arrows* exactly correspond to the structure observed by X-ray crystallography

3.2 Formation of the Continuous Planar Membrane Supported on an APS-Modified Coverglass

When an anionic vesicle solution was applied onto the positively charged APS-coverglass, a continuous planar membrane was formed. This method, therefore, produces protein-containing planar membranes from the reconstituted vesicles, as described above. The absorption spectra of LH2 and LH1-RC in the planar membranes were essentially identical to those in reconstituted membranes (Fig. 1), suggesting that these proteins were assembled into the solid-supported membranes with their structures intact. The membrane continuity and diffusivity of the constituents, lipid, LH2, and LH1-RC were determined by the fluorescence recovery after photobleaching (FRAP) method. Figure 3 shows the time-lapse images of

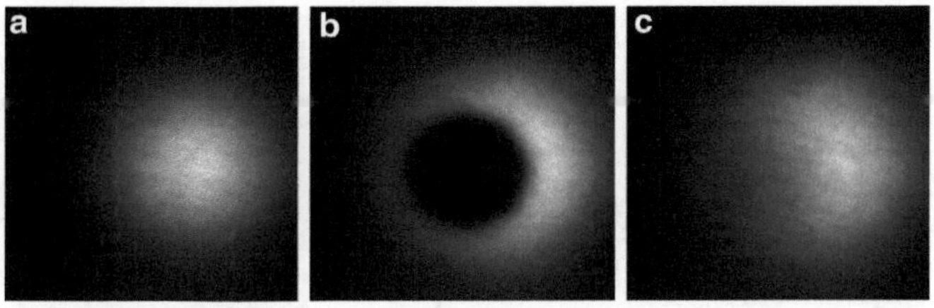

Fig. 3 Snapshots of fluorescence recovery after photobleaching measurement for the plain planar membrane. Images (**a**), (**b**), and (**c**) were before, just after, and 20 min after photobleaching, respectively. Fluorescent probe *N*-Rh-DOPE was incorporated there

such a FRAP experiment with fluorescent lipid (*N*-Rh-DOPE) in a plain lipid bilayer. The complete recovery of the fluorescence (~100% of mobile fraction) clearly indicates that the planar membrane is continuous. Mobile fractions for the proteins themselves were 90% for LH2 and 0% for LH1-RC. The difference in the mobility is likely due to the composition of proteins in the complexes. The H-subunit of LH1-RC protrudes from membrane surface (~3 nm), and so the friction between the subunit and the coverglass would be stronger than that of LH2, that has minimal hydrophilic domains [6]. Therefore, LH2 are mobile; however, LH1-RC is not mobile in the solid-supported membrane. Membranes containing proteins also exhibited such membrane continuity, i.e., 95% and 26% of mobile fraction of the *N*-Rh-DOPE for LH2- and LH1-RC-containing membranes, respectively. The difference in the diffusivity of lipid in these protein-incorporating membranes likely results from that in the mobility of these proteins there.

3.3 Lipid-Domain-Selective Assembly of LH2 and LH1-RC

Figure 4 shows the stepwise assembly of an LH2/LH1-RC-coexisting planar membrane (A-D). An anionic DMPG giant vesicle suspension was applied onto an APS-coverglass to provide a planar bilayer patch via electrostatic interaction (step A–B). Subsequently, on the addition of a cationic EDOPC vesicle containing LH2 on the surface, the vesicle selectively fused with the DMPG-membrane patch (C). Finally, a DOPG/LH1-RC proteoliposome solution was applied onto an uncovered area of the APS-coverglass, resulting in planar bilayer formation (D). The stepwise assembly was observed with TIRF microscopy. The formation of the bilayer patch (DMPG in image E) was detected by *N*-Rh-DOPE fluorescence. After photobleaching, EDOPC/LH2 vesicles selectively fused with this area (C), giving fluorescence from the incorporated LH2 molecules (image F). The outside of the patch is an area of the bare APS-coverglass, on which the anionic DOPG/LH1-RC vesicles formed the LH1-RC domain (LH1-RC in D and image G). The outside area of the LH2-patch was imaged with the fluorescent-labeled LH1-RC. These observations clearly indicate that both LH2 and LH1-RC are assembled into a solid-supported lipid bilayer in a "lipid-domain-selective" manner.

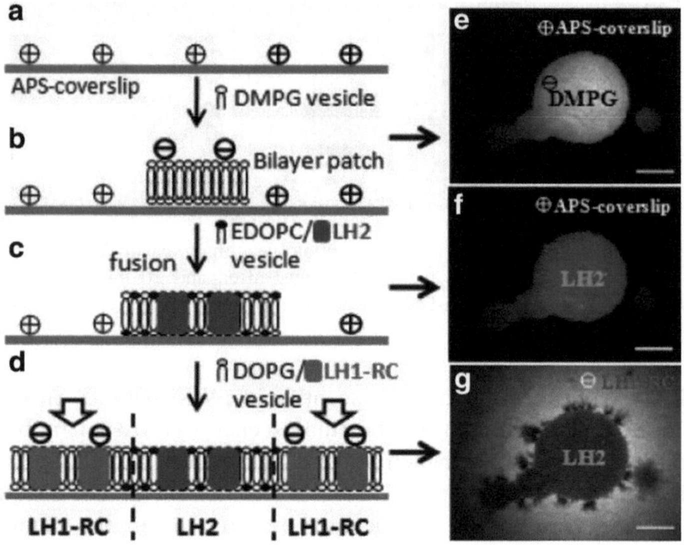

Fig. 4 Lipid-domain-selective assembly of LH2 from *Rb. sphaeroides* 2.4.1. and LH1-RC from *Rb. sphaeroides* puc705BA into the planar membrane (*scale bar:* 10 μm). Laterally separated protein domain is formed using the vesicular delivery system

4 Conclusion

We demonstrated a novel strategy for assembling LH2 and LH1-RC into a solid-supported lipid bilayer in a "lipid-domain-selective" manner. The present method can be applied to investigate the relationship between protein arrangement and function as well as to construct artificial photoenergy conversion devices with highly-efficient reaction systems.

Acknowledgments The authors would like to thank Dr. Shinichi Kitamura and Mr. Katsuyuki Suzuki (JEOL, Japan) for AFM measurement.

References

1. Svetlana et al (2004) Nature 430:26
2. Dewa et al (2006) Langmuir 22:5412
3. Angelova et al (1992) Colloid Polym Sci 89:127
4. Fang Y et al (2002) J Am Chem Soc 124:2394
5. Mcdermott et al (1995) Nature 374:517
6. Motomu et al (2005) Nature 437:29

Light-Induced Transmembrane Electron Transfer Catalyzed by Phospholipid-Linked Zn Chlorophyll Derivatives on Electrodes

Yoshito Takeuchi, Hongmei Li, Shingo Ito, Masaharu Kondo, Shuichi Ishigure, Kotaro Kuzuya, Mizuki Amano, Takehisa Dewa, Hideki Hashimoto, Alastair T. Gardiner, Richard J. Cogdell, and Mamoru Nango

Abstract Phospholipid-linked ZnPChlide *a* derivatives separated by spacer methylene groups (C_n) (PE-C_n-ZnPChlide *a*; $n=0$, 5, 11) were synthesized. When PE-C_n-ZnPChlide *a* was assembled onto an indium tin oxide (ITO) electrode modified with lipid bilayers, the assembly showed light-induced transmembrane electron transfer when illuminated at 430 nm. Interestingly, the action spectrum of the photocurrent of PE-C_n-ZnPChlide *a* on the electrode was dependent on the length of the spacer methylene groups (C_n) and the fluidity of the lipid.

Keywords Light-induced electron transfer • Lipid bilayers • Chlorophyll derivatives

1 Introduction

Synthetic porphyrin models can be very helpful in studying the effect of distance and orientation on electron transfer reactions in a biological membranes [1]. To develop a model for such electron transfer reaction where porphyrin pigments play a key role, porphyrin derivatives that are capable of electron transfer were prepared [1]. However, intramembrane electron transfer between porphyrin complexes in a lipid membrane have not been studied in depth. In our previous studies, to

Y. Takeuchi, H. Li, S. Ito, M. Kondo, S. Ishigure, K. Kuzuya, M. Amano, T. Dewa, and M. Nango
Nagoya Institute of Technology, Gokiso-cho, Showa-ku, Nagoya, 466-8555, Japan

H. Hashimoto
Osaka City University, 3-3-138 Sugimoto, Sumiyoshi-ku, Osaka, 558-8585, Japan

A.T. Gardiner and R.J. Cogdell
University of Glasgow, Glasgow G12 8TA, Scotland, UK

H. Hashimoto and M. Nango (✉)
CREST/JST, Saitama, Japan
e-mail: nango@nitech.ac.jp

T. Yao (ed.), *Zero-Carbon Energy Kyoto 2009*, Green Energy and Technology, DOI 10.1007/978-4-431-99779-5_19, © Springer 2010

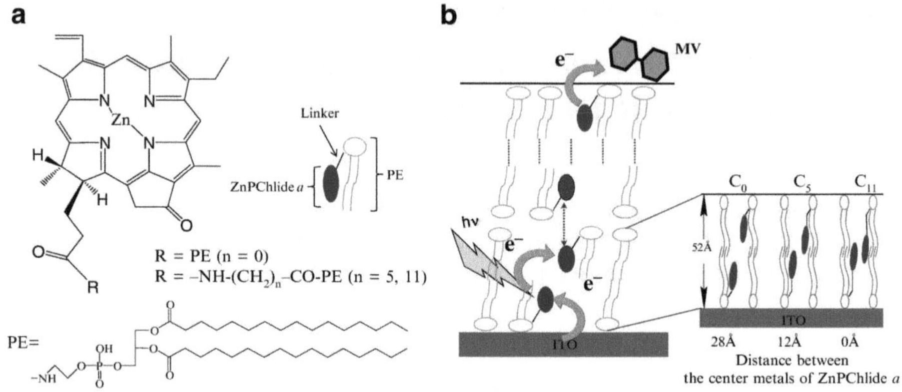

Fig. 1 (**a**) Structure of PE-C_n-ZnPChlide *a*. (**b**) The mechanism of photo-electric current from ITO electrode to methylviologen (MV) through photo-excited PE-C_n-ZnPChlide *a* lipid bilayers (left). The distance between the center metals of chlorins in the lipd bilayers on the ITO electrode is shown on right

determine the effect of distance and orientation on electron transfers in lipid bilayers, we studied the ground-state transmembrane electron transfer activity catalyzed by phospholipid- or polymer-linked manganese tetraphenyl porphyrin and mesoporphyrin derivatives [2]. In the current study, we analyze light-induced transmembrane electron transfer catalyzed by phospholipid-linked chlorophyll derivatives with spacer methylene groups (C_n) (PE-C_n-ZnPChlide *a*; n=0, 5, 11) assembled on an electrode modified with lipid bilayers (Fig. 1a). The aim of this study was to gain insight into the photoinduced of electron transfer distance between model chlorophyll complexes in lipid bilayers under illumination, as shown in Fig. 1b.

2 Experimental Section

2.1 *Materials*

Pyropheophorbide *a* methyl ester (H_2PPheide *a* ME) was provided by TAMA Biochemical Co. Ltd., Japan. Zinc acetate dehydrate was obtained from Wako Pure Chemical Industries, Ltd., Japan. High-purity egg yolk phosphatidylcholine (egg PC), dioleoylphosphatidylglycerol (DOPG) and dipalmitoylphosphatidylcholine (DMPC) were provided by Nippon Fine Chemical Co. Ltd., Japan.

2.2 *Synthesis of PE-C_n-ZnPChlide a*

H_2PPheide *a* ME was hydrolyzed with 5 N HCl to obtain pyropheophorbide *a* (H_2PPheide *a*). Then, H_2PPheide *a* was treated with N-hydroxysuccinimide and reacted with dipalmitoylphosphatidylethanolamine (PE) to yield PE-C_0-H_2PPheide *a*.

H_2PPheide *a* was reacted with 6-aminohexanoic acid using N-hydroxysuccinimide to yield H_2PPheide *a*-CONH$(CH_2)_5$COOH. In a similar manner, H_2PPheide *a*-CONH $(CH_2)_{11}$COOH was prepared using 12-aminododecanoic acid. H_2PPheide *a*-CONH$(CH_2)_5$COOH and H_2PPheide *a*-CONH$(CH_2)_{11}$COOH were reacted with PE to obtain PE-C_5-H_2PPheide *a* and PE-C_{11}-H_2PPheide *a*, respectively. Finally, PE-C_n-ZnPChlide *a* ($n=0$, 5, 11) were prepared from PE-C_n-H_2PPheide *a* ($n=0$, 5, 11) and zinc acetate dehydrate by the method described in a previous paper [2].

2.3 Preparation of PE-C_n-ZnPChlide a on an Electrode Modified with Lipid Bilayers

PE-C_n-ZnPChlide *a* was assembled on an electrode by a cast method: A mixture of PE-C_n-ZnPChlide *a* and egg PC or DMPC was dissolved with $CHCl_3$ and the solution was then cast on indium tin oxide (ITO) electrodes (surface area of 1 cm^2) at room temperature. This level of coverage yields an egg PC membrane having a thickness of approximately 8–10 μm, as measured using an Elecont micrometer (Mitsutoyo Co.).

2.4 Photocurrent Measurement

Photocurrents were measured at −0.2 V vs. Ag/AgCl in a home-made cell (100 cm^3) that contained three electrodes: an ITO electrode incorporating the PE-C_n-ZnPChlide *a* in lipid bilayers as the working electrode; an Ag/AgCl (saturated KCl) as the reference electrode; and a platinum flake as the counter electrode. The working electrode was illuminated using a xenon lamp unit (SM-25, Bunkokeiki, Japan) where the light was pared through a monochromator. An aqueous solution consisting of 0.1 M phosphate buffer (*p*H 7.0) and 0.1 M $NaClO_4$ was used as the electrolyte, and 5 mM methylviologen was used as the electron acceptor [3].

3 Results and Discussion

3.1 Synthesis of PE-C_n-ZnPChlide a

The sequence followed in the synthesis of PE-C_n-ZnPChlide *a* (Fig. 1a) is as described in the Experimental section. Each compound was purified by chromatographic separation (silica gel, methanol/chloroform 10:90, v/v) and HPLC. The results of ^{1}H NMR and MALDI-TOF-MASS analyses of the PE-C_n-H_2PPheide *a* unambiguously confirm the validity of assigned structure. The chemical shifts of NMR spectra for all peaks were as expected. The MASS spectra of PE-C_n-H_2PPheide *a*, indicated the presence of one porphyrin per phosopholipid. The absorption spectra of PE-C_n-ZnPChlide *a* in CH_2Cl_2-10% EtOH were identical to those in egg PC or DMPC vesicles, and they all showed the presence of a normal Zn chlorin chromophore.

3.2 Light-Induced Electron Transfer of PE-C_n-ZnPChlide *a* in Egg PC Bilayers on ITO Electrodes

Figure 2a shows the photocurrent response of PE-C_n-ZnPChlide *a* in egg PC bilayers on ITO electrodes when the electrodes are illuminated by pulsed light at 430 nm. Cathodic photocurrents were observed for all the PE-C_n-ZnPChlide *a* derivatives when the circuit was operated at –0.2 V, indicating that one-way electron transfers occurs from PE-C_n-ZnPChlide *a* in egg PC bilayers to methylviologen. The photocurrent density of PE-C_n-ZnPChlide *a* increased in the order: $C_5 > C_{11} \approx C_0$. Figure 2b shows the action spectra of the photocurrent of PE-C_n-ZnPChlide *a*. For example, the action spectrum of PE-C_5-ZnPChlide *a* (green line) is identical to the UV-visible absorption spectrum (solid line) of the ITO electrode modified with egg PC bilayers containing the PE-C_5-ZnPChlide *a*. Similar results were observed for PE-C_0-ZnPChlide *a*, and PE-C_{11}-ZnPChlide *a* (data not shown). The photocurrent density of these action spectra also increased in the same order as that of the photocurrent response shown in Fig. 2a: $C_5 > C_{11} \approx C_0$. These results

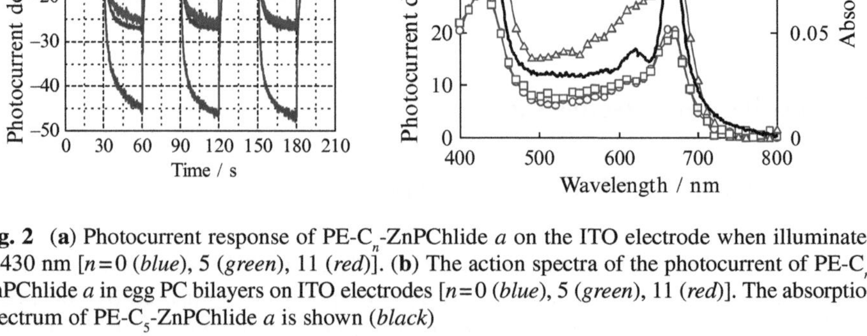

Fig. 2 (**a**) Photocurrent response of PE-C_n-ZnPChlide *a* on the ITO electrode when illuminated at 430 nm [$n=0$ (*blue*), 5 (*green*), 11 (*red*)]. (**b**) The action spectra of the photocurrent of PE-C_n-ZnPChlide *a* in egg PC bilayers on ITO electrodes [$n=0$ (*blue*), 5 (*green*), 11 (*red*)]. The absorption spectrum of PE-C_5-ZnPChlide *a* is shown (*black*)

indicated that the photocurrent density of PE-C_n-ZnPChlide *a* on the ITO electrode was dependent on the length of C_n. Further, similar results in terms of photocurrent activities were observed for all the PE-C_n-ZnPChlide *a* derivatives above the phase transition of the lipid when DMPC was used. However, little photocurrent activity, if at all, was observed below the phase transition of the lipid.

The mechanism of light-induced transmembrane electron transfer is presumed to be as follows (Fig. 1b) [2]: First, electron transfer occurs from the ITO electrode to an excited ZnPChlide *a* complex when illuminated at 430 nm; then, following this electron transfer from ZnPChlide *a* to a ZnPChlide *a* tethered to a PE molecule on the opposite side of the bilayer can occur. Finally, electron transfer from ZnPChlide *a* to methylviologen is occured. Thus, the distances between the electrode and a ZnPChlide *a* complex as well as the distances between individual ZnPChlide *a* complexes are important factors in controlling the electron transfer, as shown in the right -hand side of Fig. 1b [2]. In this system, it is assumed that ZnPChlide *a* complexes cannot bring about rapid electron transfer between the electrode and ZnPChlide *a* and nor between individual ZnPChlide *a* complexes when $n=0$. Free movement of ZnPChlide *a* is strongly limited when the methylene length is large; thus, electron transfer between the electrode and ZnPChlide *a*, as well as between individual ZnPChlide *a* complexes, is very slow when $n=11$. Therefore, $n=5$ is probably suitable for electron transfer on the electrode because the ZnPChlide *a* -moieties can freely move and approach each other for electron transfer both between the electrode and ZnPChlide *a* and between individual ZnPChlide *a* complexes. These results are consistent with the results of transmembrane electron transfer when catalyzed by phospholipid-linked manganese mesoporphyrin; in both cases, electron transfer was dependent on the length of C_n, and the optimum electron transfer was observed when $n=5$ [2].

4 Conclusions

Phospholipid-linked ZnPChlide *a* derivatives (PE-C_n-ZnPChlide *a*; $n=0, 5, 11$) were synthesized to gain insight into the dependence of the distance between chlorin complexes during light-induced chlorin-mediated electron transfer in lipid bilayers. Photocurrent action spectra of the ITO electrodes modified with PE-C_n-ZnPChlide *a*/lipid bilayers indicated that electron transfer was dependent on the length of the spacer methylene groups and the fluidity of the lipid bilayer. Thus, by selecting an appropriate membrane component and suitable chlorophyll derivatives, electron transfer in lipid bilayer systems can be analyzed systematically, and the data from the analysis can be used to construct artificial light-energy conversion systems.

Acknowledgment M.N. acknowledges the support of the AOARD-06-4084.

References

1. Drain CM, Varotto A, Radivojevic I (2009) Chem Rev 109:1630
2. Nango M, Hikita T, Nakano T, Yamada T, Nagata M, Kurono Y, Ohtsuka T (1998) Langmuir 14:407
3. Kondo M, Nakamura Y, Fujii K, Nagata M, Suemori Y, Dewa T, Iida K, Gardiner AT, Cogdell RJ, Nango M (2007) Biomacromolecules 8:2457

Raman Spectroscopic Studies on Silicon Electrodeposition in a Room-Temperature Ionic Liquid

Yusaku Nishimura, Toshiyuki Nohira, and Rika Hagiwara

Abstract The electrochemical reduction of $SiCl_4$ to Si in a room-temperature ionic liquid, trimethyl-*n*-hexylammonium bis(trifluoromethylsulfonyl)amide (TMHATFSA) has been investigated by Raman spectroscopy. For the electrolyte solution itself, most of silicon chloride species exist as $SiCl_4$ molecules in TMHATFSA. It is considered that the $SiCl_4$ molecules are stabilized by the induced dipole-induced dipole interaction with the hexyl group in $TMHA^+$ cations. In situ Raman spectroscopy revealed that $SiCl_4$ is electrochemically reduced to form amorphous Si (a-Si) as well as silicon chloride species containing Si networks as represented by Si_mCl_n ($n/m<4$). The amount of the formed a-Si increases as the electrolysis proceeds. Besides, there seems to be an incubation period in which Si_mCl_n are generated, followed by the intensive production of a-Si. This information implies that a-Si may be produced by further reduction of Si_mCl_n to form larger Si networks or by disproportionation reactions of Si_mCl_n.

Keywords Electrodeposition • Silicon • Room-temperature ionic liquid • Raman Spectroscopy • In situ measurement

1 Introduction

Propagation of solar power generation is indispensable to establish the CO_2 zero-emission energy system throughout the world. Silicon is the most favorable material for photovoltaic cells due to its abundance in the earth's crust. It is anticipated that the demands for silicon and its thin films will continuously increase in the future. In order to meet such great demands, we have to seek for low-cost processing

Y. Nishimura (✉), T. Nohira, and R. Hagiwara
Department of Fundamental Energy Science, Kyoto University, Yoshida-hommachi, Sakyo-ku, Kyoto, 606-8501, Japan
e-mail: nishimura.yusaku@t01.mbox.media.kyoto-u.ac.jp

T. Yao (ed.), *Zero-Carbon Energy Kyoto 2009*, Green Energy and Technology, DOI 10.1007/978-4-431-99779-5_20,

of silicon. Then, we have considered that an electrochemical process has a potential to be one of the low-cost production methods for silicon thin films. In general, this process has an advantage to fabricate large-area thin films with high productivity because it does not require any ultrahigh vacuum apparatuses. This is the reason why we have been engaged in the electrodeposition of silicon in non-aqueous solvents [1–4].

So far, we have reported that amorphous silicon or Si networks is formed by the electrolysis at −2.0 V vs. Pt quasi-reference electrode (QRE) in trimethyl-*n*-hexylammonium bis(trifluoromethylsulfonyl)amide (TMHATFSA) containing 0.1 mol L^{-1} $SiCl_4$ [2–4]. One of the most serious problems on this process is that the electrodeposition mechanism has not been clarified yet although comprehension of the mechanism is significant to take measures for purification of the obtained silicon and for direct control of its structure.

In this study, we attempted to clarify the electrochemical formation mechanism of Si networks by Raman spectroscopy.

2 Experimental

The electrolyte solution was comprised of TMHATFSA and $SiCl_4$. Silicon tetrachloride (Wako Pure Chemical Industries, Ltd., purity: > 99%) was dissolved into TMHATFSA as received from Stella Chemifa Corporation. The water content of the electrolyte solution supplied to the electrolysis was verified to be below 10 ppm by Karl-Fischer titration (Hiranuma Sangyo Co., Ltd., Aquacounter AQ-200). The electrolyte solutions containing different concentrations of $SiCl_4$ were supplied to a Fourier transform Raman spectrometer (Bio-Rad Laboratories, Inc., BIO-RAD FTS-175C) with a Nd:YAG laser (1,064 nm) to examine the dissolved state of $SiCl_4$ in TMHATFSA.

A conventional three-electrode cell was employed for all the electrolysis. It consisted of a Ni plate (5 mm × 5 mm in area) as a working electrode, a Pt wire (0.5 mm in diameter) coated by a heat-shrinkable tube as a QRE, and a graphite plate as a counter electrode. Each electrode surface was mechanically polished to a mirror finish. All the electrodes were ultrasonically washed in deionized (DI) water, and then rinsed out with ethanol and DI water. Before assembling the electrolytic cell, the Ni substrate except for the effective surface area was covered and insulated with Teflon® tape.

Potentiostatic electrolysis was performed at −2.0 V vs. Pt QRE in TMHATFSA containing 0.1 mol L^{-1} $SiCl_4$ at room temperature (298 K) with a potentiostat-galvanostat (Hokuto Denko Co., Ltd., model HA-301). Simultaneously, the interface between the electrolyte solution and a Ni substrate was analyzed using a micro-Raman spectrometer (Horiba Jobin Yvon, LabRAM300) with a He–Ne laser (632.8 nm). A Raman spectrum was acquired without a pause in the electrolysis.

3 Results and Discussion

3.1 Raman Spectroscopy of $SiCl_4$-TMHATFSA Electrolytes

First of all, the dissolved state of silicon chloride species in TMHATFSA was investigated by Raman spectroscopy. Figure 1 summarizes the results of peak separation for the ν_1 symmetric stretch mode of $SiCl_4$. It exhibits almost no influence of TMHATFSA on the peak position and width for the ν_1 mode of $SiCl_4$ in the electrolyte solutions. It means that silicon chloride species exists mainly as $SiCl_4$ molecules in TMHATFSA without making bonds with other molecules to form complexes, polymers, or different silicon compounds. Furthermore, it is expected on the basis of the conventional law of chemistry that $SiCl_4$ should be stabilized by the induced dipole-induced dipole (van der Waals) interaction especially with the hexyl group of $TMHA^+$ cations. This speculation is consistent with the results of density functional theory calculations. From the Raman spectroscopic study on TMHATFSA containing $SiCl_4$, there is no doubt that $SiCl_4$ is the silicon compound in the initial state for the electrolysis in TMHATFSA.

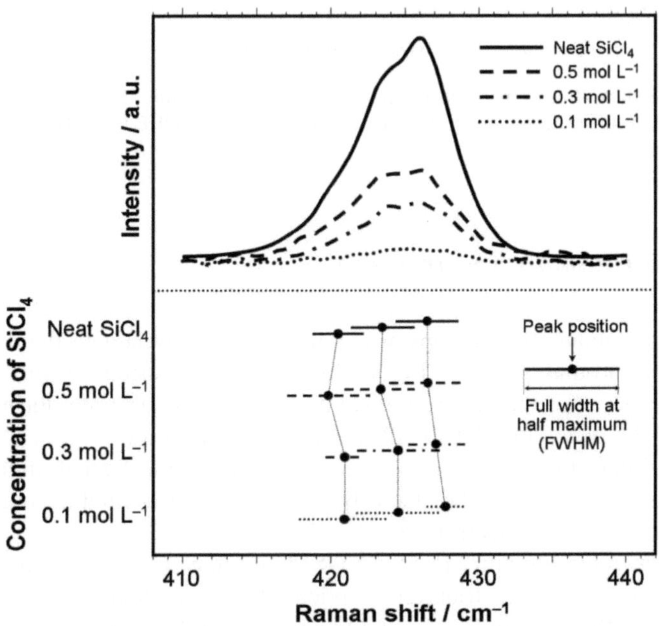

Fig. 1 Raman spectra of the ν_1 mode of $SiCl_4$ in TMHATFSA. The results of peak separation (the position and the width of each peak) are also shown

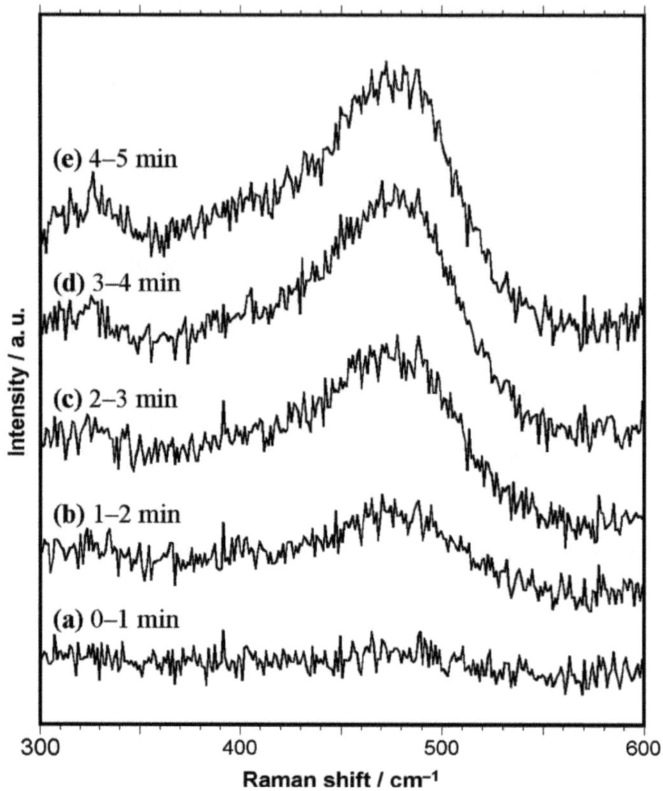

Fig. 2 Time evolution of Raman spectra in situ measured at the interface between the electrolyte solution and a Ni substrate during the electrolysis at −2.0 V vs. Pt QRE in TMHATFSA containing 0.1 mol L^{-1} $SiCl_4$

3.2 *In situ Raman Spectroscopy*

Next, in situ Raman spectroscopy was performed to investigate the transition of bonding states around Si atoms. Figure 2 shows Raman spectra in situ recorded at the interface between the electrolyte solution and a Ni substrate during the electrolysis at −2.0 V vs. Pt QRE in TMHATFSA containing 0.1 mol L^{-1} $SiCl_4$. In order to facilitate the comparison of each spectrum, the background spectrum, which was measured before the electrolysis, has been subtracted from all the spectra measured during the electrolysis.

According to the literature [5,6], a Si–Si bond in Si_2Cl_6 has a peak centered at 624 cm^{-1} while a-Si for the transverse-optical phonon at about 480 cm^{-1}. The peak position for Si–Si bonds in Si_mCl_n should shift from 624 cm^{-1} toward lower wavenumbers as Si networks become larger, assuming that the force constant for the stretching vibration does not change. The enlargement of Si networks corresponds to the decrease of the (virtual) oxidation number of Si (*n*/*m*) and the increase of the *m* value in the reduced silicon compound, Si_mCl_n. Concerning the electrolysis for the first 1 min (see Fig. 2a), the spectrum does not have any apparent peaks assigned to crystalline or amorphous Si. However, it seems to have a broad and tiny peak at wavenumbers ranging from 500 to 624 cm^{-1}. It implies that $SiCl_4$ may be gradually reduced to form Si_2Cl_6 and other reduced silicon chloride species with lower oxidation numbers in the initial stage of the electrolysis. After the 1-min electrolysis, a broad peak emerged at about 480 cm^{-1} and intensified with the progress of electrolysis. Such a trend is seen in the spectra (b)–(d) in Fig. 2. This confirms the evolution of a-Si during the electrolysis at −2.0 V vs. Pt QRE in TMHATFSA containing 0.1 mol L^{-1} $SiCl_4$.

The in situ Raman spectroscopic study invokes the electrochemical formation mechanism of Si networks as follows. Electrochemical reduction of $SiCl_4$ to Si is divided roughly into two stages. The first stage, which continues about 1 min after the electrolysis begins, is an incubation period. In this period, the electrolytic reduction of $SiCl_4$ occurs to form not a-Si but reduced silicon chloride species with Si–Si bonds as represented by Si_mCl_n ($n/m<4$). On the other hand, in the second stage, a-Si starts to form and grows intensively. It is considered that a-Si may be generated by further reduction of Si_mCl_n to form larger Si networks, or by disproportionation reactions of Si_mCl_n due to the difference in reaction rates.

4 Conclusion

Fourier transform Raman spectroscopy confirmed that most of the silicon chlorides exist as $SiCl_4$ in TMHATFSA, stabilized by the induced dipole-induced dipole interaction with the hexyl group in $TMHA^+$ cations. In situ Raman spectroscopy was also conducted for the interface between the electrolyte solution and a Ni substrate during the electrolysis at −2.0 V vs. Pt QRE in TMHATFSA containing 0.1 mol L^{-1} $SiCl_4$. It suggested that the electrochemical reduction of $SiCl_4$ to Si proceeds via two stages. In the first stage (for the first 1 min), the electrochemical reduction of $SiCl_4$ mainly produces not a-Si but reduced silicon chloride species as represented by Si_mCl_n ($n/m<4$). In the second stage, on the other hand, a-Si starts to form and grows intensively. The evolution of a-Si may be caused by the further reduction of Si_mCl_n and the formation of larger Si networks, or by the disproportionation reactions of Si_mCl_n.

References

1. Nishimura Y, Fukunaka Y (2007) Electrochemical reduction of silicon chloride in a non-aqueous solvent. Electrochim Acta 53:111–116
2. Nishimura Y, Fukunaka Y, Nohira T, Hagiwara R (2007) Electrochemical processing of nano-scale Si thin film in a hydrophobic room-temperature molten salt. ECS Trans 11:13–24

3. Nishimura Y, Fukunaka Y, Nishida T, Nohira T, Hagiwara R (2008) Electrodeposition of Si thin film in a hydrophobic room-temperature molten salt. Electrochem Solid-State Lett 11: D75–D79
4. Nishimura Y, Fukunaka Y, Nohira T, Goto T, Hachiya K, Nishida T, Hagiwara R (2008) XPS study and optical properties of Si films electrodeposited in a room-temperature ionic liquid. ECS Trans 13:37–52
5. Höfler F, Sawodny W, Hengge E (1970) Schwingungsspektren und kraftkonstanten von halogendisilanen. Spectrochim Acta A 26:819–823
6. Brodsky MH, Cardona M, Cuomo JJ (1977) Infrared and Raman spectra of the silicon–hydrogen bonds in amorphous silicon prepared by glow discharge and sputtering. Phys Rev B 16: 3556–3571

DC Connected Hybrid Offshore-Wind and Tidal Turbine Generation System

Mohammad Lutfur Rahman and Yasuyuki Shirai

Abstract "Hybrid Offshore-wind and Tidal Turbine" (HOTT) generation system (Rahman and Shirai, HOTT energy Conversion I [6-pulse GTO rectifier and inverter], 2008; HOTT energy conversion II [6-Pulse GTO Rectifier DC connection and Inverter], 2009) interconnecting method for a DC side cluster of wind and tidal turbine generators system are proposed. This method can be achieved using wind and tidal turbine generating system. Four tidal and a wind turbines generator can be interconnected easily with the proposed method, and high reliability and electric output power with high quality are also expected. This method is able to send generated power through a long-distance DC transmission line. The configuration of the hybrid wind and tidal turbine generator system is explained first, and a dynamic model of the hybrid system is developed. The dynamic performances of the HOTT system when the wind and tidal velocity is changing are then discussed. Finally, a control system to keep the DC voltage constant of the HOTT system is introduced.

Keywords Hybrid system • Offshore wind turbine • Tidal turbine

1 Introduction

The utilization of natural energy such as offshore-wind and tidal power are one of the effective answers to the global environmental problems. In general, a offshore-wind turbine generator system supplies electric power to the utility, and stable power supply to the grid is not feasible since the output power of the offshore-wind generator system fluctuates at all times with wind conditions [2,5]. Hence, we have proposed a hybrid generator system on the basis of the offshore-wind turbine generator system using a tidal generator system, and the steady-state characteristics of the hybrid system have been discussed in this paper.

M.L. Rahman (✉) and Y. Shirai
Graduate School of Energy Science, Department of Energy Science and Technology, Kyoto University, Kyoto, Japan
e-mail: lutfur@pe.energy.kyoto-u.ac.jp; shirai@energy.kyoto-u.ac.jp

T. Yao (ed.), *Zero-Carbon Energy Kyoto 2009*, Green Energy and Technology, DOI 10.1007/978-4-431-99779-5_21, © Springer 2010

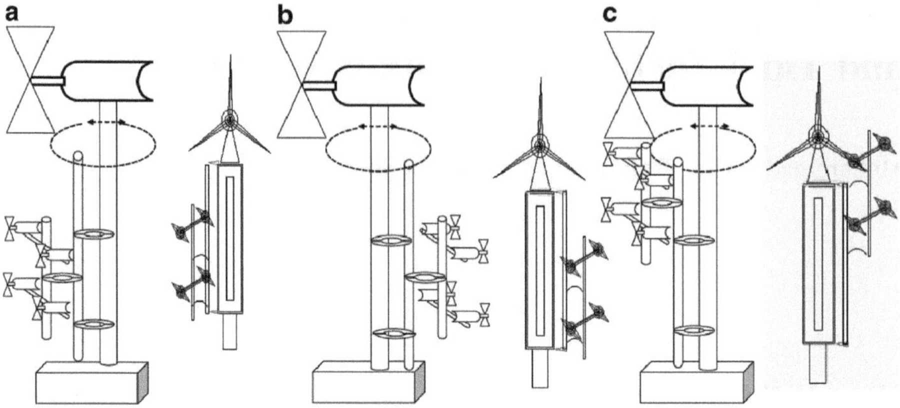

Fig. 1 Hybrid offshore-wind and tidal turbines (HOTT) (**a**) first situation (**b**) after swing (**c**) maintenance situation

The proposed HOTT generator system is given in Fig. 1. As in this figure, the system consists of an offshore-wind and tidal turbine first situation, swing situation and maintenance situation [1,4]. The hybrid model is designed to simulate a realistic situation of the power fluctuation during continuous operation of the tidal and offshore-wind turbine. It is built into combination of three steps. The first step is the offshore-wind model, which simulates the wind power. The second step is the tidal turbine model, which simulates the tidal power. The third step is to investigate hybrid system between tidal turbine and offshore wind turbine with 6-pulse GTO (Gate Turn-Off thyristor) rectifier DC side connections and inverter. HOTT energy can have a number of benefits in both environmental and socioeconomic areas. Unenclosed HOTT can avoid many of the detrimental environmental effects and CO_2 emission which is becoming a key issue, while providing significant amounts of distributed renewable energy.

2 Wind Turbine and Result

Figure 2 shows that the offshore-wind turbine individual simulation output using PSCAD/EMTDC [3]. It shows the real power (P_{wind}MW) and the mechanical torque (Tm_{wind}pu) of the offshore-wind turbine, also the AC voltage line to line ($V_{wind\text{-}L\text{-}L}$ (RMS) kV). The per-unit machine speed is controlled to be 1.014 per-unit constant throughout the simulation. The design parameters of the wind turbine are listed in Table 1.

As shown Fig. 2, the wind generator is in starting up condition until 0.9 s. Wind speed noise is given throughout the simulation period. The noise amplitude controlling parameter is 1 rad s^{-1}, a number of noise components is 30, surface drag coefficient is 0.0192, random wind speed is 8 $m s^{-1}$, and time interval of random

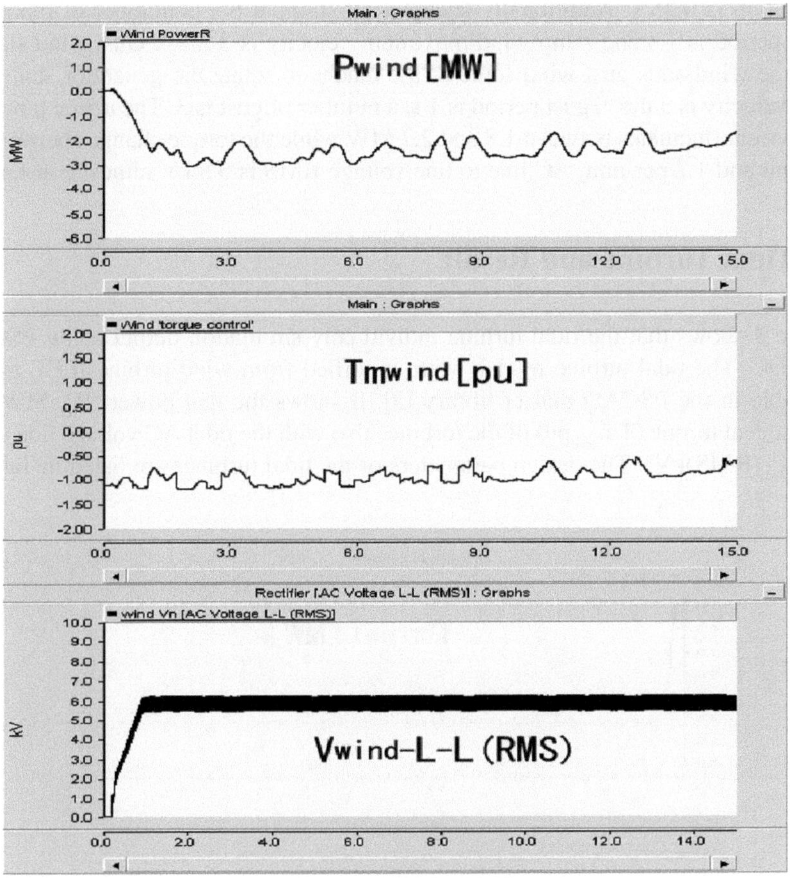

Time (s) 0.00–15.00

Fig. 2 Simulation results for offshore-wind turbine with mechanical torque

Table 1 Parameter of offshore-wind turbine

Parameters	Value
Rated power	2.3 MW
Rated line voltage	5.8 kV
Rotor radius	60 m
Air density	1.229 kg m^{-3}
Rated wind speed	8–11 m s^{-1}
Maximum power of coefficient	0.33
Stator resistance	0.066 p.u.
First cage resistance	0.018 p.u.
Second cage resistance	0.046 p.u.
Stator unsaturated leakage reactance	3.86 p.u.
Rotor unsaturated mutual reactance	0.122 p.u.
Second cage unsaturated reactance	0.105 p.u.
Moment of inertia	5 s

generation is 0.35 s. Additionally Ramp wind starts at 6 s, a number of ramp is 3, ramp period is 1 s and ramp wind maximum velocity is 3 $m s^{-1}$. Gust wind starts at 10 s the wind adds gust wind force to the blades to rotate the generator shaft, gust peak velocity is 3 $m s^{-1}$, gust period is 1 s, a number of gust is 3. The actual power for this system fluctuates between 1.8 and 2.7 MW while the torque changes between 0.7 per-unit and 1.2 per-unit. AC line to line voltage RMS is 5.8 kV almost constant.

3 Tidal Turbine and Result

Figure 3 shows that the tidal turbine individually simulation output using PSCAD/EMTDC. The tidal turbine models were modified from wind turbine IEEE models available in the PSCAD master library [3]. It shows the real power (P_{tidal}MW) and mechanical torque (Tm_{tidal}pu) of the turbine, also with the tidal AC voltage line to line ($V_{tidal\text{-}L\text{-}L}$(RMS)kV). The design parameters of the tidal turbines are listed in Table 2.

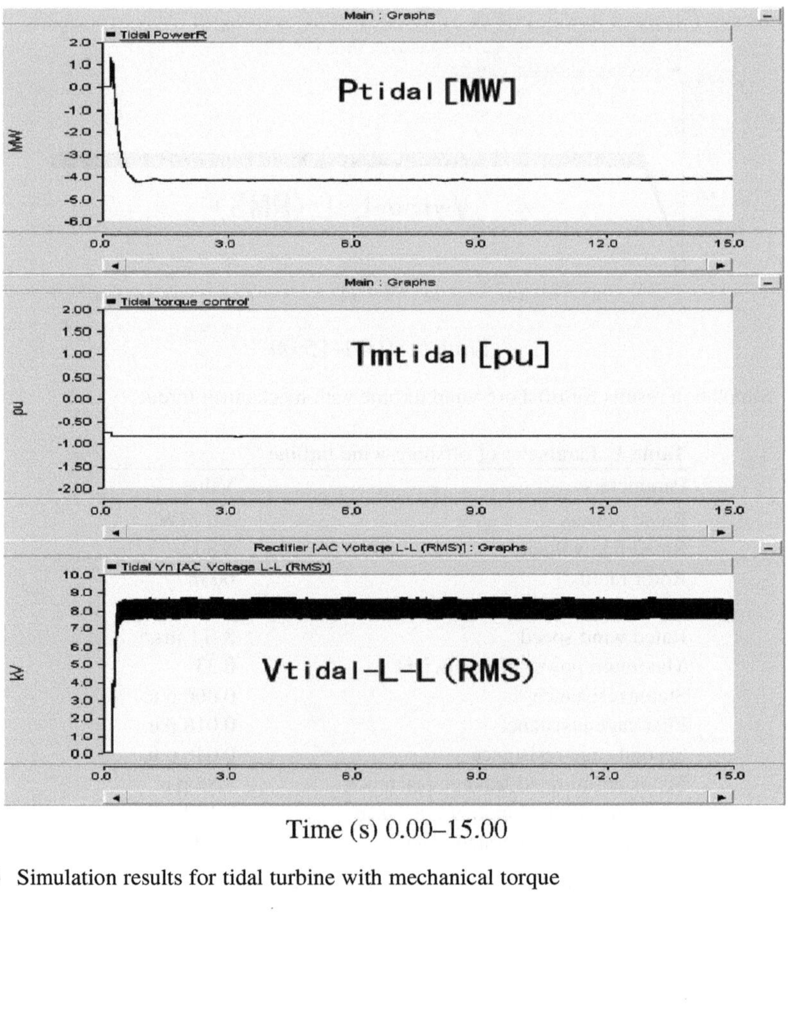

Fig. 3 Simulation results for tidal turbine with mechanical torque

Table 2 Parameters of tidal turbine

Parameters	Value
Rated power (4 turbine ×1 MW)	4 MW
Rated line voltage	8 kV
Rotor radius	15 m
Sea water density	1,026 kg m^{-3}
Rated tidal speed	2.88 m s^{-1}
Maximum power of coefficient	0.33
Stator resistance	0.066 p.u.
First cage resistance	0.018 p.u.
Second cage resistance	0.046 p.u.
Stator unsaturated leakage reactance	3.86 p.u.
Rotor unsaturated mutual reactance	0.122 p.u.
Second cage unsaturated reactance	0.105 p.u.
Moment of inertia	5 s

As shown Fig. 3, the tidal generator is in starting up condition until 0.8 s. The generator condition is assumed to be in steady-state while the tidal speed is between 2 m s^{-1} and 3 m s^{-1}. Figure 3 shows that the tidal power delivers 4.2 MW to the system at around 1.48–15 s. The input torque also has almost steady value of 1.0 per-unit. Tidal AC line to line RMS voltage is 8 kV almost constant.

4 HOTT System

Figure 4 shows that the AC power generated by the wind and tidal turbine generator are converted once into DC power with the rectifier and transmitted onto land via an underwater power cable. It is converted again into AC power through the inverter. The hybrid DC link unit is used to convert the output AC–DC–AC through a 6-pulse GTO converter and inverter [4].

Figure 5 shows the hybrid turbine simulation output using PSCAD/EMTDC, from top to bottom, the converter DC current ($I_{CON\text{-}DC}$pu) and inverter DC current ($I_{INV\text{-}DC}$pu), DC transmission power (P_{DC}pu), the converter DC voltage set point ($V_{CON\text{-}DC}$kV), the inverter DC voltage set point ($V_{INV\text{-}DC}$kV), AC voltage line to line (RMS) ($V_{INV\text{-}AC}$kV).

As shown the image, the generator starting situation (t=0.2 s to t=0.5 s) is caused by simulation difficulty, so we ignore explanation of that. The DC voltage ($V_{CON\text{-}DC}$kV) is kept to the setting point 22 kV from time t=0.51 s until t=15.0 s. AC voltage line to line (RMS) ($V_{INV\text{-}L\text{-}L}$kV) inverter side is 77 kV. The DC transmission line power ($P_{CON\text{-}DC}$ pu) shows that the total hybrid system line power is almost steady even in the ramp and gust wind period. The DC transmission line power is 0.42 per unit (6.3 MW). The DC current is 0.40 per unit. The design parameters of 6-pulse GTO inverter and transformer are listed in Table 3.

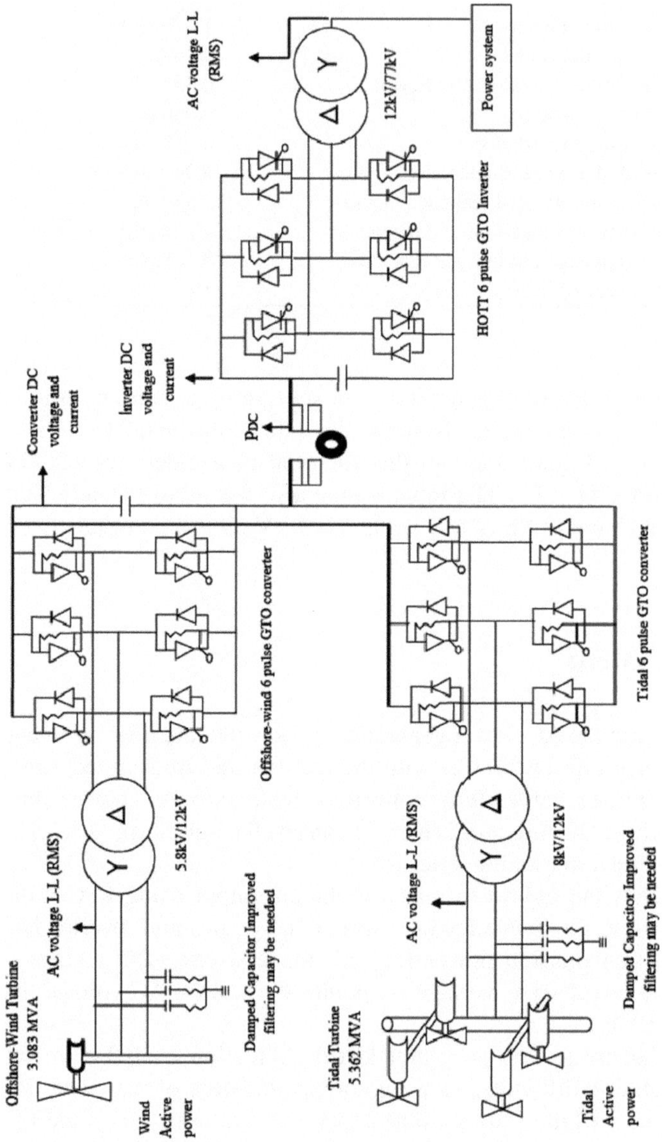

Fig. 4 HOTT system schematic

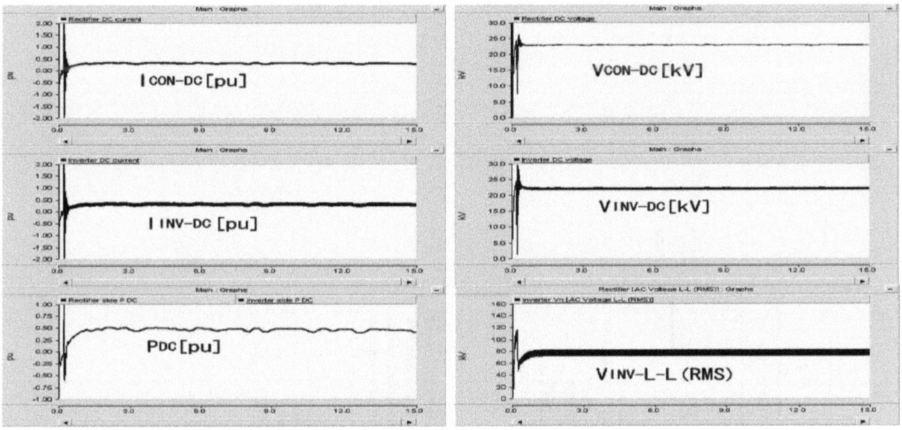

Fig. 5 Simulation results for GTO 6 pulse rectifier and inverter

Table 3 Parameters of 6-pulse GTO transformer and inverter

Parameter	Value
Base MVA	60 MVA
Base voltage	77 kV
Winding I Y	77 kV
Winding II Δ	12 kV
Frequency	60 Hz
DC voltage output	22 kV

5 Control System

Figure 6 shows the HOTT total control system, which controls the 3-phase AC voltages at the wind and the tidal generator terminals and also at the load system, and the DC voltage of the transmission line. That is, the HOTT total system must control the amplitude and the phase angle of the 3-phase voltage references for the PWM control of three GTO converters. The amplitudes *WindMR* and *TidalMR* of the 3 phase PWM sinusoidal reference signal at each generator terminal are determined to meet the reactive power at each terminal to the references *WQref* and *TQref,* respectively. The phases *WindPS* and *TidalPS* of those are determined by signals *WindPS* and *TidalPS*, that are given as a function of the voltage phase difference between the sending ends *WSEPh*, *TSEPh* and receiving end *REPh* to keep the DC transmission power P_{DC}. The PWM control at the inverter end controls the Load side AC voltage V_{AC} and the DC transmission voltage V_{DC}. The amplitude *InvMR* of the PWM sinusoidal reference signal is controlled to follow the reference *Vref* and the phase *InvPS* is given to keep the dc voltage to the reference *DCVref.*

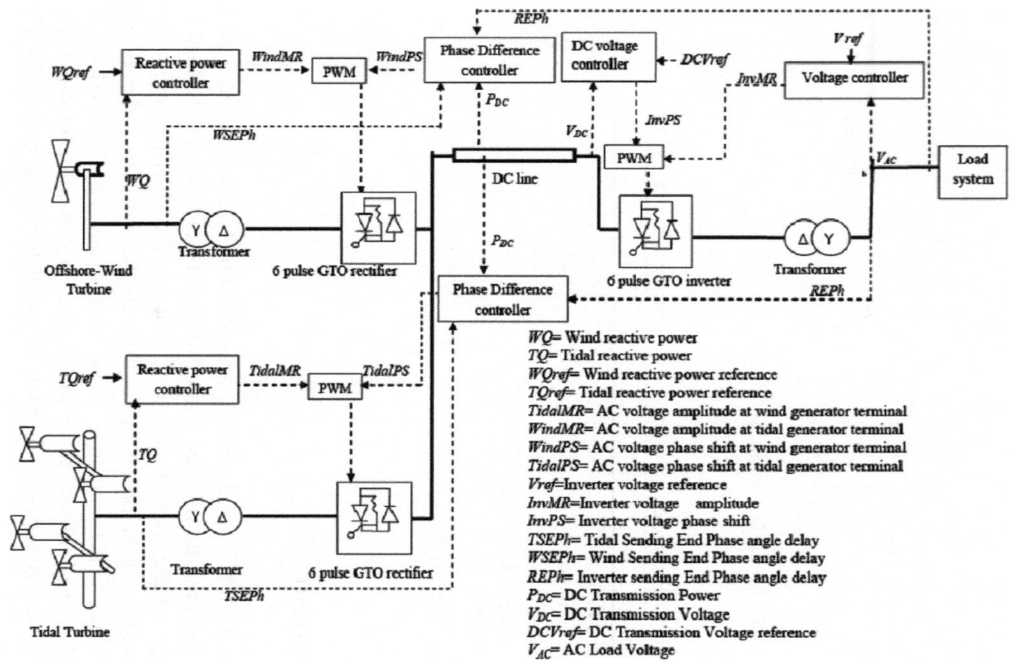

Fig. 6 HOTT control system

6 HOTT Advantage

The proposed HOTT is more flexible than the single system, so that the stable generation ranges of the wind/tidal conditions can extended by adequate system control strategy. Since the rotational speed of the wind turbine is changed irregularly with fluctuations in natural wind, the output voltage and frequency of the wind turbine generator vary widely. However, the dynamic performances of the system have been reported little [1]. In order to maintain the load power in high quality, it is desirable for a HOTT system connected to the load to keep its output voltage and power always constant even when the wind velocity changes. So, in order to realize effective control systems for the hybrid system, it is essential to develop the dynamic model of the system. In this paper, the dynamic model of the hybrid wind and tidal turbine generator system is developed. The dynamic performances of the system when the wind velocity is changing are then investigated to realize a useful control system for the hybrid system. Based on the discussion on the system performances, we propose here a control system to keep the output voltage and power of the whole system constant. Hybrid system method is considered as one of the best techniques converting tidal energy and wind energy into electricity

7 Conclusion

The PSCAD simulation results with a HOTT 6.3 MW+ test system demonstrate satisfactory operation for a range of wind and tidal speeds using 6-pulse GTO rectifier and inverter, it was successfully simulated by PSCAD/EMTDC. The key techniques of offshore-wind and tidal power are design, electric transmission and connection, system and stability operation, system investigation, reactive power and voltage control strategy, the interaction between offshore-wind and tidal turbine. The advance simulation study has to be carried out to ensure stability and also gives a better understanding of the control aspects required to make it more efficient.

Finally an output voltage control system to keep the voltage as well as output power constant for the cases when the wind velocity is changing has been proposed. It has been clarified that the output voltage can be kept almost constant with the proposed control system.

References

1. Rahman ML, Shirai Y (2008) Hybrid offshore-wind and tidal turbine (HOTT) energy conversion I (6 pulse GTO rectifier and inverter). IEEE Xplore/ICSET.2008.4747087, ISBN:978-1-4244-1887-9, 9 January 2009
2. Example tidal turbine developers. Tidal Generation Limited UK, Soil Machine Dynamic Ltd. (SMD) UK, Marine Current Turbines Ltd UK. Tidal energy. Accessed 19 October 2007

3. PSCAD/EMTDC Master library. Wind energy associated models. https://pscad.com. Accessed 25 October 2007
4. Rahman ML, Shirai Y (2009) Hybrid offshore-wind and tidal turbine (HOTT) energy conversion II (6-Pulse GTO Rectifier DC connection and Inverter). In: Proceeding, 16–19 March 2009, Europe's premier wind energy event (ewec2009, Marseille). Online conference proceedings paper: http://www.ewec2009proceedings.info
5. Marine power ocean tidal stream energy. http://www.johnarmstrong1.pwp.blueyonder.co.uk. Accessed 17 April 2008

Primary Pyrolysis and Secondary Reaction Behaviors as Compared Between Japanese Cedar and Japanese Beech Wood in an Ampoule Reactor

Mohd Asmadi, Haruo Kawamoto, and Shiro Saka

Abstract The purpose of this work is to understand the primary pyrolysis and secondary reaction behaviors of two distinct groups of woody biomass, i.e., softwood and hardwood by using Japanese cedar wood (a softwood) and Japanese beech wood (a hardwood). Their demineralized samples were also used in order to understand the influences of inorganic substance. Heat-treatment was conducted under the conditions of N_2/600°C/40–600 s in an ampoule reactor. In this paper, some characteristic features of these wood samples are reported and discussed with the different chemical structures of lignin and hemicellulose as their composing polymers.

Keywords Pyrolysis • Secondary reaction • Hardwood • Softwood • Tar composition • Char reactivity

1 Introduction

Primary pyrolysis and secondary reactions of the primary products are the fundamental steps in various thermochemical conversion processes. In wood gasification, primary pyrolysis forms primary tar, char and gas, and these are further decomposed in the secondary reaction stage. Hosoya et al. have reported the primary pyrolysis behaviors [1], influences of the inorganic substances on primary pyrolysis [2], gasification reactivities of primary products [3], focusing on softwood and its constituent polymers.

There are two major types of wood species, i.e., softwood and hardwood. Hemicellulose and lignin structures are known to be different between these groups. Therefore, such different chemical structures are expected to affect the primary pyrolysis and secondary reaction characteristics. Accordingly, in this work, primary pyrolysis and

M. Asmadi, H. Kawamoto (✉), and S. Saka
Graduate School of Energy Science, Kyoto Univesity, Kyoto, Japan
e-mail: kawamoto@energy.kyoto-u.ac.jp

T. Yao (ed.), *Zero-Carbon Energy Kyoto 2009*, Green Energy and Technology,
DOI 10.1007/978-4-431-99779-5_22, © Springer 2010

secondary reaction behaviors were compared between Japanese cedar (*Cryptomeria japonica*) wood (a softwood) and Japanese beech (*Fagus crenata*) wood (a hardwood).

2 Experimental

Extractive-free wood flour (<80 mesh) was prepared from Japanese beech wood, Japanese cedar wood and their demineralized (acid-washed) samples. Samples were used just after oven-drying at 105°C for 24 h. Ten mg of sample was put in the bottom of a Pyrex glass ampoule. After exchanging the air inside the reactor with N_2, the ampoule was closed and heated in a muffle furnace (600°C) for 40–600 s. After heat-treatment, the ampoule was cooled by flowing air for 1 min, and the non-condensable gases produced were analyzed by gas chromatography (GC) with thermal conductivity detector (TCD). Then, inside of the ampoule was extracted with methanol (1.0 ml, twice), and the soluble portion was analyzed by GC-MS to determine the lignin-derived low molecular weight (MW) products. The DMSO-d_6-soluble-portions, which were prepared with another set of the pyrolysis, were analyzed with ^{1}H-NMR to quantify the polysaccharide-derived products. The insoluble residues were obtained at the bottom of the reactor and on the glass wall at the upper side of the reactor. These dark colored residues are defined as "primary char" and "secondary char", respectively. The amounts of gaseous, tar and char factions were determined by weight difference after gas collection or extraction.

3 Results and Discussion

3.1 Gas, Tar, Char and Water Formation Behaviors

As shown in Fig. 1, a major difference was observed for the char gasification reactivities in the secondary reaction stage (120–600 s). The char yields from the cedar wood were almost constant (~30 wt%) during 120–600 s, while those from the beech wood decreased significantly from ~20 to 3 wt% in this period with the formation of gaseous products. Accordingly the char gasification reactivity was found: beech >> cedar, and hence, the beech wood formed much more gas and less char in 600 s. Demineralization reduced the reactivity of the beech wood of char significantly, while increased the cedar wood char reactivity slightly. Even after demineralization, the beech wood char still had higher reactivity.

Demineralization enhanced the tar gasification reactivities significantly. With this trends, the water formed in the primary pyrolysis stage (< 80 s) tend to be consumed more. Water was suggested to be used for tar gasification in the demineralized samples. Changes in the primary tar compositions through demineralization

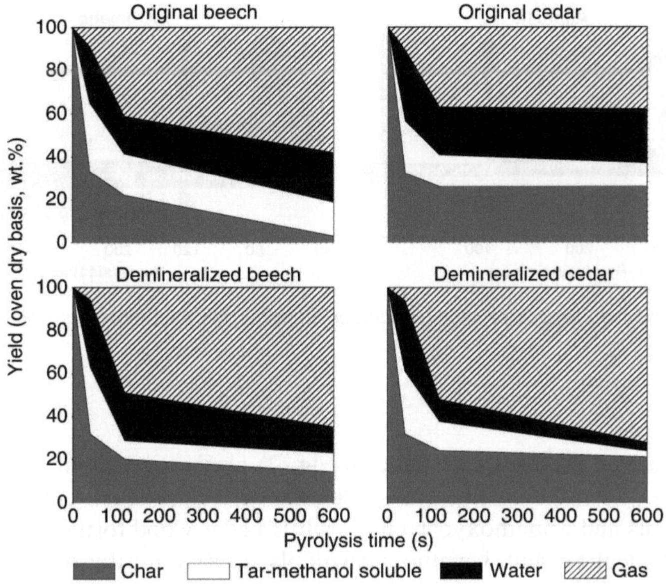

Fig. 1 Yields of gas, tar, char, and water in heat treatment of the original and demineralized wood samples (oven-dried/N_2/600°C)

are considered as a reason for this enhanced tar gasification reactivity. This tendency was much greater in the cedar wood than the beech wood. Demineralization did not increase the net gas yield from the beech wood. The enhanced tar gasification was compensated with the reduced char gasification reactivity.

3.2 *Polysaccharide- and Lignin-Derived Tar Analysis*

Figure 2 shows the time-course changes of the identified tar components from polysaccharide and lignin, which were analyzed by ^{1}H-NMR and GC-MS, respectively. The compositions of the polysaccharide-derived products (aliphatic) drastically changed depending on the heating time. All polysaccharide-derived tar components except for acetic acid and methanol disappeared within 120 or 200 s, mainly due to conversion into the non-condensable gases. Acetic acid and methanol were comparatively stable and these were the important low MW components in the tar fractions after long heating time. Demineralization increased the yield of polysaccharide-derived tar components with high gasification reactivities at primary pyrolysis stage. This would be a reason for high gas yield from the demineralized cedar wood. Tar gasification reactivity in beech wood is lower than cedar wood. Such lower tar gasification reactivity in beech wood arises from the higher yields of acetic acid, hydroxyacetone and methanol, which were more resistant for gasification.

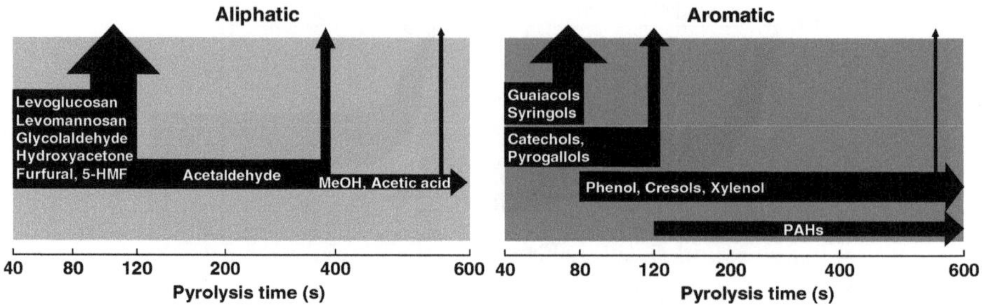

Fig. 2 Time-course changes in chemical compositions of tar fractions (oven dried/N_2/600°C/40–600 s)

Lignin-derived products (aromatic in Fig. 2) also varied significantly in short heating time (40–120 s) as like polysaccharide-derived products. The beech wood gave syringols and 3-methoxycatechols, while cedar wood formed guaiacols and catechols. Catechols and 3-methoxycatechols in 40 s are the products through the O–CH_3 bond homolysis. Structural change in the direction of guaiacols and syringols → catechols, pyrogallols, cresols, phenols and xylenol → PAHs (naphtalenes, phenanthrene, and anthracene) was observed with increasing the heating time. Although tar compositions in 40–80 s were quite different between these species, the tar compositions became similar after long heating time (120–600 s). Demineralization changed the yields of these products, while did not alter the compound types and the direction of the change in their chemical composition.

3.3 *Gas Analysis*

In primary pyrolysis stage (40 s), demineralization significantly increased the CH_4 yield. Since this CH_4 formation is suggested to arise from lignin methoxyl group [3], these results indicate that inorganic substances contained in the wood samples reduce the CH_4 formation from lignin primary pyrolysis. Contrary to this, the H_2, CO and CH_4 yield from cedar wood in the secondary reaction stage (120–600 s) were jumped up by demineralization. Enhanced steam-tar cracking reactions by demineralization are suggested to produced H_2, CO and CH_4 more preferentially than CO_2.

4 Conclusions

The following characteristic features were clarified for Japanese cedar and Japanese beech wood samples:

1. Char gasification reactivity was beech wood >> cedar wood. Demineralization reduced the reactivity of the beech wood char.
2. Aliphatic tar components (mainly from wood polysaccharides) were not different between these species except for their yields and influence of demineralization. Demineralization enhanced the formation of most of the components and this resulted in the higher gas yield from cedar wood in the secondary reaction stage. In case of beech wood, such influence was smaller and the enhanced gas formation reactivity of aliphatic tar was compensated with the reduced char gasification reactivity.
3. Aromatic tar components were quite different only in the primary pyrolysis stage. In longer heating time, the compositions became very similar between these two species.
4. Inorganic substances in wood samples increased the gas yield (H_2, CO and CH_4) in the secondary reaction stage, while decreased the gas yield (except for CH_4) in primary pyrolysis stage.

Acknowledgments This work was supported by the Kyoto University Global COE program of "Energy Science in the Age of Global Warming", and a Grant-in-aid for scientific research (B)(2) (No. 20380103, 2008.4-2011.3)

References

1. Hosoya T, Kawamoto H, Saka S (2007) Pyrolysis behaviors of wood and its constituent polymers at gasification temperature. J Anal Appl Pyrol 78:328–336
2. Hosoya T, Kawamoto H, Saka S (2007) Influence of inorganic matter on wood pyrolysis at gasification temperature. J Wood Sci 53(4):351–357
3. Hosoya T, Kawamoto H, Saka S (2008) Pyrolysis gasification reactivities of primary tar and char fractions from cellulose and lignin as studied with closed ampoule reactor. J Anal Appl Pyrol 83:71–77

Some Low-Temperature Phenomena of Cellulose Pyrolysis

Seiji Matsuoka, Haruo Kawamoto, and Shiro Saka

Abstract Pyrolysis behavior of cellulose was studied at relatively low pyrolysis temperature (below 280°C, under nitrogen). This paper focuses on the reducing end-group in cellulose as a potential reactive site. Number of the reducing end-group could be reduced down to 17% in Avicel PH-101 by heat treatment with glycerol. Thermal glycosylation occurred between the reducing end-groups and hydroxyl groups of glycerol. With this cellulose (named as G-cellulose) as a model with less reducing end-groups, influences of the reducing end-groups on color formation and weight-loss behavior were studied. Color formation was significantly inhibited in G-cellulose. Initial weight-loss in thermogravimetric (TG) analysis was also reduced. Thus, these low temperature phenomena were suggested to arise from the reducing end-groups in cellulose. Furthermore, pyrolysis studies of the cellulose – methyl-β-D-glucopyrnoside (Me-β-Glc) mixture implied that the reducing end-groups activate the transglycosylation reactions for Me-β-Glc. This may be involved in the initial weight-loss in TG analysis.

Keywords Cellulose • Pyrolysis • Color formation • Weight loss • Thermal glycosylation • Reducing end-group

1 Introduction

Cellulose pyrolysis is known as a temperature-dependent process; cellulose decomposes rapidly at >300°C to form volatile products such as levoglucosan, glycolaldehyde and furans, although slow formation of H_2O, CO, and CO_2, reduction in degree of polymerization (DP) into the leveling-off DP (LODP) and color formation proceed at <300°C. In kinetic studies of cellulose, these high and low temperature processes are considered as two competitive pathways following the "active cellulose" formation (1). This type of kinetic model is known as "Broido-Shafizadeh" model [1].

S. Matsuoka, H. Kawamoto (✉), and S. Saka
Graduate School of Energy Science, Kyoto University, Kyoto, Japan
e-mail: kawamoto@energy.kyoto-u.ac.jp

T. Yao (ed.), *Zero-Carbon Energy Kyoto 2009*, Green Energy and Technology,
DOI 10.1007/978-4-431-99779-5_23,

However, the reactions included in these pathways are still unknown, especially for the low temperature processes such as active cellulose formation and low temperature char and gas formation steps. Cellulose is not a chemically homogeneous polymer. One end of cellulose polymer exhibits an aldehyde-character, hence this is called as "reducing end-group". In monosaccharides, reducing sugars have generally higher reactivity than the non-reducing sugars. For example, glucose decomposes at much lower temperature than the corresponding glucosides. Based on these lines of information, the reducing end-groups in cellulose are expected to have higher pyrolysis reactivity than other parts of cellulose, although the content of the reducing end-group is usually very small. In this paper, the influences of the reducing end-groups on low temperature phenomena of cellulose pyrolysis were investigated by using glycerol-treated cellulose. We found that large part of the reducing end-groups was converted to the non-reducing ends through thermal glycosylation occurring between cellulose reducing end-groups and hydroxyl groups of glycerol.

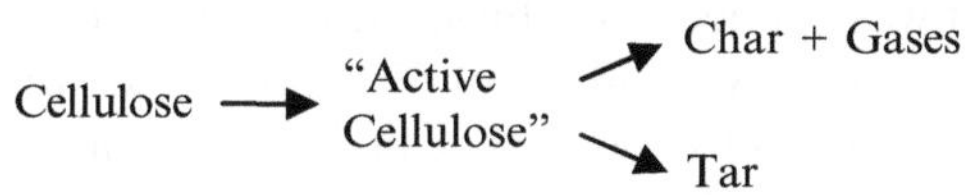

(1)

2 Experimental

Avicel PH-101 (Asahi Kasei Co. Ltd.) and filter paper cellulose (Kiriyama Co. Ltd.) was used. Content of the reducing group was determined by BCA (bicinchoninic acid) method as described in the literature [2], which is based on the reduction of Cu^{2+} to Cu^{+}. Thermogravimetry analysis (TGA) was performed with Shimadzu TGA-50 by using 1 mg of samples. Dynamic TGA was performed at various heating rate (1, 2.5, 5, 10, 20, 40°C min^{-1}). Isothermal TGA was performed at 240–280°C for 600 min after heating up to the desired temperatures at a heating rate of 10°C min^{-1}. Co-pyrolysis of cellulose with Me-β-Glc was carried out as described below. A 1:2 mixture of filter paper cellulose and Me-β-Glc was heated in a muffle furnace under the conditions of 200–280°C / 10–120 min / N_2. The pyrolysates were extracted with D_2O containing 2-furancarboxylic acid as an internal standard. The D_2O-soluble portions were analyzed by ^{1}H-NMR and GPC.

3 Results and Discussion

3.1 Preparation of Alcohol-Treated Cellulose as a Model with Less Number of Reducing Ends

The reducing end-groups in cellulose were found to be converted to the non-reducing glycosides through heat treatment with various alcohols such as glycerol, glucitol and even methanol vapor at 180–240°C. This reaction proceeded without addition of any catalysts. Formation of the glycosidic bonds between cellulose reducing ends and alcohols was confirmed with the successive extraction with water, an alcohol with UV-absorption and measurement of reducing end-groups.

A cellulose–mannitol mixture (1:2, w/w) was heated in N_2 at 240°C for 10 min, and the resulting mixture was washed with water repeatedly. The amount of mannitol included in the mixture decreased with increasing the washing time, down to a certain level. However, one mol of mannitol per about 1,000 mol of cellulose–glucose unit could not be removed from the cellulose sample even after repeated extraction. These results indicate the binding of mannitol to cellulose. Use of phenyl ethanol with UV-absorption character gave more direct proof for chemical bonding between alcohol and cellulose. The phenyl ethanol-treated cellulose (after repeated extraction with methanol) showed UV-absorption in high molecular weight (MW) region in GPC analysis of the acetate derivatives.

The BCA method was applied to evaluate the reducing end contents of cellulose samples. Heat treated Avicel with glycerol (3 equivalents, w/w) were prepared at 240°C (10 min in N_2) and subsequent repeated extraction with water. Hereafter this treated cellulose is called as "G-cellulose" in this paper. G-cellulose had 5.1 nmol mg^{-1} of reducing groups (glucose equivalent) and this value is only about 17% of that of original Avicel (30.5 nmol mg^{-1}). Thus, the content of reducing end-groups is dramatically reduced in G-cellulose, although the reducing end-groups are not completely eliminated. These results are consistent with the previous proposal that cellulose reducing end-groups are effectively converted to glycosides by heat treatment with alcohol. In this paper, G-cellulose was used as a model cellulose with less number of reducing end-group in order to investigate the role of the reducing end-groups on low temperature pyrolysis of cellulose.

3.2 Color Formation

Color formation behaviors were compared between G-cellulose and Avicel PH-101 under the conditions of 200–240°C/10–30 min/N_2. Under all experimented conditions, color formation of G-cellulose was effectively inhibited, although color of Avicel PH-101 changed to light brown to brown under the 10 min heating conditions. These results suggest that low temperature color formation of cellulose is related to the reactions occurring at the reducing end-groups as depicted in Fig. 1. The aldehyde character may play an important role in this color formation, although the detail mechanisms are unknown presently.

3.3 Weight-Loss Behavior

Figure 2 shows the weight-loss behaviors in isothermal TG analysis of G-cellulose and Avicel PH-101 in N_2 flow (10 ml min^{-1}) at 260, 270 and 280°C. At 240°C, no

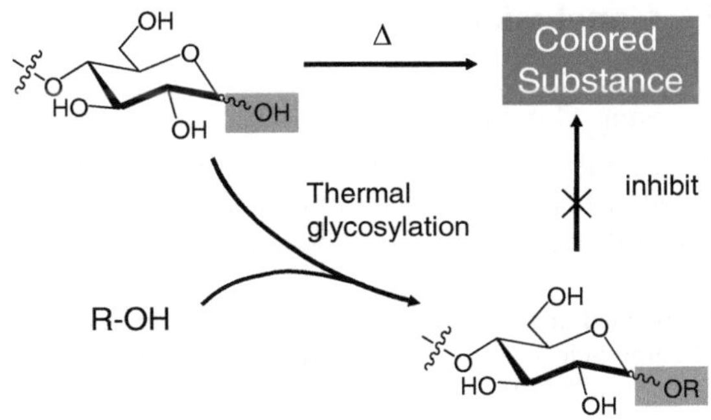

Fig. 1 Role of the reducing end-groups of cellulose on color formation at relatively low temperature

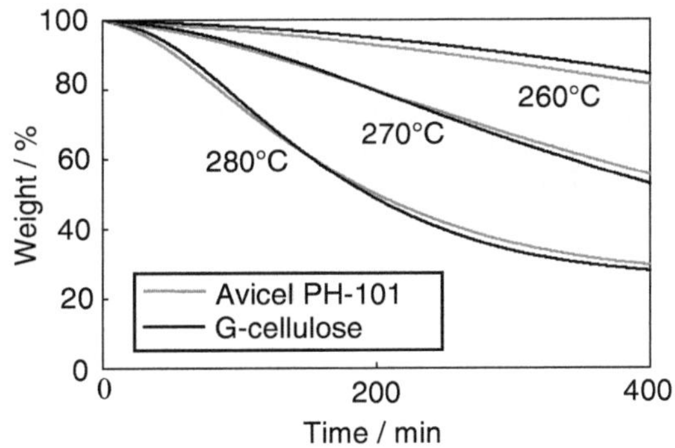

Fig. 2 Isothermal TG analysis of G-cellulose and Avicel PH-101

detectable weight loss was observed up to 600 min heating. Although the effects are comparatively small, the weight loss behavior of G-cellulose was different from Avicel PH-101 in the rates of initial and main weight-loss stages. The observed influences on these rates were opposite with each other. The initial weight-loss rate became slow in G-cellulose, while the main weight-loss rate was rather larger than Avicel PH-101. The same tendency was observed also in dynamic TG analysis. These results indicate that the initial slow weight-loss is related to the reactions of the reducing end-groups and that this conversion decreases the main devolatilization rate.

An interest result was also obtained from the co-pyrolysis experiment of cellulose with Me-β-Glc. Thermalglycosylation reaction of Me-β-Glc was significantly enhanced in the presence of cellulose. This indicates a possibility that the reducing end-groups in cellulose activate the transglycosylation of the co-existing glycoside. Although further study is necessary to extend this proposal to low temperature weight-loss behavior of cellulose, these results suggest that such activation for transglycosylation may be involved in this low temperature devolatilization.

4 Conclusions

By using G-cellulose with less reducing end-groups, some roles of the reducing end-groups on low temperature color formation and weight-loss behavior were suggested. These results will be useful to understand the low temperature pyrolysis of cellulose, which has not been focused yet.

Acknowledgments This work was supported by the Kyoto University Global COE program of " Energy Science in the Age of Global Warming", and a Grant-in-aid for scientific research (B) (2) (No. 20380103, 2008.4-2011.3)

References

1. Bradbury AGW, Sakai Y, Shafizadeh F (1979) A kinetic model for pyrolysis of cellulose. J Appl Polym Sci 23(11):3271–3280
2. Zhang Y-HP, Lynd LR (2005) Determination of the number-average degree of polymerization of cellodextrins and cellulose with application to enzymatic hydrolysis. Biomacromolecules 6(3):1510–1515

Rotational Temperature Measurements in a Molecular Beam with High-Order Harmonic Generation

Kazumichi Yoshii, Godai Miyaji, and Kenzo Miyazaki

Abstract We have developed a new method to measure molecular rotational temperature in a supersonic gas beam, using nonresonant pump and probe femtosecond laser pulses, where the pump forms a rotational wave packet of molecules, and the probe produces high-harmonic radiation from coherently rotating molecules. The rotational temperature can accurately be derived with high spatial and temporal resolutions from the Fourier spectrum of time-dependent harmonic signals. The validity of this method was demonstrated for an expanding supersonic flow of N_2 beam with a rapid temperature decrease.

Keywords Femtosecond laser • Rotational temperature • Molecular beam • High-order harmonic generation

Supersonic molecular beams have been used extensively in material processing, as well as in a variety of fields of molecular science and technology [1]. In the experiment using a molecular beam, the rotational temperature T_{rot} is one of the most important parameters to characterize the molecular system concerned. The most powerful tool to measure T_{rot} in a thin supersonic beam has been coherent anti-Stokes Raman scattering (CARS), where tunable pulsed lasers are usually used to observe the spectrum in resonance to the rotational levels [2, 3]. In addition to the frequency domain measurement, the time-domain CARS spectroscopy has also been developed using femtosecond (fs) laser pulses [4, 5], which has the advantage of high temporal resolution. However, the CARS spectroscopy, as well as its related nonlinear optical methods such as degenerate four-wave mixing (DFWM), often meets some difficulties arising from the complexity of molecular rotational

K. Yoshii
Graduate School of Energy Science, Kyoto University, Kyoto, Japan

K. Yoshii, G. Miyaji, and K. Miyazaki (✉)
Advanced Laser Research Section, Institute of Advanced Energy, Kyoto University, Kyoto, Japan
e-mail: miyazaki@iae.kyoto-u.ac.jp

T. Yao (ed.), *Zero-Carbon Energy Kyoto 2009*, Green Energy and Technology,
DOI 10.1007/978-4-431-99779-5_24,

transitions, e.g., nonresonant background and signal frequencies close to those of the pump laser pulses [5, 6].

In this paper we report a sensitive and versatile method to measure T_{rot} in a supersonic molecular beam, which has high spatial and temporal resolutions in a broad experimental range of molecular beams. The method employs a pump and probe technique with fs laser pulses, where the pump pulse creates a rotational wave packet that leads to temporal alignment of molecules, and the probe pulse generates high-order harmonic radiation from aligning molecules [7, 8]. In contrast to CARS spectroscopy and DFWM, the present method detects the harmonic radiation in a vacuum ultraviolet spectral region far from the fundamental wavelength concerned. The harmonic signal measured as a function of time delay Δt between the pump and probe pulses provides us a definite information of T_{rot} in the supersonic molecular beam.

The experimental setup was the same as in our previous experiments [7–9]. Briefly, the linearly polarized, 40 fs, 800 nm, pump and probe pulses with a time delay Δt are produced from the output of a Ti:sapphire laser system. These pulses were collinearly focused with a 500 mm focal-length lens into a pulsed N_2 beam, where the spot size was measured to be 70 μm. The pump pulse induces transient alignment of molecules through the formation of a rotational wave packet, which is recurrent under field free conditions, and the delayed probe pulse generates high harmonic radiation from coherently rotating molecules as a function of Δt. The pump and probe intensities were optimized for the measurement of T_{rot} in the N_2 beam and fixed at 0.75 and 2.2×10^{14} W cm^{-2}, respectively. The harmonic radiation produced was detected by an electron multiplier mounted on a vacuum ultraviolet monochromator.

The pulsed supersonic N_2 beam of 160 μs duration is jetted into vacuum from the 1-mm diameter nozzle of a piezoelectric valve. The pulsed valve was synchronously operated with the fs laser at a repetition rate of 10 Hz. The pulsed valve attached on a vacuum chamber was mounted on a positioning stage, and the distance x between the nozzle exit and the optic axis was changed by translating the stage (see the inset in Fig. 2).

In the present study we assume that the rotational distribution in the gas jet was in a thermal equilibrium with a Boltzmann distribution. Then the recently developed theory provides the n-th harmonic signal per molecule as a function of Δt [10, 11]. The validity of theoretical expressions for N_2 and O_2 has been verified with our recent experimental study [9]. For N_2 used in this work, the harmonic signal is given as

$$S^{(n)}(\Delta t) \sim c_1+c_2+ \langle\langle\cos^2\theta\rangle\rangle(\Delta t) + c_3\langle\langle\cos^2\theta\rangle^2\rangle(\Delta t) + c_4\langle\langle\cos^4\theta\rangle\rangle(\Delta t) + \text{---}, \quad (1)$$

where c_i is the coefficient depending on the intensity and duration of the probe pulse [10, 11]. The inner bracket of $\langle\langle\cos^2\theta\rangle\rangle$ and subsequent terms stands for the expectation value for the rotational wavepacket consisting of coherently populated rotational states at $\Delta t=0$, and the outer bracket corresponds to the thermal average over the Boltzmann distribution at T_{rot}. It is noted that the time-dependent harmonic signal $S^{(n)}$ is dominated by the initial rotational temperature T_{rot} included in such

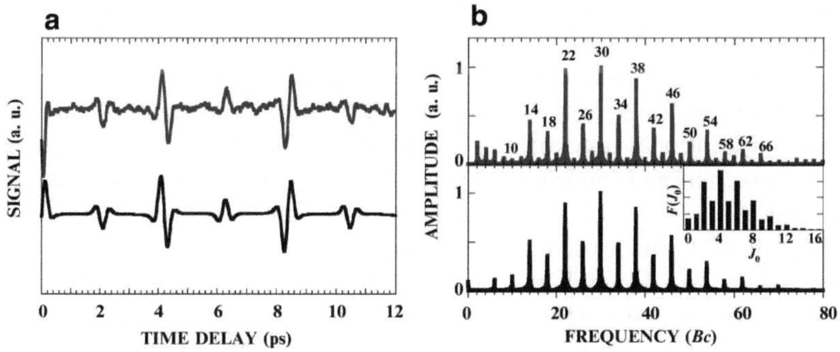

Fig. 1 (**a**) Time-dependent 19th harmonic signals observed at $x/D=0.1$ for $P_0=1$ atm (upper) and calculated (lower) for $T_{rot}=110$ K, and (**b**) their frequency spectra. The *inset* in (**b**) indicates the rotational distribution at $T_{rot}=110$ K

thermally averaged terms as $<<\cos^2\theta>>$, once the rotational wave packet is formed at $\Delta t \approx 0$ with the pump pulse. The evaluation of T_{rot} in the experiment can accurately be made by comparing the frequency spectrum of the observed time-dependent harmonic signal with the theoretical, because the spectrum is very sensitive to the distribution of rotational states at the unique fitting parameter T_{rot}, as shown below.

Figure 1 shows (a) typical examples of the time-dependent harmonic signal observed at the relative position $x/D=0.1$ for $P_0=1$ atm (upper trace) and the signal ϕ (Δt) calculated for the same pump and probe intensities (lower trace), and (b) their corresponding frequency spectra.[1] In the calculation for N_2, we used only the single dominant term $\phi(\Delta t)\equiv c_2<<\cos^2\theta>>(\Delta t)$ to reproduce the signal and its frequency spectrum, since the other terms provide the identical rotational distribution to derive T_{rot} [9]. In Fig. 1, the best agreement between theory and experiment is obtained for $T_{rot}=110$ K. As discussed in detail in the previous studies [8, 10], the time-dependent behavior of harmonic signal is dominated by the beat frequency $\Omega\equiv(E_{J+2}-E_J)/2\pi=(4J+6)Bc$ between any pair of rotational states populated through the Raman transition $\Delta J=\pm 2$, where E_J is the eigenenergy of a state with the angular momentum J, B is the rotational constant, and c is the speed of light. The frequency spectrum represents a distribution of coherently rotating molecules, and their coherent superposition produces the characteristic time-dependent signal including the revival structure of molecular alignment. The inset in the lower spectrum of Fig. 1b represents the initial rotational distribution at $T_{rot}=110$ K. Since N_2 is a homonuclear molecule with nuclear spin of 1, the population $F(J_0)$ of even and odd-J_0 states has a ratio of 2:1. This population alternation is clearly seen in the distribution of Ω. As shown in the inset in Fig. 1b, the initial population is peaked at $J=4$. This population peak is shifted to the $J=6$ state for $\Omega=30Bc$ with the pump pulse. The experimental result shown in Fig. 1b includes the additional small

[1]In producing the frequency spectrum of the observed time-dependent signal, we ignore the initial rapid change at $\Delta t \approx 0$ that is induced by the high intensity of superimposed pump and probe pulses [8].

frequency components that originate from the higher terms in $S^{(n)}$, as discussed in our previous studies [8, 10].

T_{rot} in the free expansion of a supersonic molecular beam decreases with an increase in x/D and/or P_0 [12, 13]. In particular, the decrease in T_{rot} is so rapid in a region of non-isentropic gas expansion, which is usually observed for a small value of x/D [13]. To see this and confirm the applicability of the present method, we measured T_{rot} as a function of x/D (<2) for different P_0, using the 19th harmonic signals.

The result of T_{rot} measured is plotted as a function of x/D as shown in Fig. 2. As expected, a rapid decrease in T_{rot} is observed in the region of $x/D=0.1-1.0$, e.g., $T_{rot}=110$ K decreases to 50 K at $P_0=1$ atm. For comparison, we calculated the translational temperature T in the same region of x/D, assuming the isentropic expansion of molecular gas [13], which is usually less than T_{rot}. The calculation provided $T=245$ K at $x/D=0.1$ and its monotonous decrease to $T=94$ K at $x/D=1.5$. As shown in Fig. 2, the measured value of T_{rot} is much lower than the calculated value of T. This characteristic property of T_{rot} in a region of small x/D reconciles with those reported so far for the pulsed supersonic free jets of N_2 [14] and other molecules [15, 16].

The present technique provides a versatile way to measure T_{rot}, since the nonresonant fs laser can commonly be applied to different kinds of molecules. We have confirmed that T_{rot} in pulsed O_2 and CO_2 beams can accurately be estimated with this method. The temporal resolution in this measurement is on a ps scale with a spatial resolution in the focused beam size. In addition, the time domain measurement has allowed us to evaluate T_{rot} of an ensemble of low-density molecules by observing multiphoton ionization of aligned molecules with a time-of-flight mass spectrometer [17].

In summary, we have demonstrated a sensitive and versatile method to measure rotational temperature in a pulsed supersonic molecular beam, which is based on the observation of time-dependent HHG signal from coherently rotating molecules. The results obtained are consistent with those reported so far.

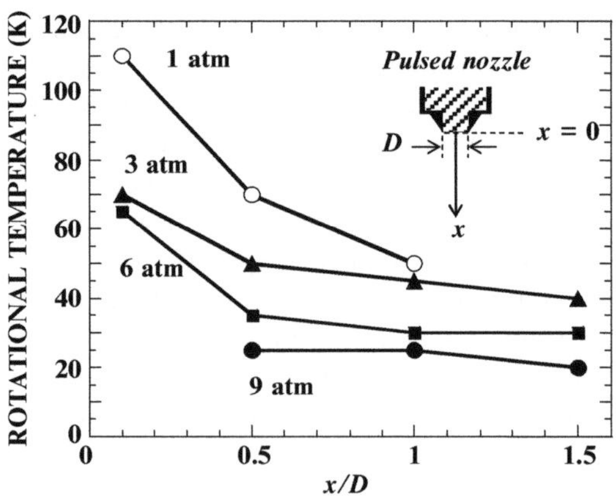

Fig. 2 Rotational temperature in the supersonic N_2 gas jet measured as a function of the distance x/D from the nozzle for different source pressures

Acknowledgments The authors thank Abdurrouf for helpful discussion. This work is partially supported by the Grant-in-Aid for Scientific Research (A) 18206010.

References

1. Scoles G (ed) (1988) Atomic and molecular beam methods. Oxford University Press, New York
2. Huber-Wälchli P, Guthals DM, Nibler JW (1979) Chem Phys Lett 67:233–236
3. Duncan MD, Österlin P, Byer RL (1981) Opt Lett 6:90–92
4. Lang T, Kompa KL, Motzkus M (1999) Chem Phys Lett 310:65–72
5. Lang T, Motzkus M, Frey HM, Beaud P (2001) J Chem Phys 115:5418–5426
6. Hornung T, Skenderović H, Kompa KL, Motzkus M (2004) J Raman Spectrosc 35:934–938
7. Kaku M, Masuda K, Miyazaki K (2004) Jpn J Appl Phys 43:L591–L593
8. Miyazaki K, Kaku M, Miyaji G, Abdurrouf A, Faisal FHM (2005) Phys Rev Lett 95:243903
9. Yoshii K, Miyaji G, Miyazaki K (2008) Phys Rev Lett 101:183902
10. Faisal FHM, Abdurrouf A, Miyazaki K, Miyaji G (2007) Phys Rev Lett 98:143001
11. Faisal FHM, Abdurrouf A (2008) Phys Rev Lett 100:123005
12. Gallagher RJ, Fenn JB (1974) J Chem Phys 60:3487–3491
13. Miller DR (1988) In Scoles G (ed) Atomic and molecular beam methods. Oxford University Press, New York, pp 14–53
14. Barth HD, Huisken F, Ilyukhin AA (1991) Appl Phys B 52:84–89
15. Barth HD, Huisken F (1990) Chem Phys Lett 169:198–203
16. Huisken F, Pertsch T (1986) Appl Phys B 41:173–178
17. Miyazaki K, Shimizu T, Normand D (2004) J Phys B 37:753–761

Chemical Conversion of Lignocellulosics as Treated by Two-Step Hot-Compressed Water

Natthanon Phaiboonsilpa, Xin Lu, Kazuchika Yamauchi, and Shiro Saka

Abstract Chemical conversion of lignocellulosics in two-step semi-flow hot-compressed water treatment was investigated for Japanese beech (*Fagus crenata*) as one of the hardwoods and Japanese cedar (*Cryptomeria japonica*) as one of the softwoods, with the first step at 230°C / 10 MPa for 15 min to decompose hemicellulose and lignin, and the second step for cellulose at 270°C / 10 MPa / 15 min and 280°C / 10 MPa / 30 min for Japanese beech and Japanese cedar, respectively. As a result, totally 97.2% of Japanese beech and 87.8% of Japanese cedar could be solubilized by the hot-compressed water with 2.8% and 12.2% water-insoluble residue composed mainly of lignin, respectively. In addition to the hydrolyzed products, the dehydrated, fragmented and isomerized products as well as organic acids were additionally recovered in the water-soluble portion. The differences observed in lignin between Japanese beech and Japanese cedar would be due mainly to the inherent differences in lignin structure between hardwoods and softwoods.

Keywords Cellulose • Hardwood • Hemicellulose • Hot-compressed water • Hydrolysis • Lignin • Lignocellulosics • Softwood

---

N. Phaiboonsilpa, K. Yamauchi, and S. Saka (✉)
Graduate School of Energy Science, Kyoto University, Kyoto, 606-8501, Japan
e-mail: saka@energy.kyoto-u.ac.jp

X. Lu
On the Leave from College of Food Science and Technology, Nanjing Agricultural University, Nanjing 210095, China

X. Lu
Presently, College of Food Science and Engineering, Northwest Agriculture and Forestry University, Yangling 712100, China

T. Yao (ed.), *Zero-Carbon Energy Kyoto 2009*, Green Energy and Technology, 166
DOI 10.1007/978-4-431-99779-5_25, © Springer 2010

1 Introduction

In an attempt to alleviate both of the environment and energy security problems, bioethanol produced from various biomass resources has long been proposed as a promising biofuel for several decades. Lignocellulosics, one of the most abundant cellulosic biomass on the earth, can be utilized to obtain fermentable saccharides for the ethanol production [1]. However, it is difficult to attain high yield of saccharides from lignocellulosics due to the difference in reactivity of cellulose and hemicelluloses [2]. In this work, therefore, two-step hot-compressed water treatment of lignocellulosics, with the first step performed at low severity to hydrolyze hemicellulose and decompose lignin, and the second step at a higher for the hydrolysis of cellulose, was investigated. In addition, to gain better insights into the two-step hot-compressed water treatment of various kinds of lignocellulosics, a comparative study on the two-step treatment of hardwoods and softwoods was performed.

2 Experimental

Extractive-free wood flour of Japanese beech and Japanese cedar were used [3]. The two-step hot-compressed water treatment was carried out by the semi-flow biomass conversion system developed in our laboratory. All the experiments and analyses were conducted in compliance with the procedures described in our previous papers [3, 4].

3 Results and Discussion

3.1 Effect of Treatment Temperatures on Hydrolysis of Cellulose and Hemicellulose

Prior to the two-step treatment, one-step semi-flow hot-compressed water treatment on the wood samples was preliminarily conducted in order to determine the treatment temperatures at 10 MPa for the first and second steps of the treatment. The results are shown in Fig. 1. The maximum yields of hemicellulose-derived saccharides could be attained at the treatment temperature of 230°C for both kinds of wood, whereas the cello-saccharides from cellulose were recovered in high yield at 270°C and 280°C in case of Japanese beech and Japanese cedar, respectively. Moreover, the crystalline structure of cellulose was found to be obviously destroyed over 270°C [3].

Based on these preliminary experiments, 230°C was thus set at 10 MPa as the temperature for the first-step treatment, while the temperature for the second-step treatment was selected at 270°C for Japanese beech and 280°C for Japanese cedar at 10 MPa to highly recover cello-saccharides from cellulose.

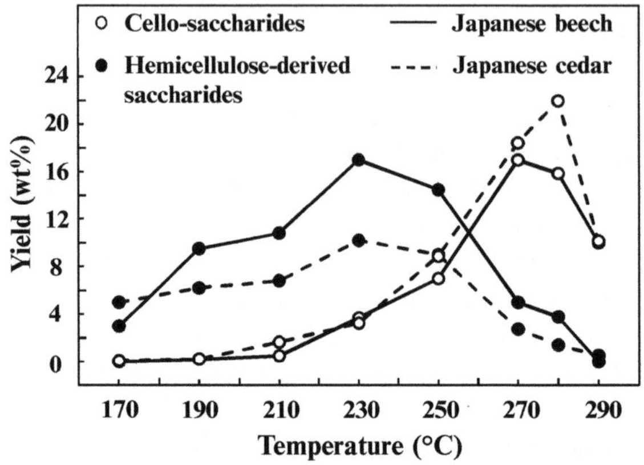

Fig. 1 Yield of the saccharides produced from Japanese beech and Japanese cedar wood flours as treated by hot-compressed water at 170–290°C/10 MPa/21 min. *Open and filled circles* correspond to cello-saccharides and hemicellulose-derived saccharides; *solid and dash lines* represent Japanese beech and Japanese cedar, respectively

3.2 Two-Step Semi-Flow Hot-Compressed Water Treatment

Figure 2 shows the temperature profiles at 10 MPa and various products obtained from Japanese beech and Japanese cedar as treated by two-step hot-compressed water. In the first step, hydrolyzed products of Japanese beech were xylo-saccharides such as xylose and xylo-oligosaccharides, glucuronic acid and acetic acid from the major hardwood hemicellulose, *O*-acetyl-4-*O*-methylglucuronoxylan. In addition, lignin-derived products mainly recovered were sinapyl alcohol and coniferyl alcohol from syringyl and guaiacyl units of hardwood lignin (Fig. 2a).

On the contrary, those from Japanese cedar were glucomanno-saccharides such as mannose, glucose and oligomeric glucomannan, galactose and acetic acid from the major softwood hemicellulose, *O*-acetyl-galactoglucomannan; xylo-saccharides such as xylose and xylo-oligosaccharides, arabinose and glucuronic acid from the minor softwood hemicellulose, arabino-4-*O*-methylglucuronoxylan; and coniferyl alcohol from guaiacyl unit of softwood lignin (Fig. 2b). In the second step, on the other hand, the hydrolyzed products for both kinds of wood were cello-saccharides such as glucose and cello-oligosaccharides from cellulose.

Dehydrated, fragmented and isomerized products as well as organic acids were also recovered in the water-soluble portion (data not shown). Totally, 97.2% of Japanese beech and 87.8% of Japanese cedar could be solubilized with 2.8% and 12.2% water-insoluble residue composed mainly of lignin, respectively. In comparison with the hardwood, softwood showed higher resistance to be hydrolyzed. This might be due to a greater condensed type lignin in softwood which can prevent the cell wall from swelling and thus results in its higher resistance to be hydrolyzed.

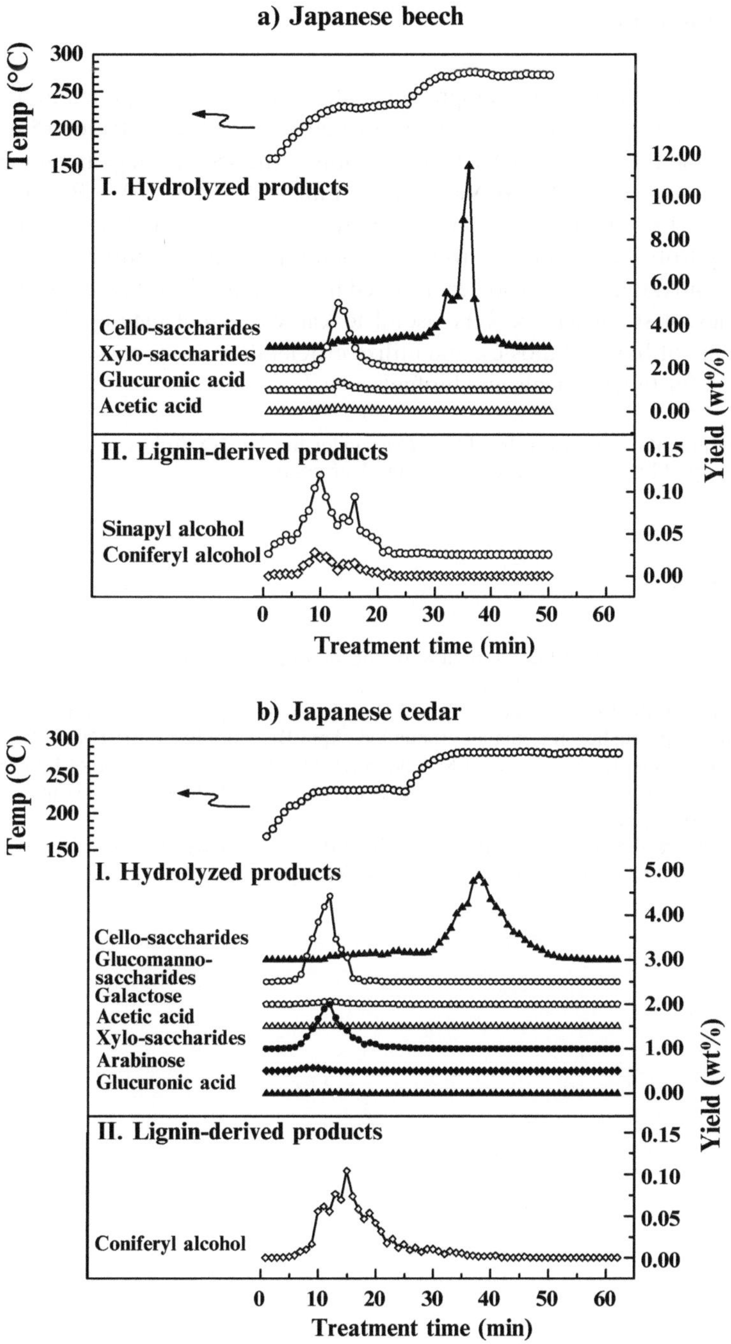

Fig. 2 Hydrolyzed and lignin-derived products from (**a**) Japanese beech and (**b**) Japanese cedar as treated by two-step semi-flow hot-compressed water. *Left axis* corresponds to temperature (*open circles*); *right axis* corresponds to product yields

4 Conclusions

The two-step semi-flow hot-compressed water treatment was applied to hydrolysis of lignocellulosics. Japanese beech was treated at 230°C / 10 MPa for 15min as the first stage and then 270°C / 10 MPa for 15 min as the second stage treatment, while Japanese cedar at 230°C / 10 MPa for 15 min and 280°C / 10 MPa for 30 min, respectively. In comparison with one-step, the two-step method is very effective to obtain hydrolyzed products from hemicellulose and cellulose, separately. Concomitantly, lignin was also decomposed in part in the first stage of the treatment. These lines of study will be very useful to know inherent nature of the chemical components of lignocellulosics, and utilize efficiently various products from lignocellulosics for biochemicals and biofuels.

Acknowledgments This work has been done under the NEDO project and under a partially financial support by the GCOE Program, Kyoto University.

References

1. Zacchi G, Galbe M (2002) A review of the production of ethanol from softwood. Appl Microbiol Biotechnol 59:618–628
2. Söderström J, Pilcher L, Galbe M, Zacchi G (2002) Two-step steam pretreatment of softwood with SO_2 impregnation for ethanol production. Appl Biochem Biotechnol 98:5–21
3. Phaiboonsilpa N, Lu X, Yamauchi K, Saka S (2009) Chemical conversion of lignocellulosics as treated by two-step semi-flow hot-compressed water. In: World Renewable Energy Congress 2009 – Asia, pp 235–240
4. Lu X, Yamauchi K, Phaiboonsilpa N, Saka S (2009) Two-step hydrolysis of Japanese beech as treated by semi-flow hot-compressed water. J Wood Sci 55(5):367–375

Method for Improving Oxidation Stability of Biodiesel

Jiayu Xin and Shiro Saka

Abstract In order to find the influences of antioxidant addition and storage temperature on stability, induction period of safflower biodiesel stabilized with propyl gallate whose concentration spreads from 0 to 5,000 ppm was studied by Rancimat method at the temperature range of 100–120°C. Kinetics on its oxidation was described by the first order rate law. In order to improve the oxidation stability of biodiesel economically, Klason lignin was introduced to supercritical treatment during biodiesel production. It was consequently found that the rapeseed biodiesel prepared by supercritical methanol method at 300°C / 20 MPa for 20 min with a small amount of lignin addition had the induction period longer than 6 h at 110°C in Rancimat test.

Keywords Biodiesel • Oxidation stability • Antioxidant • Kinetics • Supercritical methanol method • Lignin • Rancimat test

1 Introduction

Biodiesel has lower oxidation stability compared with petroleum diesel because biodiesel has high content of unsaturated methyl esters, especially poly-unsaturated methyl esters easily oxidized such as methyl linoleate ($C_{18:2}$) and methyl linolenate ($C_{18:3}$). The unsaturated methyl esters lead to the formation of decomposed compounds such as acids, aldehydes, esters, ketones, peroxides and alcohols. These products not only affect the properties of biodiesel, but also bring the problems of engine operation. As a result, the European Committee for Standardization established a standard (EN 14214) for biodiesel in 2003, which requires that biodiesel must reach a minimum induction period of 6 h as tested by Rancimat method at 110°C.

Therefore, the purpose of this study was to establish an oxidation reaction law of the antioxidant preventing oxidation of biodiesel. Rancimat test which

J. Xin and S. Saka (✉)
Graduate School of Energy Science, Kyoto University, Kyoto, Japan
e-mail: saka@energy.kyoto-u.ac.jp

T. Yao (ed.), *Zero-Carbon Energy Kyoto 2009*, Green Energy and Technology,
DOI 10.1007/978-4-431-99779-5_26, © Springer 2010

can accelerate oxidation was carried out at various temperatures for safflower biodiesel stabilized with propyl gallate. Furthermore, a method to improve the oxidation stability of biodiesel as prepared by supercritical methanol method with inexpensive additives was introduced.

2 Materials and Methods

Oxidation stability of safflower biodiesel samples with different concentrations of propyl gallate (Sigma, Japan, >99%) was studied according to EN 14112 [1] in Rancimat equipment model 743 (Metrohm, Herisau, Switzerland).

Lignin was prepared by Klason method from Japanese beech wood flour. The given amount of the prepared Klason lignin, rapeseed oil and methanol were charged into the reaction vessel (Inconel-625; 5 ml in volume) and the biodiesel was prepared by the batch-type reactor at 300°C/20 MPa as described in the previous work [2].

3 Results and Discussion

3.1 Effect of Temperature and Antioxidant Concentration on Oxidation Stability

Figure 1 shows the influence of antioxidant concentrations on the oxidation stability of safflower biodiesel for the test temperatures as set at 100, 105, 110, 115 and 120°C. We can see clearly that the higher the temperature, the shorter the induction period. In addition, the higher the propyl gallate concentration, the longer the induction period. The effect of propyl gallate concentration on the induction period is more evident when the antioxidant concentration is less than 1,000 ppm.

3.2 Kinetics on Oxidation of Biodiesel as Stabilized with Propyl Gallate

Oxidation reaction of organic compound is a chain reaction process and rather complicated, consisting of numerous elementary steps. The process is an autoxidation process if chain propagation reaction is faster than chain termination; if the antioxidant is active and its concentration is high enough, chain propagation will be broken through the reaction of transfer hydrogen atom from antioxidant to intermediate peroxyl radicals.

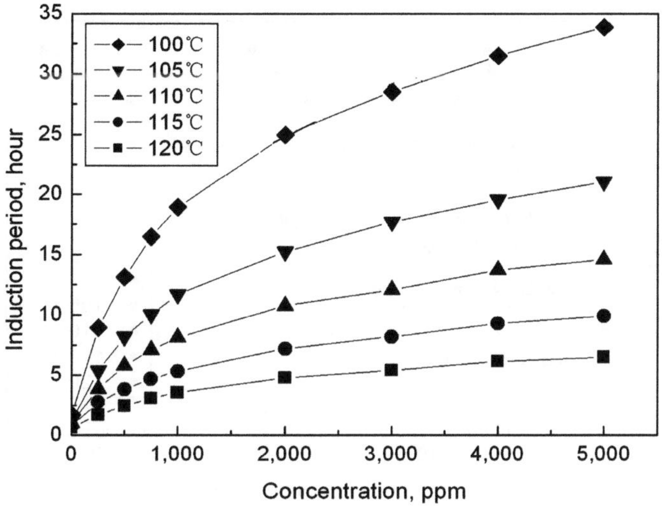

Fig. 1 Influence of propyl gallate concentration on the oxidation stability of safflower biodiesel for test temperatures as set at 100, 105, 110, 115 and 120°C

Antioxidant in biodiesel plays an important role in determining length of induction period. At the beginning of Rancimat test, the concentration of propyl gallate is C_0. With the increase of oxidation time, concentration of propyl gallate decreases and at the end of induction period t_i, its concentration becomes to be C_{cr}. The C_{cr} refers to the concentration of propyl gallate below such value to have no effect on retarding the oxidation of biodiesel. Therefore, the rate of consumption of antioxidant is considered on the basis of the kinetics of the first order reaction for which the rate equation is

$$\frac{dc}{dt} = -kc, \tag{1}$$

where c is the concentration of propyl gallate, t is the oxidation time tested by Rancimat method, while k is the reaction constant of propyl gallate consumption. Integration of (1) within the concentration range from C_0 to C_{cr} and time limits $t=0$ to $t=t_i$ results in the dependence as following equation:

$$\ln C_0 = k(t_i - t_{i0}) + \ln C_{cr}. \tag{2}$$

This shows a linear relation between initial concentration of antioxidant C_0 and induction period t_i. Figure 2 shows a dependence of the induction period of safflower biodiesel on the Napierian (natural) logarithm (ln) of propyl gallate concentration at 100, 105, 110, 115 and 120°C. Straight lines were determined to fit the data in order to adopt the first order rate equation. As expected, the lines show a high degree of correlation.

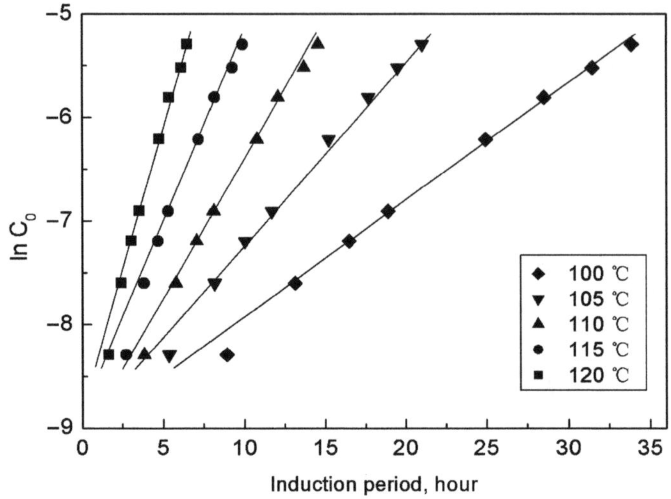

Fig. 2 Dependence of the induction period of safflower biodiesel on the logarithm (ln) of propyl gallate concentration at 100, 105, 110, 115 and 120°C

3.3 Improvement of the Oxidation Stability of Biodiesel as Prepared by Supercritical Methanol Method with Lignin

Among natural phenolic polymers, lignin is the most abundant one, composing up to one-third of the products found in plant cell walls such as trees and agricultural crops. Commercial lignin is currently produced as a co-product of the pulp and paper industry, separated from fibers by a chemical pulping process. Minami and Saka [3] reported that the wood components such as lignin, hemicelluloses and cellulose could be decomposed and liquefied in supercritical methanol at 270°C / 27 MPa. Lignin was prepared by concentrated sulfuric acid method from Japanese beech (*Fagus crenata* Blume) flour as Klason lignin. The sulfuric content determination also showed that the obtained Klason lignin is sulfur free. The lignin was then subjected to the supercritical methanol treatment simultaneously with rapeseed oil, to be liquefied to obtain lignin-derived phenolic products as antioxidants.

Figure 3 shows an effect of lignin addition on oxidation stability of rapeseed biodiesel prepared by supercritical methanol method at 300°C / 20 MPa for 20 min. It is apparent that the course of the oxidation stability is correlated with lignin addition; but the increasing trend of oxidation stability at high lignin addition is not as evident as that at low addition. Under this reaction condition, 15 mg lignin added in the system is enough for the lowest requirement to satisfy the Rancimat test to be longer than 6 h.

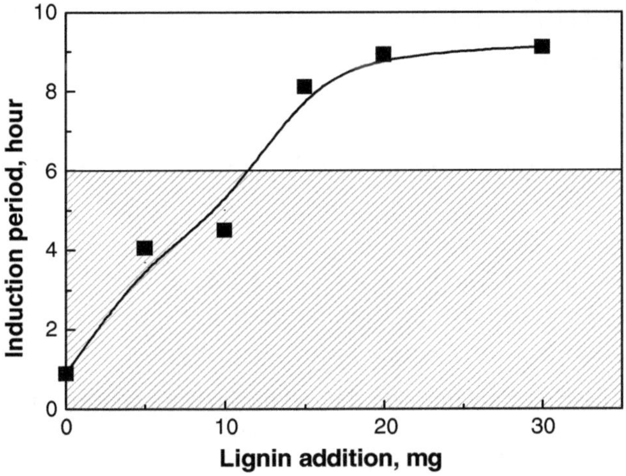

Fig. 3 Effect of lignin addition on oxidation stability of rapeseed biodiesel prepared by supercritical methanol method at 300°C/20 MPa for 20 min. (*Shadowed area* is a range unsatisfactory in EN14214 standard. Given condition is 1.8 ml rapeseed oil and 3.2 ml methanol with lignin designated)

4 Conclusions

In a wide range of concentration of propyl gallate as antioxidant from 250 to 5,000 ppm and temperatures from 100°C to 120°C, the process of oxidation of safflower biodiesel could be described by the first order kinetic law. The results obtained in this work suggested that biodiesel fuel stored at lower temperature is favorable for long time storage of biodiesel without degradation. Oxidation stability of biodiesel prepared by supercritical methanol method with lignin addition was also studied. It was consequently found that the oxidation stability was improved for biodiesel prepared by supercritical method with lignin addition at a temperature of 300°C/20 MPa and molar ratio between rapeseed oil and methanol of 1:42. These data demonstrate that producing biodiesel by supercritical methanol with lignin addition provides an inexpensive and technically acceptable way to improve its oxidation stability.

References

1. European Committee for Standardization (2003) EN 14112, Determination of oxidation stability (accelerated oxidation test). European Committee for Standardization, Berlin
2. Saka S, Kusdiana D (2001) Biodiesel fuel from rapeseed oil as prepared in supercritical methanol. Fuel 80:225–231
3. Minami E, Saka S (2003) Comparison of the decomposition behaviors of hardwood and softwood in supercritical methanol. J Wood Sci 49:73–78

Construction of the Artificial Enzyme for Using Solar Energy

Shun Nakano, Masatora Fukuda, Kazuki Tainaka, and Takashi Morii

Abstract We have recently introduced novel methods for constructing a wide variety of ligand-binding receptors and fluorescent biosensors based on a framework of ribonucleopeptide (RNP) composed of an RNA subunit and a peptide subunit. The modular RNP complex would be potentially applicable to the fabrication of the artificial enzyme because each subunit could be functionalized separately. In this study, we focused on the material conversion process in the photosynthesis and aimed at the development of the artificial photodriven enzymes based on the RNP complex. The strict control of the configuration and orientation of functionalized motifs is needed for the rational design of the RNP-based enzyme. To clarify the relationship between the structure and the function of RNP receptor, secondary structure of the ATP-binding RNP receptor was investigated. As a result, the model for the ATP-binding RNP complex could be proposed based on the chemical structure analyses.

Keywords RNA • Peptide • Biomolecular assembly

1 Introduction

The living things have very highly effective energy utilization and material conversion system by using the biopolymer including the enzymes. To realize sustainable society, one of the important goals is to construct a novel functional biopolymer and to establish new energy and material conversion system by imitating material recycling system in nature. Currently, efficient utilization of solar energy which is permanent,

S. Nakano, M. Fukuda, K. Tainaka, and T. Morii
Graduate School of Energy Science, Kyoto University, Yoshidahonmachi, Sakyo-ku, Kyoto 606-8501, Japan

K. Tainaka and T. Morii (✉)
Institute of Advanced Energy, Kyoto University, Gokasho, Uji, Kyoto 611-0011, Japan
e-mail: t-morii@iae.kyoto-u.ac.jp

T. Yao (ed.), *Zero-Carbon Energy Kyoto 2009*, Green Energy and Technology, DOI 10.1007/978-4-431-99779-5_27, © Springer 2010

inexhaustible and clean "renewable energy source" has been strongly desired. The fossil resources used now are originated in the carbonic acid fixation by the photosynthesis of the plants. If the system that can convert the targeting materials to the industrially useful compounds by using solar energy based on usage of artificial biopolymers can be constructed, it is helpful for the decrease of the negative environmental impacts and the realizing of highly effective energy utilization system. In this study, we focused on the chemical reaction which proceeds at the photosynthetic reaction center and aimed the construction of artificial photodriven enzymes that uses ribonucleopeptide (RNP) as the scaffold and the development of the novel methodology of construction of the artificial enzymes.

We have developed the method of constructing the RNP-based receptors and fluorescent sensors [1, 2]. In this method, we inserted the randomized sequence in the RNA subunit of RNP which the three-dimensional structure and the interaction mode of the RNA and the peptide subunits have already been established [3] (Fig. 1). RNA receptors which bind to targeting small molecules including ATP can be selected from this library by applying in vitro selection method [4]. Then, the modification of the peptide subunit of RNP receptors by a fluorophore enabled to convert RNP receptors into fluorescent biosensors that exerted fluorescent spectral changes upon binding to ATP.

Thus, functional biopolymers can be constructed without loss of the function of each subunit by using the RNA-peptide complex for a basic framework and functionalizing each subunit step by step. In this study, we aim to construct an artificial reductase that reduces the electron acceptor by applying this method. Initially, the substrate-binding motif which is specifically bound with the targeting electron carrier could be constructed at the RNA subunit of the RNP. Subsequently, the light antenna motif which absorbs sunlight efficiently, undergoes the photoinduced charge injection into an electron carrier and achieves a long-lived charge-separated state could be constructed at the peptide subunit of the RNP. Thus, our RNP-based enzyme would be expected to act as the photodriven redox enzyme for the targeting electron carrier.

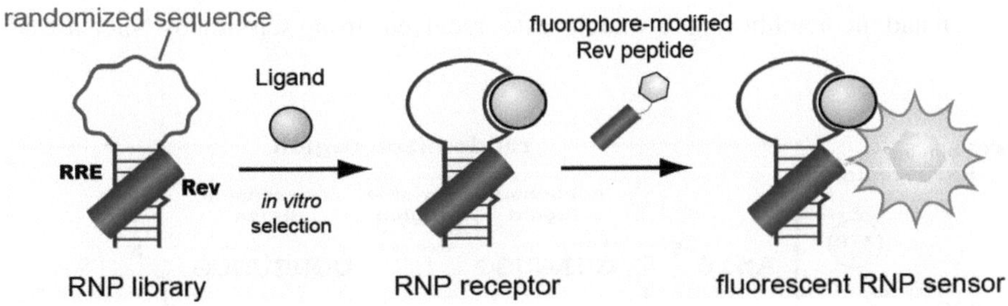

Fig. 1 Schematic illustration shows the stepwise molding strategy to construct ribonucleopeptide (RNP) receptors and sensors. At first, the randomized sequences are introduced to RNA subunit of Rev-RRE complex to construct RNP library. Then, RNP receptors were selected from the library by applying the in vitro selection method. Moreover, the peptide subunit of RNP receptors can be modified by fluorophore to convert to the RNP sensors that can detect the change of the concentration of the ligand

Since the activity of the enzyme and its three-dimensional structure have the close relationship, it is thought that the geometry of each subunit of RNP is an important element that decides the activity of the artificial enzyme in this methodology. Therefore, there is a necessity for knowing the relationship between the structure and the function of RNP receptor to the rational design and strict control of the orientation of each subunit. Consequently, first of all, we aimed to presume the substrate interaction mode of ATP binding RNP receptors based on the secondary structure analyses.

2 The Secondary Structure Analyses of An16/Rev Complex

The mapping by hydrolytic enzymes, the chemical probing with DMS [5] and the in-line probing [6] are performed to solve the part which participates in the substrate-binding of ATP-binding receptors. The ATP-binding receptors that are used to the secondary structure analyses were selected from the RNP library which is consists of 7 to 40 randomized nucleotides and RRE sequence. The various length of RNA subunits were selected and their consensus sequence was 5′-GUAGUGG-UGUGUGUG-3′ that is divided into two regions. The region named variable region that is different among each clones is exist in the middle of these consensus region. Besides, there is the region which consists of complementary sequences in both ends of the consensus region. Among these clones, dissociation constant against ATP of An16 is relatively low ($K_D = 1.9\ \mu M$) and its randomized sequence includes 19 nucleotides which is relatively shorter than other RNA clones of ATP-binding receptors (Fig. 2). Furthermore, the quantitative formation of the RNP complex with Rev peptide of An16 has been already evaluated. From these reasons, it is thought that An16 is moderate for the structure analyses. Therefore, it is thought that we can clarify the interaction mode and the model of ATP-binding RNP receptor based on the secondary structure analyses of An16 receptor.

At first, the mapping by using hydrolytic enzymes was performed. The variable region and the neighboring few nucleotides received strong scission by Nuclease S1

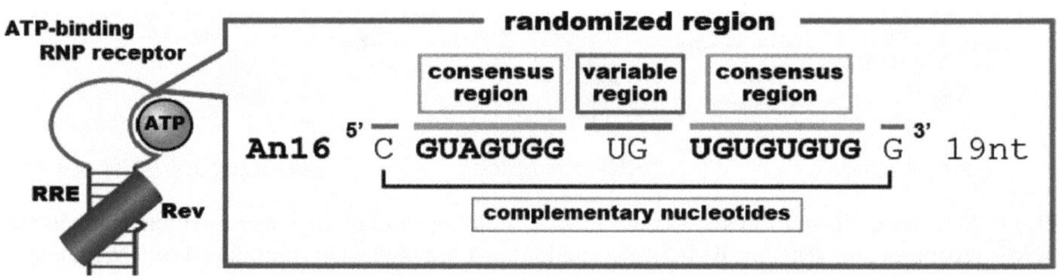

Fig. 2 The nucleotide sequence of An16 in randomized region selected by in vitro selection for ATP. It seems to consist of the three regions. The nucleotides shown by bold faces are the consensus sequences

that recognizes specifically the single-stranded nucleotides. On the other hand, the RRE sequence region was cleaved by RNase V1 that recognizes the double-stranded nucleotides. Because the tendency of band intensity and its changes are corresponded to that of Rev-RRE complex which has already reported [7], it is thought that the RRE sequence region takes the structure which is similar to that of the Rev-RRE complex. Interestingly, the major part of the consensus region was cleaved by both Nuclease S1 and RNase V1. So, this region is difficult to assign to the specific structure. It is assumed that these regions directly relate to ATP-binding.

Then, we attempted the chemical probing with DMS to evaluate the accessibility of the N7 of guanine to the solution. Because the nucleotides in the variable region and the neighboring ones show the strong band intensity, it is thought that these nucleotides are easy to access to the solution. Meanwhile the major of the nucleotides in the consensus region were hard to be cleaved. Besides, the band intensity of these nucleotides further decreased with addition of ATP. These results agree with the result of the enzyme mapping well, and show the possibility that the part of the consensus region takes part in the substrate binding.

In addition, the structural freedom of the RNA chain at each nucleotide was evaluated by the in-line probing. The band intensity of A34 and U38 bulge in RRE region, two nucleotides of variable region and U15 were strong. It implies that the structural freedoms of the RNA chain of these nucleotides are high. Especially, the strong band intensity of two bases of the variable region shows the possibility that this region takes single-stranded structure as seen up to now in the mapping by hydrolytic enzyme and the chemical probing with DMS. In contrast, the band intensity of other bases decreased along with the substrate binding and the tendency in which a specific structure was formed was able to be observed. From the above-mentioned results, ATP-binding RNP receptors took the loop structure around the variable region, and it was shown that the part of the consensus region that positioned at the both ends of the variable region might be the substrate binding core.

3 Conclusions

In this study, we have analyzed the secondary structure of ATP-binding RNP receptor through the nucleotide mapping in the solution in the absence or presence of the substrate. As a result, we have succeeded in deducing a possible part that bind directly to the substrate using An16/Rev of the ATP-binding RNP. Further investigations of the secondary structure mapping and other structural analyses of the ATP-binding RNP, such as the NMR measurement, would allow construction of the three dimensional structural model of the ATP-binding RNP receptor. Such structural information would lead to a precise design of functional RNPs.

References

1. Morii T, Makino K et al (2002) In vitro selection of ATP-binding receptors using a ribonucleopeptide complex. J Am Chem Soc 72:4617–4622
2. Hagihara M, Morii T et al (2006) A modular strategy for tailoring fluorescent biosensors from ribonucleopeptide complexes. J Am Chem Soc 128:12932–12940
3. Williamson JR et al (1996) αHelix-RNA major groove recognition in an HIV-1 Rev peptide-RRE RNA complex. Science 273:1547–1551
4. Ellington AD, Szostak JW (1990) In vitro selection of RNA molecules that bind specific ligands. Nature 346:818–822
5. Peattie DA, Gilbert W (1980) Chemical probes for higher-order structure in RNA. Proc Natl Acad Sci USA 77:4679–4682
6. Breaker RR et al (1999) Relationship between internucleotide linkage geometry and the stability of RNA. RNA 5:1308–1325
7. Kjems J et al (1992) Specific binding of a basic peptide from HIV-1 Rev. EMBO J 11:1119–1129

Development of Fluorescent Ribonucleopeptide-Based Sensors for Biologically Active Amines

Fong Fong Liew, Masatora Fukuda, and Takashi Morii

Abstract Development of functional biomacromolecules is essential for a sustainable energy technology that is based on the principle of energy-utilization and energy-transformation pathways in the living systems. Though there still remain technological barriers, use of biomacromolecules for energy-utilization and transformation will contribute to realize a society that would be harmless to the nature. Biosensor transduces a binding event between a "reporter" and its target ligand into an optical signal, thereby enables monitoring the interaction of biologically active ligands such as amines. Previously, we have established the strategy for the construction of receptors and biosensors based on ribonucleopeptide (RNP) scaffold. RNP receptors that bind to a specific target molecule were constructed by the combination of RNA aptamer technology with the Rev Responsive Element (RRE)-HIV Rev peptide complex. Modification of Rev peptide with a fluorophore enables construction of histamine-binding fluorescent RNP library. Eventually, RNP sensors for histamine that show higher selectivity to histamine over histamine analogs included imidazole, ethylamine and L-histidine were identified through the screening of the histamine-binding fluorescent RNP library.

Keywords RNA • Peptide • Biosensor • Histamine

1 Introduction

Biologically active amines, such as histamine, catecholamine, and serotonin, have a close relevance to the regulation of movement in the central nervous system and are presumably associated with the pathophysiology of Parkinson's and Huntington's

F.F. Liew, M. Fukuda, and T. Morii
Graduate School of Energy Science, Kyoto University, Yoshidahonmachi, Sakyo-ku, Kyoto 606-8501, Japan

T. Morii (✉)
Institute of Advanced Energy, Kyoto University, Gokasho, Uji, Kyoto 611-0011, Japan
e-mail: t-morii@iae.kyoto-u.ac.jp

T. Yao (ed.), *Zero-Carbon Energy Kyoto 2009*, Green Energy and Technology,
DOI 10.1007/978-4-431-99779-5_28,

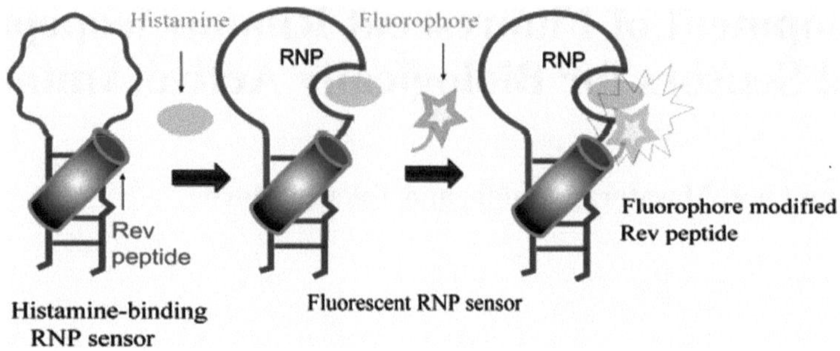

Fig. 1 A stepwise molding strategy for the construction of RNP fluorescent sensors. Combination of the RNA subunits as a histamine-binding receptor and a fluorophore-modified Rev peptide provided a histamine-specific RNP fluorescent sensor

diseases, psychosis, and drug addiction [1]. Quantitative and high-sensitive diagnostic methods for biologically active amines were persistently required for exploration of a new class of drugs against such neurological disorders. Our strategy for the stepwise molding based on a framework of ribonucleopeptide (RNP) provides simultaneously a wide variety of fluorescent biosensors with diverse functions, i.e., high signal-to-noise ratios, different wavelengths and various concentration ranges for the ligand detection (Fig. 1) [2]. In the first step to fabricate RNP based fluorescent sensors, an RNA-derived RNP pool was prepared by a structure-based design of the Rev Responsive Element (RRE)-HIV Rev peptide complex [3] appended with a randomized nucleotides region as a ligand binding domain, adjacent to the RRE segment. *In vitro* selection method [4] was applied to a randomized nucleotides region to afford a series of ATP-binding RNP receptors with high selectivity and affinity [3]. In the second step, ATP-binding RNP receptors were successfully functionalized as ATP binding fluorescent sensors by the chemical modification of the *N*-terminal of the Rev peptide with a various kind of fluorophore. RNP-based fluorescent sensors are also applicable to the other class of small molecules, such as biologically active amines.

Herein we report on the fluorescent RNP sensors for histamine with diverse functions. RNA subunits as a ligand binding domain were prepared by in vitro selection method, as previously reported [3, 5]. Rev peptide motifs as a function of signal transducer were afforded by the fluorescent labeling to the *N*-terminal of peptide. Histamine selective RNP fluorescent sensors were conveniently screened by the RNP pool composed of RNA subunits and Rev peptide motifs, and as a result, displayed expedient optical and binding properties.

2 Result and Discussion

RNP receptors for histamine were isolated from an RNA-derived pool through in vitro selection method [5]. After each round of selection, an RNP pool was incubated with an immobilized histamine-agarose resin and the resin-bound fraction of

RNP was recovered with buffer. Those bound RNA fractions were collected, through reverse transcription and followed by PCR amplification to generate new DNA pool. After 15 rounds of in vitro selection, RNP that has binding ability to histamine RNA library was obtained. Through DNA sequencing analysis, sequences of RNA subunit that predicted to form the histamine-binding site was identified. RNP receptors with specific binding affinity to histamine were shown in Fig. 2. These clones did not have consensus sequence.

Figure 3 shows relative ratios of fluorescence intensity (I/I_0) in the absence (I_0) and the presence (I) of histamine for fluorescent RNPs with 7mC-Rev, Pyr-Rev, 6FAM-Rev and NBD-Rev monitored at 390, 390, 535 and 535 nm. RNP receptors with fluorophore-modified Rev peptide showed an increase in the fluorescence emission intensity upon addition of histamine [3, 5]. The H05 RNA complexed with 7mC-Rev, 6FAM-Rev and NBD-Rev with different I/I_0 value of 1.68, 1.24 and 1.23,

clone	Nucleotide Sequences
H01	AGACGGCCUUACGUGGUGUAAAG
H02	GUCUGCCAUUAAAUGUUCAGACCGGAAUCCGGCUAAGU
H05	UGGGAAUAACAAUAGUCCUAACUGGCAACU

Fig. 2 Nucleotide sequences obtained for the randomized region of histamine-binding RNP receptors

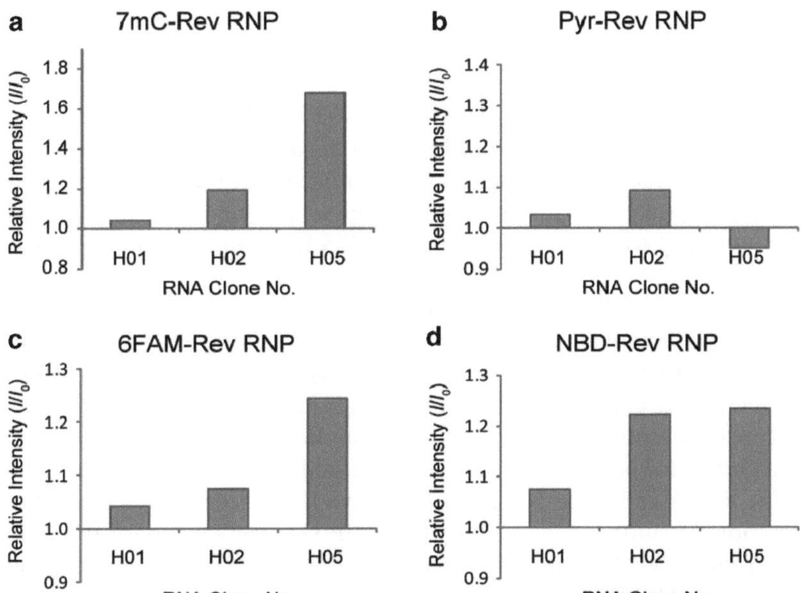

Fig. 3 Relative fluorescence intensity changes (I/I_0) of RNPs with existence of 1 mM histamine are shown in the bar graphs for (**a**) 7mC-Rev RNP, (**b**) Pyr-Rev RNP, (**c**) 6FAM-Rev RNP, and (**d**) NBD-Rev RNP

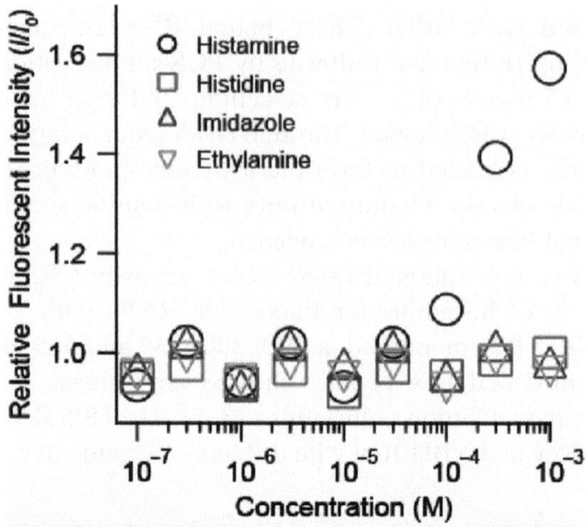

Concentration [1mM]	Histamine	Histidine	Imidazole	Ethylamine
Relative Fluorescent Intensity (I/I_o)	1.57	1.00	0.97	0.97

Fig. 4 Direct titration of a fluorescent RNP complex (1 mM) of the H05 RNA subunit and 7mC-Rev (H05/7mC-Rev) with histamine, histidine, imidazole and ethylamine. (0.1, 0.3, 1, 3, 10, 30, 100, 300 μM and 1 mM)

respectively as shown in Fig. 3a,c,d. The H05-Pyr Rev peptide (RNP) exhibited the slight fluorescence changes by the addition of histamine. On the other hands, H05 RNP with other fluorophores showed the significant changes in fluorescent intensity in the presence of 100 μM to 1 mM histamine. It is suggested that the dissociation constant (K_D) for the complex between H05 RNA and histamine is more than 100 μM.

Characteristic of H05/7mC-Rev peptide binding with different histamine analogues was investigated by measuring the fluorescence intensity changes with presence of various histamine analogues (Fig. 4). Titration curves for H05/7mC-Rev by increasing concentrations of histamine analogues were judged by relative fluorescence intensity changes. From the nonlinear regression analysis of the titration curve, dissociation constant for the complex between H05/7mC-Rev and histamine was found to be 502.5 μM. Relative fluorescence intensity ratio (I/I_0) at saturation of substrate was 2.28. H05 RNP showed selectivity in binding of histamine analogues. Both the imidazole moiety and the aliphatic chain of histamine were recognized by H05 RNP. It was assumed that presence of carboxyl group at histidine would inhibit the formation of ligand-fluorescent RNP complex.

3 Conclusion

In conclusion, fluorescent histamine sensors were successfully constructed by utilizing the modular strategy for tailoring fluorescent RNP sensor. The complex of histamine-binding RNA aptamer and the fluorescent Rev peptides afforded fluorescent RNP sensors that showed distinct changes in the fluorescence emission intensities upon binding to histamine. H05/7mC-Rev revealed a distinct selectivity for histamine over the structurally related analogs of histamine. Development and construction of histamine biosensor is valuable in early stages of preclinical development as a therapeutic or as diagnostic tool. It would be possible to construct RNP-based fluorescent biosensors for other biologically active amines.

References

1. Nestler EJ (1994) Hard target: understanding dopaminergic neurotransmission. Cell 79: 923–926
2. Hagihara M, Morii T et al (2006) A modular strategy for tailoring fluorescent biosensors from ribonucleopeptide complexes. J Am Chem Soc 128:12932–12940
3. Williamson JR et al (1996) αHelix-RNA major groove recognition in an HIV-1 Rev peptide-RRE RNA complex. Science 273:1547–1551
4. Ellington AD, Szostak JW (1990) In vitro selection of RNA molecules that bind specific ligands. Nature 346:818–822
5. Hagihara M, Morii T et al (2008) Context-dependent fluorescence detection of a phosphorylated tyrosine residue by a ribonucleopeptide. J Am Chem Soc 130:8804–8812

Light Energy Induced Fluorescence Switching Based on Novel Photochromic Nucleosides

Katsuhiko Matsumoto, Yoshio Saito, Isao Saito, and Takashi Morii

Abstract Photochromic molecules are potentially applicable to the optical module for molecular switches and sensors. Herein, we report on a novel photochromic vinylpyrene-substituted 2'-deoxyguanosine analogue (^{VPy}**G**) that undergoes the reversible *E–Z* isomerization accompanied with a unique "on–off" fluorescence switching. The highly fluorescent *E*-isomer was rapidly converted to the non-fluorescent *Z*-isomer under visible light irradiation (>420 nm). Conversely, when the *Z*-isomer was illuminated with UV-light (~365 nm), *Z*- to *E*-isomerization also took place rapidly. Furthermore, such reversible photoisomerization was repeated more than 10 times without any side reaction. The drastic and reversible fluorescence change of the photochromic guanine base ^{VPy}**G** might be useful for molecular devices.

Keywords DNA • Photochromic nucleotide • Photo-isomerization

1 Introduction

Fluorescent photoswitching molecules have attracted remarkable interest for their possible application to fluorescent sensors [1–5], optical devices, and bioimaging [6–8] due to its high sensitivity and selectivity. We designed and synthesized C_8-vinylpyrene-substituted 2'-deoxyguanosine ^{VPy}**G**, which would be potentially promising as a photochromic nucleobase for fluorometric sensing, genetic analysis and photoregulation of nucleic acid structures. This molecule showed a unique "on–off" fluorescence switching owing to its rapid photoisomerization between highly fluorescent (*E*-isomer) and non-fluorescent states (*Z*-isomer) (Fig. 1).

K. Matsumoto and T. Morii (✉)
Graduate School of Energy Science, Kyoto University, Yoshidahonmachi,
Sakyo-ku, Kyoto 606-8501, Japan
e-mail: t-morii@iae.kyoto-u.ac.jp

Y. Saito and I. Saito
Department of Materials Chemistry and Engineering, School of Engineering,
Nihon University, Koriyama, Fukushima 963-8642, Japan

T. Yao (ed.), *Zero-Carbon Energy Kyoto 2009*, Green Energy and Technology,
DOI 10.1007/978-4-431-99779-5_29, © Springer 2010

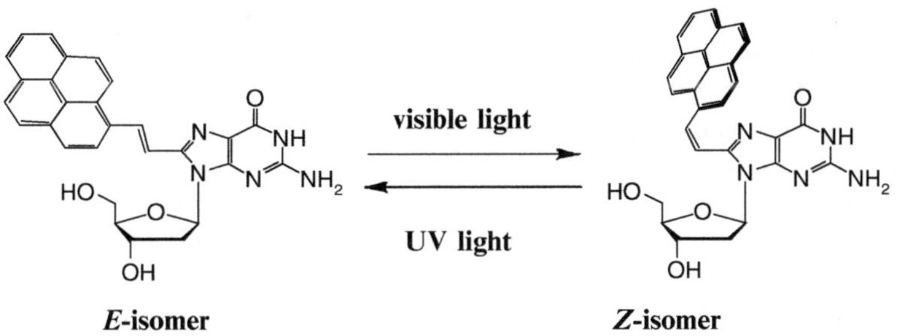

Fig. 1 Light energy induced fluorescence switching of a photochromic nucleoside **VPyG**

2 Results and Discussion

VPyG was prepared according to Scheme 1. 2'-Deoxyguanosine **2** was bromated with *N*-bromosuccinimide and then the amino group was protected with *N,N*-dimethylformamide diethylacetal in methanol at 60°C. The N^2-protected C_8-bromo-2'-deoxyguanosine **4** was then subjected to Pd(0)-mediated Still coupling [9, 10] to afford C_8-vinyl substituted 2'-deoxyguanosine analogue **5**. The compound **5** was coupled with 1-bromopyrene in the presence of sodiume acetate, to give N^2-protected **VPyG**. The N^2-protected **VPyG** was treated with NH_4OH, to afford fluorescent switchable nucleoside **1** (**VPyG**). *E*-isomer of **VPyG** was easily converted to *Z*-isomer **2** under room light. The mixture of *E*- and *Z*-isomers was separated in a pure form by using HPLC.

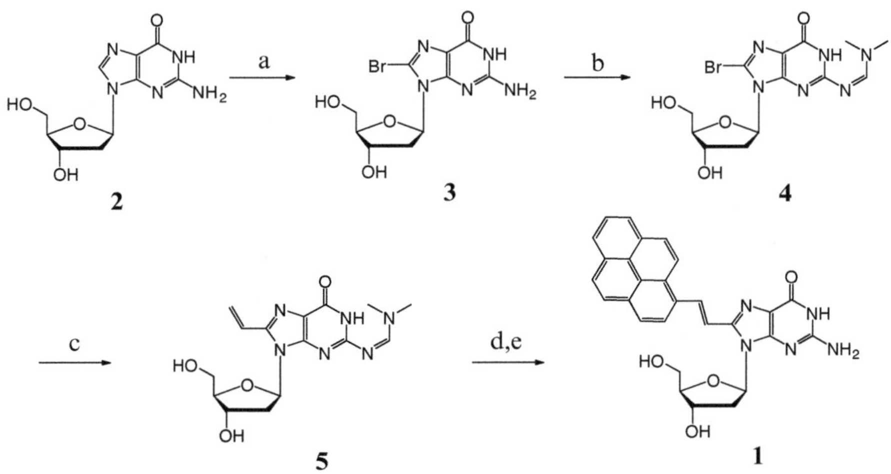

Scheme 1 Synthesis of fluorescent switchable nucleoside **VPyG**. Reagent and coditions: (**a**) *N*-bromosuccinimide, H_2O, r.t., 0.5 h, 82% (**b**) *N,N*-dimethylformamide diethylacetal, methanol, 60°C, 3 h, 98% (**c**) $Pd(PPh_3)_4$, $Sn(CHCH_2)_4$, Et_3N, DMF, 60°C, 12 h, 62% (**d**) 1-bromopyrene, $Pd(PPh_3)_4$, CH_3COONa, DMF, 80°C, 12 h, 31%, (**e**) NH_4OH/methanol, r.t., 8 h, quant

In order to evaluate the photochromic property of ^{VPy}**G**, the photoisomerization of ^{VPy}**G** in methanol was initially investigated. Upon illumination of *E*-isomer with visible light (>420 nm), the *Z*-isomer was predominantly formed (92%) in the photostationary state as determined by HPLC. *E*-isomer was then regenerated by UV irradiation (365 nm) of its *Z*-isomer, and it was afforded in 82% yield. These results suggest that highly reversible *E*- to *Z*- photoisomerization of ^{VPy}**G** can be achieved by illumination at 365 and 420 nm. Thus, the *E–Z* isomerization of photoresponsive ^{VPy}**G** was conducted by 420 nm light separated from a 100 W Xenon lamp by a filter solution and 365 nm light from a UV-transilluminator. Photoirradiation of *E*-isomer at 420 nm resulted in a rapid decrease in the absorption at 405 nm with a blue shift of ca. 40 nm and in an increase in the absorption at 280 nm with a blue shift of ca. 2 nm, indicating a quantitative *E* to *Z* photoisomerization (Fig. 2b). On the other hand,

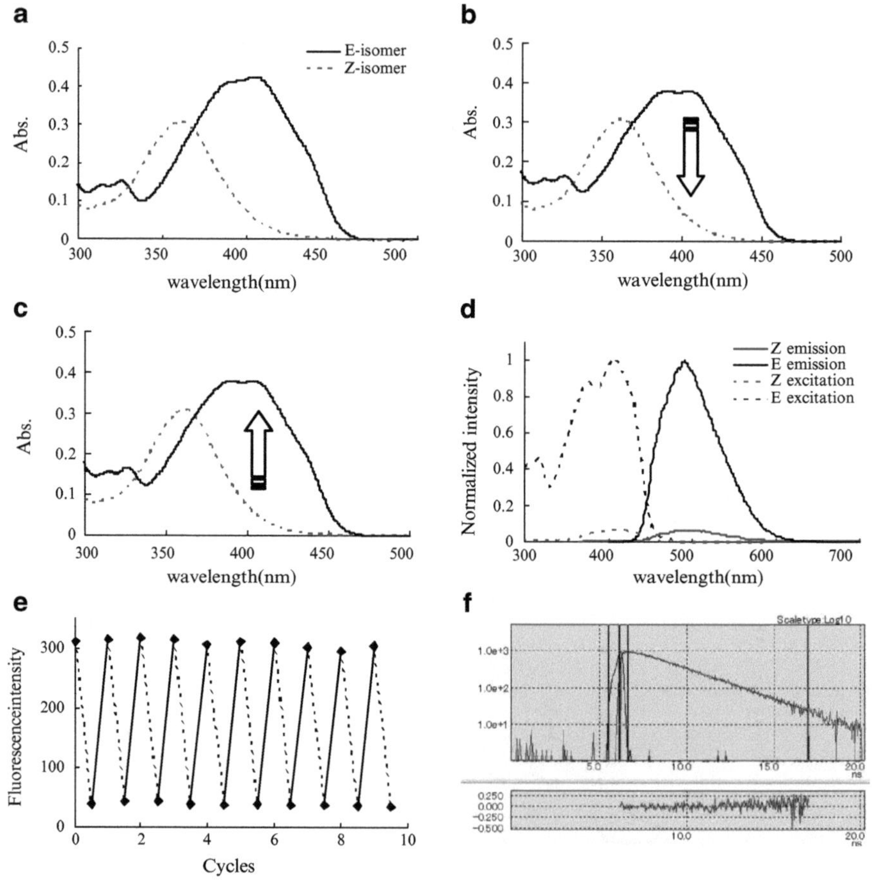

Fig. 2 (**a**) Absorption spectra of *E*- and *Z*-isomer of ^{VPy}**G**. (**b**) Photoisomerisation of *E*-isomer with illumination at 420 nm over 300 s, (**c**) *Z*-isomer with illumination at 365 nm over 60 s, (**d**) Fluorescence spectra and fluorescence excitation spectra of *E*- and *Z*-isomers. (**e**) Fluorescent switching between *E*- and *Z*-isomer alternate illumination at 420 nm during 300 s and 365 nm during 60 s, respectively. (**f**) Fluorescence decay of the *E*-isomer in methanol

when *Z*-isomer was illuminated at 365 nm, the absorbance of the peak at 280 nm decreased with a red shift of 2 nm, while the intensity of the peak at 365 nm increased with a red shift of ca. 40 nm (Fig. 2c). By using azobenzene as a standard [11], the quantum yield of *Z*- to *E*-isomerization at 365 nm was calculated to be approximately 0.46, which is significantly larger than that of *E*- to *Z*-isomerization (0.09).

Subsequently, we examined the fluorescence properties of $^{\mathbf{VPy}}\mathbf{G}$ as a photo-switching fluorescent molecule in more detail. *E*-isomer showed bright fluorescence around 490 nm but *Z*-isomer had a very weak fluorescence at the same wavelength. The very weak fluorescence of *Z*-isomer was presumably attributed to the residual fluorescence of *E*-isomer formed during fluorescence measurement. This assumption was supported by the fact that the fluorescence excitation spectra of the low intensity fluorescence peak at 490 nm were identical with that of the fluorescence excitation spectra of *E*-isomer, not for the *Z*-isomer (Fig. 2d). Finally, we monitored the periodic fluorescence response of $^{\mathbf{VPy}}\mathbf{G}$ originated from the photoisomerization process. As shown in Fig. 2e, the *E*–*Z* isomerization was repeated in 10 times without any side reaction. The fluorescence life time of the *E*-isomer was also measured in methanol at room temperature (Fig. 2f).

3 Conclusions

In conclusion, we have successfully synthesized a novel fluorescence switchable nucleoside $^{\mathbf{VPy}}\mathbf{G}$. The $^{\mathbf{VPy}}\mathbf{G}$ showed a very rapid and reversible photoisomerization without any side reaction. *E*-isomer showed strong bright bluish fluorescence but *Z*-isomer did not show any fluorescence. Therefore, $^{\mathbf{VPy}}\mathbf{G}$ was shown to be a useful fluorescence switching molecule.

References

1. Feringa BL (2007) The art of building small: from molecular switches to molecular motors. J Org Chem 72:6635–6652
2. Green JE, Choil JW, Boukai A et al (2007) A 160-kilobit molecular electronic memory patterned at 10^{11} bits per square centimetre. Nature 445:414–417
3. Balzani V, Credi A, Venturi M (2003) Molecular devices and machines: a journey into the nano world. Wiley-VCH, Weinheim
4. Irie M, Mrozek T, Daub J et al (2001) In: Feringa BL (ed) Molecular switches. Wiley-VCH, Weinheim
5. Irie M (2000) Diarylethenes for memories and switches. Chem Rev 100:1685–1716
6. Hein B, Willig KI, Hell SW (2008) Stimulated emission depletion (STED) nanoscopy of a fluorescent protein-labeled organelle inside a living cell. Proc Natl Acad Sci USA 105:14271–14276
7. Andresen M, Stiel AC, Fölling J et al (2008) Photoswitchable fluorescent proteins enable monochromatic multilabel imaging and dual color fluorescence nanoscopy. Nat Biotechnol 26:1035–1040

8. Folling J, Belov V, Kunetsky R et al (2007) Photochromic rhodamines provide nanoscopy with optical sectioning. Angew Chem Int Ed 46:6266–6270
9. Milstein D, Stille JK (1978) A general, selective, and facile method for ketone synthesis from acid chlorides and organotin compounds catalyzed by palladium. J Am Chem Soc 100:3636–3638
10. Stille JK (1986) The palladium-catalyzed cross-coupling reactions of organotin reagents with organic electrophiles [new synthetic methods (58)]. Angew Chem Int Ed Engl 25:508–524
11. Conti I, Marchioni F, Credi A et al (2007) Cyclohexenylphenyldiazene: a simple surrogate of the azobenzene photochromic unit. J Am Chem Soc 129:3198–3210

Development of Nanocrystalline Co–Cu Alloys for Energy Applications

Motohiro Yuasa, Hiromi Nakano, and Mamoru Mabuchi

Abstract Nanocrystalline Co–Cu alloys were processed by electrodeposition, and their mechanical and magnetic properties were investigated. The nanocrystalline Co alloys exhibited very high strength of about 2 GPa. Also, the activation volume for the nanocrystalline Co alloys was very low, compared with those for conventional Co alloys. Clearly, the deformation mechanisms of the nanocrystalline Co alloys are different from those of conventional Co alloys. Besides, the nanocrystalline Co alloys showed unique ferromagnetic properties, different from conventional Co alloys.

Keywords Cobalt • Nanocrystalline • Nanoscale lamellar structure

1 Introduction

Co alloys are promising energy materials because they exhibit high heat resistance, ferromagnetism and so on. For more applications, it is desirable to improve the mechanical and functional properties of Co alloys. Nanocrystallization can give rise to significant enhancement of the mechanical and magnetic properties in metallic materials. However, nanocrystalline metals tend to be very brittle with a ductility of less than a few percent in tensile tests [1, 2], due to the absence of dislocation activity [3]. Hence, it is required to develop nanocrystalline Co alloys with unique grain boundaries for enhancement of the properties. In the present work, a nanocrystalline Co–Cu alloy having nanoscale lamellar structure with a narrow spacing of 3 nm is processed by electrodeposition, and their mechanical and magnetic properties are investigated.

M. Yuasa (✉) and M. Mabuchi
Graduate School of Energy Science, Kyoto University, Kyoto, Japan

H. Nakano
Cooperative Research Facility Center, Toyohashi University of Technology, Toyohashi, Japan

T. Yao (ed.), *Zero-Carbon Energy Kyoto 2009*, Green Energy and Technology, DOI 10.1007/978-4-431-99779-5_30, © Springer 2010

2 Experimental

A nanocrystalline Co–Cu alloy was processed by electrodeposition. The electrolyte composition was $CoSO_4{\cdot}7H_2O$ (1 M) and $CuSO_4{\cdot}5H_2O$ (0.025 M). Microstructure of the Co–Cu alloy was investigated by transmission electron microscopy. Mechanical properties of the Co–Cu alloy were investigated by the tensile and hardness tests at room temperature. The hardness tests were performed with a diamond Berkovich tip at constant loading rates of 13.24, 1.324 and 0.378 mN s^{-1}. Magnetic properties were measured at room temperature by a vibrating sample magnetometer.

3 Results and Discussion

A transmission electron microscopy image of the Co–Cu alloy is shown in Fig. 1. The grain size of the Co–Cu alloy was 110 nm. Most of the grains contained a high-density fine nanoscale lamellar structure. In previous studies [4, 5], the nanocrystalline Cu with nanoscale twins with a spacing of tens of nanometers was fabricated by electrodeposition. Note that the Co–Cu alloy developed in the present work contained nanoscale lamellar structure with a much smaller spacing of 3 nm.

The yield (0.2% proof) stress and ultimate tensile strength of the Co–Cu alloy were 1,420 and 1,875 MPa, respectively, from the tensile test. The yield strength is higher than that of nanocrystalline Co with a grain size of 12 nm (=1,002 MPa) [6]. Also, the Co–Cu alloy showed an elongation to fracture of 3.3%, which is larger than those for nanocrystalline metals with a grain size of less than 10 nm and containing no nanotwins [1, 2]. It has been demonstrated in nanocrystalline Cu with nanotwins that twin boundaries can act as dislocation sources [5]. Not only do twin boundaries behave as obstacles to dislocation motion, but they also serve as dislocation sources during further deformation. Therefore, the high ductility for the Co–Cu alloy may be attributed to boundaries of the nanoscale lamellar structures acting as nucleation/accumulation sites of dislocations as well as the twin boundaries.

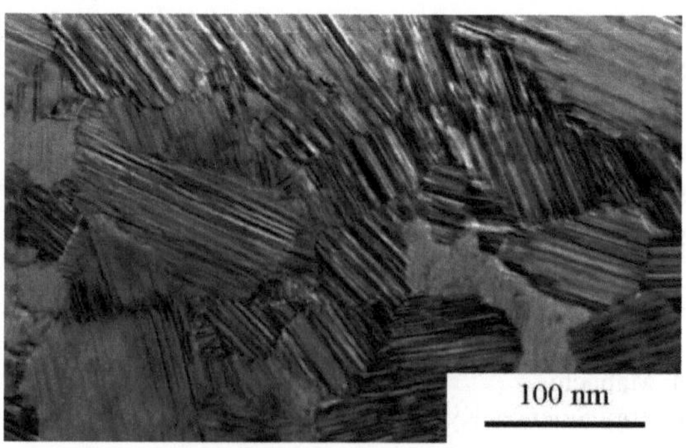

Fig. 1 A transmission electron micrograph of the Co–Cu alloy

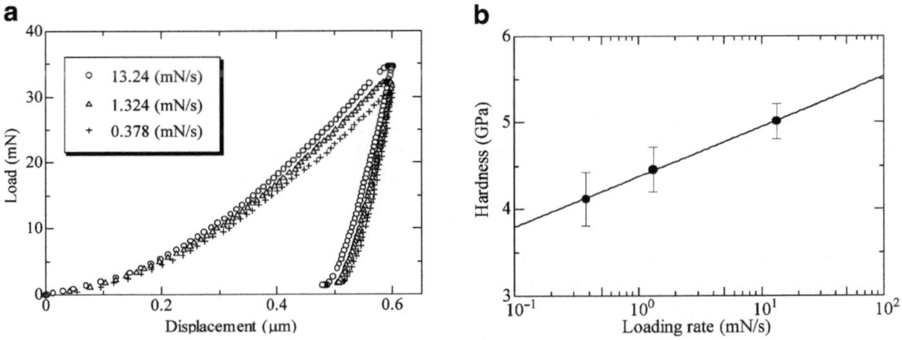

Fig. 2 The results of hardness tests for the Co–Cu alloy, (**a**) load-displacement curves at three different loading rates and (**b**) variation in hardness as a function of loading rate

Load-displacement curves obtained from the hardness tests at the three loading rates are shown in Fig. 2a. The hardnesses of the Co–Cu alloy were 4.12–5.02 GPa. As shown in Fig. 2a, a higher load was required at a higher loading rate to impose the same displacement. The variation in hardness as a function of loading rate is shown in Fig. 2b.

From the results in Fig. 2, the strain rate sensitivity and activation volume [7] were 0.055 and $3.3b^3$ for the Co–Cu alloy, respectively. The activation volume for the Co–Cu alloy is much lower than those for the nanotwin Cu [8]. Clearly, the low activation volume for the Co–Cu alloy is attributed to the nanoscale lamellar structure of 3 nm.

From the results by a vibrating sample magnetometer, the saturation magnetization and coercivity of the Co–Cu alloy were 1.85 Wb m^{-2} and 11.07 kA m^{-1}. Childress and Chein [9] investigated the magnetic properties of Co_xCu_{1-x} alloys, whose structure remained single-phase fcc up to $x=0.80$, and they showed that the saturation magnetization monotonically decreases with increasing Cu concentration. However, the Co–Cu alloy exhibited greater saturation magnetization than Co bulk (=1.82 Wb m^{-2}), although the Cu concentration was 7% in the lamellar phase [10]. Note that the presence of the nanoscale lamellar structure enhanced the saturation magnetization.

4 Conclusions

- Nanocrystalline Co–Cu alloy fabricated by electrodeposition contained a high-density fine nanoscale lamellar structure with the spacing of about 3 nm.
- The Co–Cu alloy showed the high tensile strength of about 2 GPa. Also, the activation volume for the Co–Cu alloy was very low ($=3.3b^3$).
- The Co–Cu alloy showed greater saturation magnetization than Co bulk. This is attributed to the presence of the nanoscale lamellar structure.

References

1. Wang N, Wang Z, Aust KT, Erb U (1997) Mater Sci Eng A 237:150–158
2. Iwasaki H, Higashi K, Nieh TG (2004) Scripta Mater 50:395–399
3. Schiotz J, Tolla FDD, Jacobsen KW (1998) Nature 391:561–563
4. Shen YF, Lu L, Lu QH, Jin ZH, Lu K (2005) Scripta Mater 52:989–994
5. Shen YF, Lu L, Dao M, Suresh S (2006) Scripta Mater 55:319–322
6. Karimpoor AA, Erb U, Aust KT, Palumbo G (2003) Scripta Mater 49:651–656
7. Asaro RJ, Suresh S (2005) Acta Mater 53:3369–3382
8. Lu L, Schwaiger R, Shan ZW, Dao M, Lu K (2005) Acta Mater 53:2169–2179
9. Childress JR, Chien CL (1991) Phys Rev B 43:8089–8093
10. Nakamoto Y, Yuasa M, Chen Y, Kusuda H, Mabuchi M (2008) Scripta Mater 58:731–734

Investigation of SI-CI Combustion with Low Octane Number Fuels and Hydrogen using a Rapid Compression/Expansion Machine

Sopheak Rey, Haruo Morisita, Toru Noda, and Masahiro Shioji

Abstract In order to clarify the possibility of spark assisted compression-ignition (SI-CI) combustion, experiments were made by using the rapid compression/expansion machine (RCEM) which provides advantages compared to conventional engine for its completely homogeneous charge preparation and its simplicity of the setting of engine condition. Combustion processes of low octane number fuels and hydrogen were investigated by analyzing in-cylinder pressure and heat-release rate. Primary reference fuels (PRFs) with hydrogen addition were used to alter octane number RON and hydrogen ratio r_H at a fixed equivalence ratio $\phi=0.45$, compression ratio $\varepsilon=13$ and spark timing $\theta_i=-20°$ATDC. The rates of heat release are greatly changed by RON and r_H. From the results of systematic tests for various RON and r_H, the criteria of HCCI combustion, SI combustion and SI-CI combustion were exhibited. In the case of further hydrogen addition, combustion was completed with a simple SI form. Based on the combustion characteristics for various conditions, the feasibility and advantage of SI-CI combustion are discussed.

Keywords HCCI • Hydrogen • Lean mixture • Octane number • Primary reference fuels • Spark ignition

1 Introduction

In response to the huge energy demand and stringent emission regulation, new and efficient combustion concept such as homogeneous charge compression ignition (HCCI) was introduced to exhibit higher thermal efficiency and lower NO_x emission.

S. Rey (✉), H. Morisita, and M. Shioji
Graduate School of Energy Science, Kyoto University, Kyoto, Japan

T. Noda
Nissan Research Center, Nissan Motor Co., Ltd., Kanagawa, Japan

T. Yao (ed.), *Zero-Carbon Energy Kyoto 2009*, Green Energy and Technology,
DOI 10.1007/978-4-431-99779-5_31, © Springer 2010

HCCI operations are, however, restricted at a narrow range of engine output due to knock at higher loads and misfire at lower loads [1].

In order to solve knock problem, some fuels have been used to control the combustion ignition of HCCI such as ethanol, liquefied petroleum gas (LPG) or hydrogen as the auto-ignition suppressor. Besides that, other fuels with high research octane number (RON) fuel such as gasoline or high RON primary reference fuel (PRF) have been tested with HCCI [2–5]. On the other hand, in order to solve the problem of misfire, higher intake temperature, compression ratio, equivalence ratio or low octane number fuel was used [6].

In recent, the spark assisted compression-ignition (SI-CI) combustion was proposed to confirm the ignition of lean mixture and to control the heat-release rate, in which compression-ignition (CI) combustion of the end gas may moderately follow spark-ignition (SI) combustion. Until now, only small researches were made with SI-CI with certain PRF fuels and without the addition of hydrogen. Being different from HCCI, SI-CI could operate with low intake temperature resulting in the increase of load limit, and could control combustion by its spark timing. SI-CI differs from SI for its ability of lean combustion and for having higher thermal efficiency for higher compression ratio. High flammability and high RON of hydrogen give more merits in enhancing SI-CI combustion [7, 8].

Based on the experimental results, the criteria of SI-CI combustion with RON and hydrogen ratio were clarified to emphasize the features of HCCI operation. This research was investigated in order to better understand the combustion characteristics and behavior of SI-CI combustion with low octane number fuels and hydrogen using rapid compression/expansion machine (RCEM).

2 Experimental Setup

In this experiment, RCEM was used and its specifications are shown in Table 1. It is a system which is able to change compression ratio by valve close timing. Schematic diagram of experimental set-up as shown as in Fig. 1a describes that RCEM consists of a single-cylinder engine (YANMAR NF19SK) operated by a motor (TOSHIBA VF IKK-FBK8-132S). A mixing chamber with a volume of 1,650 cm^3 was installed to prepare fuel/air mixture with a given condition.

The fuel/air mixture was introduced into the combustion chamber through a solenoid intake valve (CKD 4F110-06-CS-AC100V). A specially designed combustion chamber replaces the original head of the engine in Fig. 1b. A spark plug (NGK BCPR5ES-11) was mounted at the opposed side of combustion window.

Table 1 RCEM specifications

Bore × stroke	110 × 106 mm	IVC	−180°ATDC
Compression ratio	13	Spark ignition	−20°ATDC
Engine speed	600 rpm	Fuel	PRF, H2

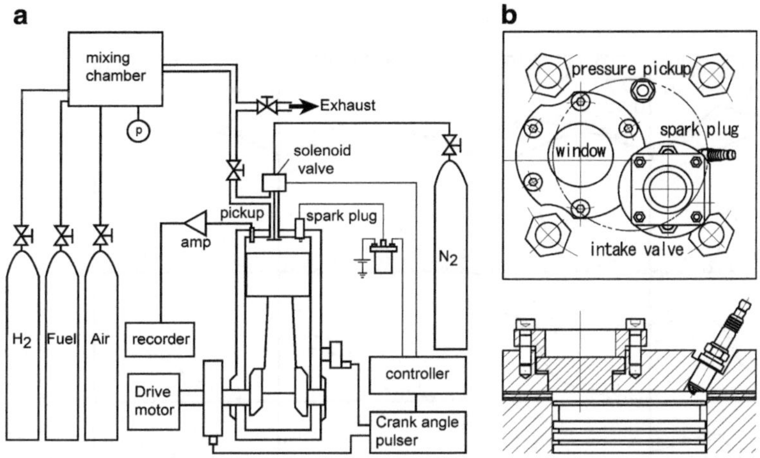

Fig. 1 (**a**) Schematic diagram of experimental set-up and (**b**) schematic diagram of the combustion chamber

The in-cylinder pressure was measured by using a piezoelectric pressure transducer (KISLTER 6052B). High precision pressure gauge sensor (KYOWA PAB-A-1MP) with an amplifier (KYOWA DPM-711B) was used for mixture preparation. Initial temperature was set to $T_i = 97°C$ at every experiment and initial pressure was measured at the time of valve closing for the compression stroke, and equals to $p_i = 0.121$ MPa. The average cylinder head temperature is 56°C. The RCEM speed is set to 600 rpm by the direct adjusting on motor inverter (TOSHIBA VF-S9) and the equivalence ratio was set to $\phi = 0.45$ and was confirmed every single experiment by using the gas chromatography (YANACO G6800 CS-FD.TC G).

3 Results and Discussions

In this research, the experiments were conducted with $\varepsilon = 13$, $\phi = 0.45$. Other parameters such as research octane numbers (RON) of PRFs and hydrogen ratio based on heat fraction (r_H) were used to determine the combustion characteristic and behavior of SI-CI.

Figure 2a shows the basic cylinder pressure data p, its calculated and derivative data such as average combustion temperature T_{ave}, rate of heat release $dq/d\theta$ and combustion efficiency χ_q. From these results, with RON0 and at $r_H = 0.1$, the combustion is in HCCI mode which cool flame combustion starts at −10°ATDC and the hot flame combustion starts at 3°ATDC. When the hydrogen ratio is increased at $r_H = 0.30$ ($\phi_H = 0.10$), the SI-CI combustion occurs. The spark ignition combustion starts a bit early compared to cool flame combustion. Due to the result of flame propagation of first stage combustion, the compression ignition starts at 7°ATDC. Further increasing of hydrogen ratio at $r_H = 0.47$, the combustion is in type of simple

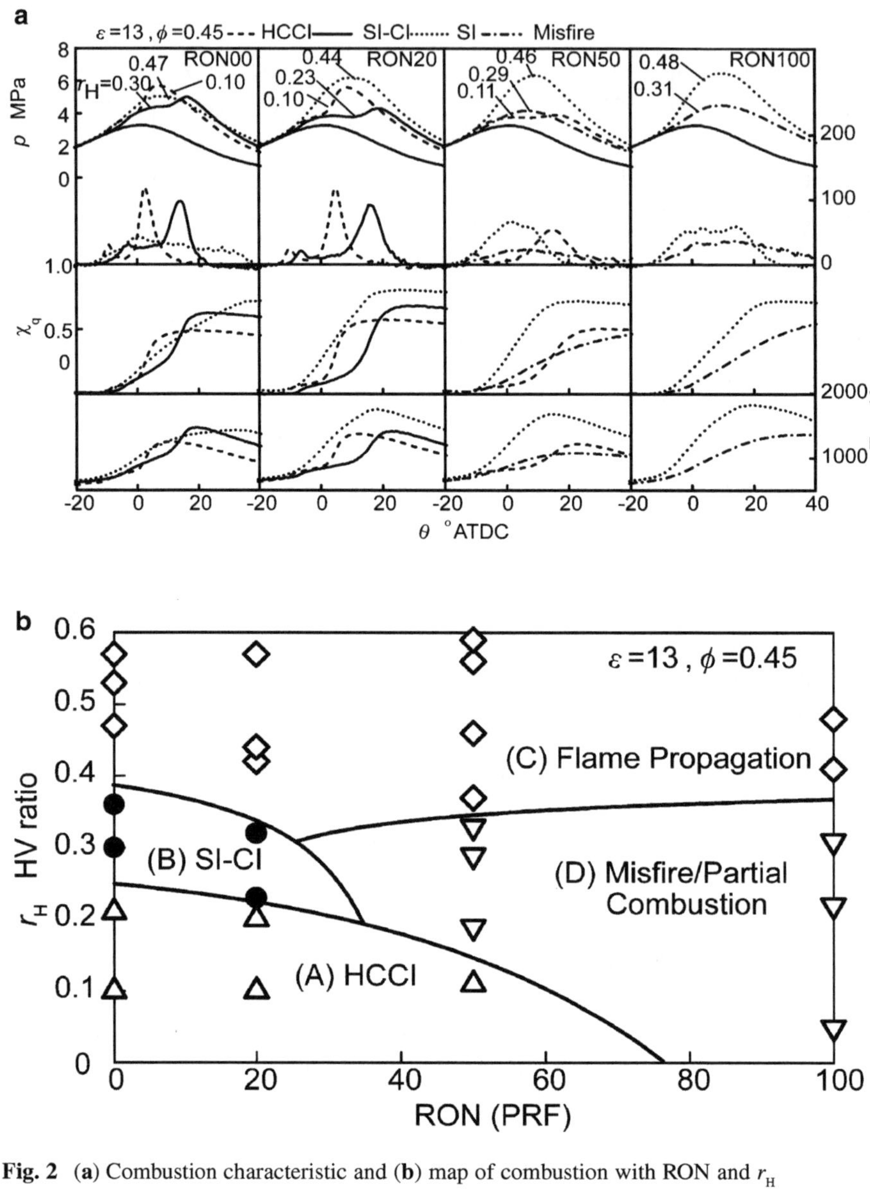

Fig. 2 (**a**) Combustion characteristic and (**b**) map of combustion with RON and r_H

flame propagation (SI) due to the high flame propagation speed and high resistant auto-ignition of hydrogen. As the combustion is changed from HCCI to SI-CI, the combustion efficiency increases due to the more complete combustion of SI-CI at lean mixture compared to HCCI.

Similar to RON0, combustion of RON20 also gives HCCI, SI-CI and SI with various ranges of r_H. In this case, the CI degree of SI-CI combustion is lower and smaller than that of RON0. The decrease of this CI is due to the increase of octane number. For RON50 at r_H=0.11, the HCCI combustion was very weak due to high octane number of fuel. When r_H increases (r_H=0.29), only partial combustion of flame propagation shows and when r_H=0.46, only SI combustion is showed. The combustion of RON100 is similar to that of RON50. However, RON100 does not have any HCCI.

The combustion map in Fig. 2b is drawn based on the data in Fig. 2a. On this map, HCCI, SI-CI, SI and Misfire/partial combustion represented by area (A), (B), (C) and (D) respectively. For RON0, area (A), area (B) and area (C) are ranged with r_H from 0 to 0.25, from 0.25 to 0.37 and over or about 0.4 respectively. For RON20, area (A), area (B) and area (C) are ranged with r_H from 0 to 0.22, from 0.22 to 0.33 and over or about 0.33 as in respective order. Area (D) locates at low r_H and with high octane number fuel. For RON50 and RON100, area (D) is at r_H lower or equal to 0.33.

In general, the intensity of HCCI increases when r_H decreases. Furthermore, observation also recognizes that SI-CI intensity has similar trend as HCCI. SI-CI range becomes narrower when octane number fuel increases, and its area moves from high to low r_H. The best r_H for making SI-CI in this case is around 0.3 for RON00 and around 0.25 for RON20.

Various combustion parameters such as χ_q, *IMEP*, T_{max}, $dq/d\theta_{max}$ and $dp/d\theta_{max}$ in the function of r_H for various RONs are compared and discussed in Fig. 3a. SI-CI is observed with low octane number fuels and shows better combustion compared to HCCI: for higher combustion efficiency, higher *IMEP*, lower pressure rise and lower heat-release rate.

From Fig. 3b, combustion efficiency is drawn against the pressure rise and reveals the combustion behavior of HCCI, SI-CI and SI. SI-CI gives higher combustion efficiency and lower pressure rise. Lower pressure rise suggests that SI-CI possibly prevent knock at high load and therefore, operation limit can be expanded. However, remained low combustion efficiency of SI-CI compared to conventional engine will be improved in the next research by increasing higher compression ratio.

4 Conclusions

From the experimental results, some conclusions are drawn as the following:

- With low octane number fuel, SI-CI combustion proves its occurrence, and shows better combustion such as having higher *IMEP* and higher combustion efficiency, and lower pressure rise and lower heat release rate than HCCI.

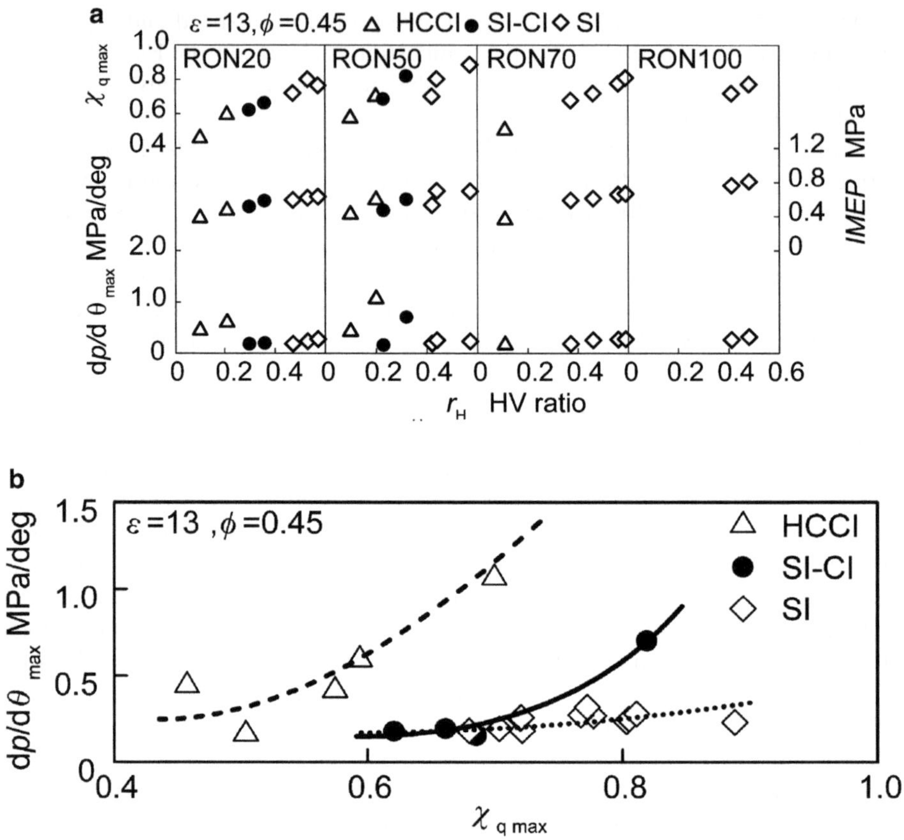

Fig. 3 (**a**) Combustion behavior of various RON and (**b**) combustion efficiency of each combustion types

– SI-CI overcomes the narrow operation range of HCCI as knock can be avoided, misfire can be secured and load can be increased.

Combustion improvement by SI-CI shows another progress in solving the problems of HCCI and helps to reduce the exhaust gas emission in the future power system by its efficient combustion.

References

1. Najt PM, Foster DE (1983) Compression-ignited homogeneous charge combustion. SAE Paper 830264
2. Antunes JG, Mikalsen R, Roskilly A (2008) An investigation of hydrogen-fuelled HCCI engine performance and operation. Int J Hydrogen Energy 33:5823–5828

3. Lu X, Hou Y, Zu L, Huang Z (2006) Experimental study on the auto-ignition and combustion characteristics in the homogeneous charge compression ignition (HCCI) combustion operation with ethanol/n-heptane blend fuels by port injection. Fuel 85:2622–2631
4. Shudo T, Yamada H (2007) Hydrogen as an ignition-controlling agent for HCCI combustion engine by suppressing the low-temperature oxidation. Int J Hydrogen Energy 32:3066–3072
5. Yeom K, Jang J, Bae C (2007) Homogeneous charge compression ignition of LPG and gasoline using variable valve timing in an engine. Fuel 86:494–503
6. Machrafi H, Cavadiasa S (2008) An experimental and numerical analysis of the influence of the inlet temperature, equivalence ratio and compression ratio on the HCCI auto-ignition process of Primary Reference Fuels in an engine. Fuel Process Technol 89:1218–1226
7. Urushihara T, Itoh T et al (2005) A study of a gasoline-fueled compression ignition engine – expansion of HCCI operation range using SI combustion as a trigger of compression ignition. SAE Trans 114(3):419–425
8. Yoshiwa K, Urushihara T et al (2006) Study of high load operation limit expansion for gasoline compression ignition engine. J Eng Gas Turbines Power 128(2):377–387

Comparison Between the Hexaboride Materials as Thermionic Cathode in the RF Guns for a Compact MIR-FEL Driver

Mahmoud Bakr, Kyohei Yoshida, Keisuke Higashimura, Satoshi Ueda, Ryota Kinjo, Heishun Zen, Taro Sonobe, Toshiteru Kii, Kia Masuda, and Hideaki Ohgaki

Abstract Thermionic RF guns are used as highly brilliant electron source for linac-driven FEL (free electron laser). They can potentially produce an electron beam with high energy, small emittance, short pulse duration, inexpensive and compact configuration in comparison with other high brightness electron sources, e.g., DC guns and photocathode RF guns. The most critical issue of the thermionic RF gun is the transient cathode heating problem due to the electron back-bombardment when the gun is used for an FEL driver. The heating property of cathode strongly depends on the physical properties of the cathode material such as electron stopping power and range. We investigated the heating property of six hexaboride materials against the back-bombarding electrons by numerical calculation of the stopping power and range. In this investigation, the emission property of the cathode was also taken into account, since high electron emission is required for generation of high brightness electron beam. As a result, calcium hexaboride material has best properties for thermionic RF gun cathode material in the backbombardment effect point of view.

Keywords Back-bombardment • Hexaborides • RF gun

1 Introduction

A mid-infrared free electron laser (MIR-FEL) facility (KU-FEL: Kyoto University Free Electron Laser) has been constructed for energy science in Institute of Advanced Energy, Kyoto University [1]. A thermionic RF gun has been chosen for

M. Bakr(✉), K. Yoshida, K. Higashimura, S. Ueda, R. Kinjo, T. Sonobe, T. Kii, K. Masuda, and H. Ohgaki
Institute of Advanced Energy, Kyoto University, Gokasho, Uji, Kyoto 611-0011, Japan
e-mail: m-a-bakr@iae.kyoto-u.ac.jp

H. Zen
UVSOR, Institute for Molecular Science, 38 Nishigo-Naka, Myodaiji, Okazaki 444-8585, Japan

T. Yao (ed.), *Zero-Carbon Energy Kyoto 2009*, Green Energy and Technology, DOI 10.1007/978-4-431-99779-5_32, © Springer 2010

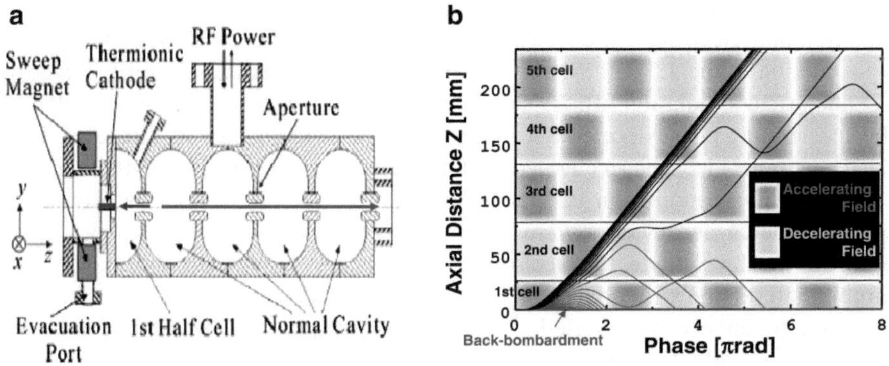

Fig. 1 (**a**) Schematic side view of thermionic RF gun, (**b**) Electrons motion in KU-RF gun by 1D numerical simulation, showing axial distances from the cathode as a function of time [6]

electron source of KU-FEL linac, because it has features of compactness, an easy-handling and high brightness of the output beam. KU-FEL thermionic gun consists of a 4.5-cell cavity with total length 30 cm, driven by a 10-MW RF power, which provides up to 10 MeV electron beam. A high-quality electron source is crucial for a compact and economical FEL device. We improved the cathode of the thermionic RF gun science 2007, which was difficult to produce high-energy electron beam with long macropulse. Successfully we have produced a long macropulse by substituting of the dispenser tungsten-base (W-BaO/CaO/Al_2O_3) with a single crystal of lanthanum hexaboride (LaB_6) with 2 mm diameter. However the back-bombardment in the thermionic cathode still affect the macropulse duration below 6 μs [2].

The purpose of this paper is to report on the comparison between the most candidate materials which used as cathodes in RF guns. Hexaboride materials are the most promising materials which may be used as emitter of thermionic cathode for RF guns; the primary compounds of hexaboride materials considered in this paper are calcium hexaboride CaB_6, lanthanum hexaboride LaB_6, cerium hexaboride CeB_6, strontium hexaboride SrB_6, barium hexaboride BaB_6 and thallium hexaboride ThB_6 in the viewpoint of the backbombardment effect in a thermionic RF gun.

2 Back-Bombardment Electrons

KU thermionic RF gun equips a thermionic cathode at one end of a cylindrically symmetric RF cavity as shown in Fig. 1a. In the RF cavity, electric field strength varies sinusoidal with the frequency of RF power fed to the gun. At the time when accelerating field exists in the cavity, electrons are extracted from the cathode and gain kinetic energy. After half period of RF cycle, the electric field changes its direction and electrons are decelerated to the cathode. Figure 1b, shows the electrons motion in the 4.5 cell S-band thermionic RF gun used for KU-FEL driver by 1D numerical simulation [3]. As shown in Fig. 1b, electrons which emitted late in the accelerating phase decelerated after the electric field changes its direction. Then the decelerated electron accelerated back towards the cathode and eventually hit the cathode. This phenomenon is unavoidable if a thermionic cathode is used as an electron source and electrons are continuously emitted from the cathode. When electrons hit the cathode, the electrons lose its energy with penetrating the cathode through interaction with bound electrons in the cathode material. Most of the electrons kinetic energy is converted to thermal energy, and then the cathode is heated up. When the cathode heated by back-bombarding electrons, the temperature of the cathode increases, hence the current density on the cathode surface J_c increases. Thus the beam current increases during a macropulse. When the beam current in the RF cavity increases the acceleration voltage of the RF cavity decreases. Eventually the electron beam energy decreases as a result of cavity voltage decrease [4]. Since electron beam with long macropulse duration and constant beam energy is mandatory for FEL lasing, from this point of view the energy decrease due to the backbombardment has to be eliminated.

3 Cathode Materials and Calculation Method

3.1 *Hexaboride Materials*

The alkaline-earth metals, rare-earth metals and thorium form of borides of the type MB_6, all these compounds have the same cubic crystal structure. The small boron atoms form a three dimensional framework structure which surrounds the large metal atom. The metallic character of these compounds is evident from their desired properties for an excellent cathode material, such as low work function, low volatility, low electrical resistivity, high mechanical strength, and high chemical resistance. The borides are characterized by high melting temperatures, and, most of them, high thermal conductivity. Further characteristics are a fair corrosion resistance, chemical inertness, good wear resistance and a thermal shock resistance much better than that of oxide ceramics [5]. Many of these properties are of great interest for technical applications as well as small spot size applications such as SEM, TEM, surface analysis and metrology, and for high current applications such as microwave tubes, lithography, electron-beam welders, X-ray sources and free electron lasers [6].

The essential physical and chemical properties for the hexaboride materials under considerations in the comparison are listed in Table 1, such as the molecular weight, density, work function, Richardson constant, melting temperature, effective atomic number and effective atomic weight.

Table 1 The physical and chemical properties for the hexaboride materials

	CaB_6	LaB_6	CeB_6	SrB_6	BaB_6	ThB_6
Molecular weight (g mol^{-1})	104.946	203.772	204.986	152.490	202.193	296.904
Density (g cm^{-3})	2.490	4.720	4.797	3.390	4.390	6.990
Melting temperature (K)	2,508	2,483	2,463	2,508	2,543	2,468
Richardson constant (A cm^{-2} K^{-2})	2.6	29	3.6	0.14	16	0.5
Work function (V)	2.86	2.66	2.59	2.67	3.45	2.92
Effective atomic number	10.728	40.447	41.228	23.962	39.639	71.429
Effective molecular weight (g mol^{-1})	22.518	94.735	96.035	53.734	93.194	176.731

3.2 Thermionic Emission Properties

The normalized rms thermal emittance of electrons emitted from a hot cathode is described by the following equation [7]:

$$\varepsilon_{n,rms} = \frac{r_c}{2}\sqrt{\frac{k_B T}{m_e c^2}}, \tag{1}$$

where r_c is the cathode radius, k_B is Boltzmann's constant (~1.38×10^{-23} J K^{-1}), T is the cathode surface temperature m_e the electron mass and c is the speed of light. From the above relation, in order to obtain small thermal emittance less than 9 π mm mrad required for the KU-FEL, the diameter of the cathode must be in the range of a few mm at the temperature of 1,500–2,500 K. On the other hand, high emission density is required to produce a few tens ampere peak current for FEL applications. Moreover, the current density of 10 and 30 Acm^{-2} is required for FEL amplification and FEL gain saturation respectively [4]. Hexaboride materials can emit such an intense current over long lifetimes. A single crystal is preferable for obtaining low emittance because of its extremely flat surface (roughness ≤1 μm) with low porosity after surface material evaporation. The electron emission behavior is well described by (2), known as the Richardson–Dashman equation which given as

$$j_c = AT^2 \exp\left(-\frac{\phi}{k_B T}\right), \tag{2}$$

where ϕ (V) is the wok function of the material used as the electron emitter. This is a material dependant property. As one can see, a higher work function requires higher temperature and ultimately more power to achieve the desired electron current density. A (A cm^{-2} K^{-2}) denotes the Richardson's constant or the emission constant, specific to each material as well. One can see the current density will increase with decreasing work function. The emission constants and the effective work functions for the hexaborides under consideration are listed in Table 1.

3.3 The Range and Stopping Power

The range, *R*, of electrons inside the material is useful for evaluation of effects associated with deep penetration of electrons, such as back-bombardment electrons. The extrapolated range usually defined as the thickness of material at which the extension of the linearly decreasing region of the transmission curve becomes zero. At low energies, *R* is frequently determined from linear extrapolation of the number energy curve measured for a given thickness of the absorber. For absorbers of high atomic number, the transmission curve often does not show the linear region, and the extrapolation is then made of the tangent at the steepest point of the curve.

The range *R* of monoenergetic electrons in the energy region 0.3 keV – 30 MeV for the absorbers of atomic number 6–92 has been found to be expressed by a single semiempirical equation of the form [8].

$$R = \frac{a_1}{\rho}\left[\frac{\ln(1+a_2\tau)}{a_2} - \frac{a_3\tau}{1+a_4\tau^{a5}}\right], \tag{3}$$

where

$$a_1 = \frac{2.335A}{Z^{1.209}}, \qquad a_2 = 0.000178\,Z, \qquad a_3 = 0.9891 - 0.000301\,Z,$$
$$a_4 = 1.468 - 0.0118\,Z, \qquad a_5 = \frac{1.232}{Z^{0.109}}, \tag{4}$$

where *R* (m) denotes to the electron range, ρ is the material density, τ is the incident kinetic energy in the units of the rest energy of the electron, and the parameters a_i (i = 1, 2, …, 5) are given by simple function of atomic number *Z*, and the atomic weight *A*. In case of mixture or compound *Z* and *A* should be replaced by the effective values of the atomic number Z_{eff} and effective atomic weight A_{eff} as shown below,

$$Z_{\mathrm{eff}} = \sum_i f_i Z_i, \quad A_{\mathrm{eff}} = \frac{Z_{\mathrm{eff}}}{(Z/A)_{\mathrm{eff}}}, \quad (Z/A)_{\mathrm{eff}} = \sum_i \frac{f_i Z_i}{A_i}, \tag{5}$$

where f_i is the fraction by weight of the constituent element with Z_i and A_i, the effective atomic number and the effective atomic weight for the materials under consideration are listed in Table 1.

4 Results and Discussions

Assuming that, all the cathodes have the same conditions of pressure, vacuum level, heating method, applied electric field in the RF cavity, area and diameter. Moreover, the back-bombardment electrons have the same range of energy 0.3 keV – 1 MeV.

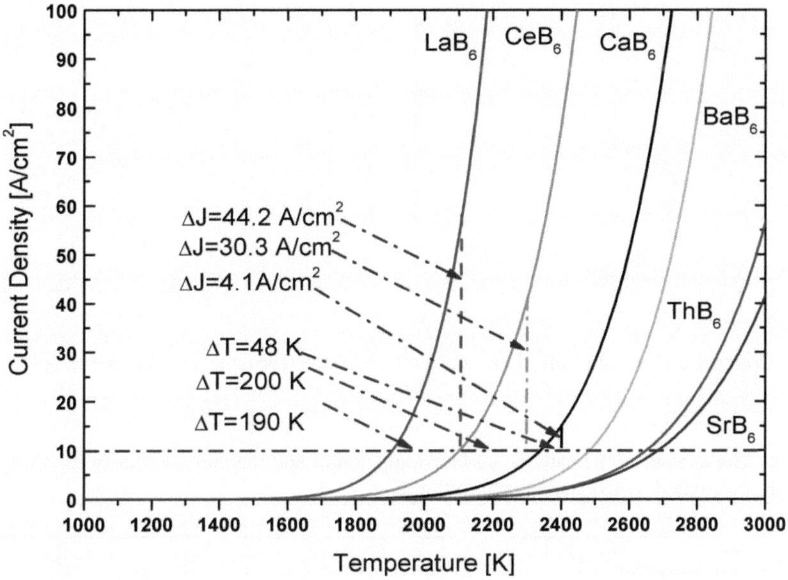

Fig. 2 The current density of the hexaboride materials as a function of temperature has been calculated by using (2), and the change in the current density due to the change in the cathode surface temperature

The typical emission characteristics from the cathodes are shown in Fig. 2. The characteristic curves are determined by using Richardson–Dushman equation, (2). As one can see from Fig. 2, that three of the hexaboride materials ThB_6, BaB_6 and SrB_6 are not satisfied to produce current density 10 $A\,cm^{-2}$ below the melting temperature (~2,500 K). This value of current density is lower than the required value to get FEL amplification.

From the current density calculations on the considered materials, three hexaboride materials (CaB_6, LaB_6, and CeB_6) could produce current density satisfied FEL amplification; these materials will be used in the next step to calculate the range and stopping power.

The range due to the back-bombardment electrons form the range 0.3 keV – 1 MeV was calculated for single crystals of CaB_6, LaB_6, and CeB_6 by using (3), and hence the stopping power was determined from the relation $\Delta E/\Delta R$ as shown in Figs. 3a and 3b respectively.

The range calculation for the back-bombardment electrons indicated that CaB_6 has longer range comparison with LaB_6 and CeB_6 as shown in Fig. 3a. Moreover, the stopping power of CaB_6 is lower than the stopping power of LaB_6 and CeB_6 as shown in Fig. 3b. The longer range of the electrons inside the CaB_6 can be explained as the low density of CaB_6, 2.49 $g\,cm^{-3}$. Both LaB_6 and CeB_6 have approximately the same range and stopping power as shown in Figs. 3a, and 3b respectively.

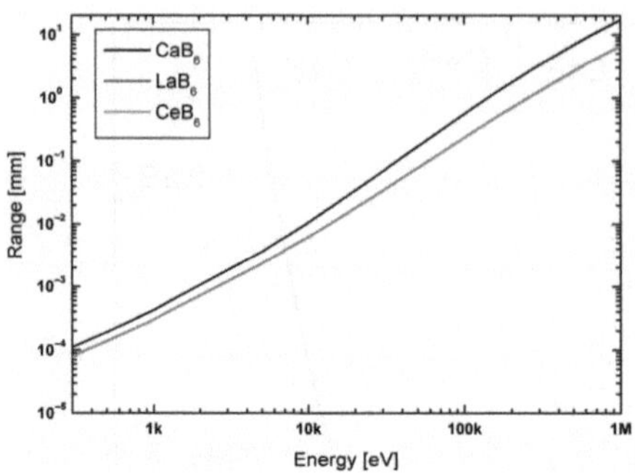

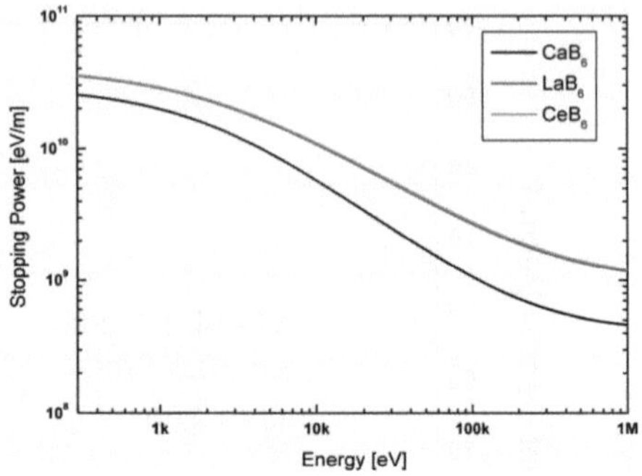

Fig. 3 Comparison between CaB_6, LaB_6 and CeB_6 in (**a**) the reachable range (**b**) the stopping power

Table 2 The change of the cathode surface temperature and the current density for CaB_6, LaB_6 and CeB_6 calculated by (6), and shown in Fig. 2

Electrons energy (keV)	CaB_6		LaB_6		CeB_6	
	ΔT (K)	ΔJ (A cm^{-2})	ΔT (K)	ΔJ (A cm^{-2})	ΔT (K)	ΔJ (A cm^{-2})
30	48	4.1	190	44.2	200	30.3
100	20	2.1	94	14.5	99	10
300	12	1.3	56	8	60	7

The deposited heat at 1 μm from the cathode surface was calculated in this analysis, after considering the that the back-bombardment electrons are monoenergetic electrons, and the number of electrons which hit the cathode material is constant (form particle simulation at KU) [4]. Under these conditions, the deposited heat in the cathode surface by back-bombardment electrons is given by

$$\frac{C\rho Sz\Delta T}{\Delta t} = Q \tag{6}$$

where C denotes the specific heat capacity, ρ the density of the cathode, ΔT the change of the cathode surface temperature within time Δt, S the cathode surface area, z the depth of cathode from the surface, and Q the heat input due to the back-bombarding electrons. The change in the cathode surface temperature for CaB_6, LaB_6 and CeB_6 due to the deposited heat at 1 μm from the surface within the macropulse of 5 μs has been calculated by using (6), for different energies of the back-streaming electrons 30, 100 and 300 keV.

Table 2, shows the change of the cathode surface temperature and the change of the current density from 10 A/cm due to the heat deposited at different back-bombardment electron energies. One can see from Table 2 and Fig. 2, that the change of the cathode surface temperature and the current density in case of CaB_6 material is lower than LaB_6 and CeB_6. This means that the effect of back-bombardment electrons in case of CaB_6 is lower than the effect on LaB_6 and CeB_6. From the previous results the CaB_6 it seems the best candidate material to be used for KU-FEL linac driver under some considerations. The first one, using CaB_6 in FEL amplification with large beam spot size, where the required current density is 10 $A\,cm^{-2}$, in this meaning, the area of the cathode surface must be increased. If the area duplicated, the thermal emittance will be introduced and will be twice according to (1). Even with the increasing of the thermal emittance it's still applicable to use CaB_6 as electron source because the emittance does not exceed the limit for emittance requirements for the KU-FEL [4]. The second consideration, the cathode surface temperature increasing due to the back-bombardment effect makes the cathode temperature near from the melting temperature; as a result the evaporation rate from the cathode surface will increase, moreover very small fragments from the cathode surface will melt up by the continues operating. Eventually, the lifetime of the cathode will decrease. Care must be taken to properly optimize cathode temperature to obtain the required emission without overheating the crystal. On the other hand, in case of applications with small beam spot size, where large total current and high current density >30 $A\,cm^{-2}$ are required for FEL gain saturation experiments, LaB_6 and CeB_6 are the material of choice for high current cathodes in a variety of advanced. Even with CeB_6 cathode has smaller current density than LaB_6 at the same temperature as shown in Fig. 2, while the change of the cathode surface temperature due to the back-bombardment electrons slightly lower than the change in surface temperature of LaB_6. Hence, the change of the current density is much lower than LaB_6 cathode. Moreover, it's reported that CeB_6 has another advantage over LaB_6 relating to lifetime, its evaporation rate at normal operating temperatures near 1,800 K is lower than that of LaB_6 [9]. From applications with high current point of view, the CeB_6 is best candidate to be used. Experiments are required to confirm which hexaboride material agrees the numerical calculations.

5 Conclusions

Six of the hexaboride materials were investigated in the way to find which material has low effect to the back-bombardment electrons in KU-FEL thermionic RF gun. The strategy started with checking the current density of the hexaboride materials, three of the materials did not match the required current density for FEL amplification (10 $A\,cm^{-2}$) below the melting temperature of these materials, SrB_6, BaB_6 and ThB_6. The range and stopping power for the other materials (CaB_6, LaB_6 and CeB_6) were calculated as function of the back-bombardment electrons energy of the range 0.3 keV – 1 MeV. The results indicated that the stopping power of CaB_6 material was less than LaB_6 and CeB_6, in other words the heat deposited at 1 μm of the cathode surface was lower than the other materials. As a result, the change in the cathode surface temperature due to the deposited heat was lower than the other materials. Furthermore, the change of the current density will be smaller than the other materials. Based upon the findings of this investigation, the calcium hexaboride cathode proved the superior electron emitter, not from the electron emission standpoint, but from the back-bombardment effect and FEL amplification with large beam spot size point of view. However, the cerium hexaboride cathode would provide longer lifetime, and could satisfy the FEL gain saturation applications with small beam spot size and wider range of MIR-FEL applications benefits over other cathodes. We will perform experiments to confirm the above mentioned results.

Acknowledgment The author wish to thank the GCOE program, (Energy Science in the Age of Global Warming) Kyoto University, for financial support.

References

1. Yamazaki T et al (2002) Free Electron Laser 2001:II13–II14.
2. Kii T, Nakai Y, Fukui T, Zen H, Kusukame K, Okawachi N, Nakano M, Masuda K, Ohgaki H, Yoshikawa K, Yamazaki T (2007) AIP Conf Proc 879:248–251
3. Masuda K, Kusukame K, Kii T, Ohgaki H, Zen H, Fukui T, Nakai Y, Yoshikawa K, Yamazaki T (2005) Particle simulations of a thermionic RF gun with gridded triode structure for reduction of backbombardment. In: Proceedings of the 27th International Free Electron Laser Conference, pp 588–591
4. Zen H (2009) Doctor Thesis, Institute of Advanced Energy, Kyoto University
5. Lafferty JM (1951) Boride cathodes. J Appl Phys 22(3):299–309
6. Lundström T (1985) Pure Appl Chem 57(10):1383–1390
7. Kobayashi H (1992) Emittance measurement for high-brightness electron guns. In: 1992 Linear Accelerator Conference, Ottawa, Canada
8. Tabata T, Ito R, Okaba S (1972) Generalized semiempirical equations for the extrapolated range of electrons. Nuclear Instrum Methods 103:85–91
9. http://www.emsdiasum.com/microscopy/products/microscope/lab6_ceb6.aspx?mm=8

Indicators for Evaluating Phase Stability During Mechanical Milling

Kosuke O. Hara, Eiji Yamasue, Hideyuki Okumura, and Keiichi N. Ishihara

Abstract Mechanical milling (MM) is a non-equilibrium processing technique, which can provide new materials with metastable structures. However, plausible explanation and prediction of the phase formation during MM is still difficult due to its dynamic nature. In this study, to develop the criteria of the phase stability during MM, two indicators, the molar atomic volume (V_{atom}) and the number of atoms in the reduced unit cell (Z_{atom}), were proposed, and the validity of them was investigated by plotting milling-induced polymorphic transformations reported in literatures for the variation of V_{atom} and Z_{atom}. As a result, phases with the smaller V_{atom} or Z_{atom} values were found preferable during MM.

Keywords Mechanical milling • Polymorphic transformation • Crystal structure • Phase stability

1 Introduction

Highly-functional materials are required for effective energy use and conversion. Since the function of a material largely depends on its crystal structure, a new material with a different structure from ordinary one might provide superior functions.

Mechanical milling (MM) is one of the processing techniques to produce non-equilibrium phases such as amorphous, high-temperature/high-pressure and disordered phases [1]. However, explanation of the phase formation is difficult only in terms of the Gibbs free energy due to the dynamic nature of MM, where the phase stability during MM is thought to be different from an ambient condition [2].

There have been some reports noting that the phases produced by MM have larger density or smaller volume [3, 4]. However, the validity of volume as an indicator has not been investigated so far. Furthermore, any parameter other than volume has not been proposed. Criterion of the phase stability using parameters related to the crystal structure would be useful for material design by MM. Thus, the aim of

K.O. Hara(✉), E. Yamasue, H. Okumura, and K.N. Ishihara
Graduate School of Energy Science, Kyoto University, Kyoto 606-8501, Japan

T. Yao (ed.), *Zero-Carbon Energy Kyoto 2009*, Green Energy and Technology,
DOI 10.1007/978-4-431-99779-5_33, © Springer 2010

this study is to propose a set of indicators of the phase stability during MM and investigate the validity of them. In this paper, only polymorphic systems are focused at the beginning of this kind of study because it is more complicated to define the indicators for multiple phases present.

2 Candidate Indicators

The milling-induced polymorphic transformations were picked up from literatures as much as possible, which are listed in Table 1. In this table, the transformations are categorized in terms of whether they proceed completely or not, and the sign of the Gibbs free energy change (ΔG). As a preliminary investigation, several parameters related to the crystal structure such as the unit cell volume and density were calculated for these systems. As a result, the molar atomic volume (V_{atom}) and the number of atoms in the reduced unit cell (Z_{atom}) were found probable to serve as the indicators. Thus, these two parameters were investigated in detail. The values of the changes of V_{atom} (ΔV_{atom}) and Z_{atom} (ΔZ_{atom}) are included in Table 1.

Table 1 Studied polymorphic transformations with the values of ΔV_{atom} and ΔZ_{atom}. In the column "Mark", *circles and triangles* represent, respectively, the transformations which proceed almost completely and those which proceed partially (leading to two-phase coexistence). *Open circles* represent the transformations with positive ΔG, while *solid circles* represent those with negative ΔG

Chemical formula	Reported transformation	Mark	ΔV_{atom} ($cm^3 mol^{-1}$)	ΔZ_{atom}	Refs.
AgI	Hexagonal → cubic	•	−0.025	−2	[5]
$CaCO_3$	Calcite → aragonite	Δ	−0.54	10	[5]
$CaCO_3$	Vaterite → calcite	•	−0.13	−50	[5]
$CuFe_2O_4$	Tetragonal → cubic	O	−0.082	0	[5]
Er_2S_3	Monoclinic → cubic	O	−1.3	−16.7	[4]
Eu_2O_3	Cubic → monoclinic	O	−0.81	−25	[5]
Fe_2O_3	Maghemite → hematite	•	−0.51	−43.4	[5]
FeS_2	Marcacite → pyrite	•	−0.21	6	[5]
Lu_2S_3	Rhombohedral → cubic	Δ	−1.8	3.3	[4]
$MgSiO_3$	Orthoenstatite → clinoenstatite	•	−0.21	0	[5]
$MoSi_2$	Tetragonal → hexagonal	Δ	−0.069	6	[6]
Ni_3Sn_2	Orthorhombic → hexagonal	O	0.020	−5	[1]
PbO	Litharge → massicot	Δ	−0.36	4	[5]
PbO_2	$\alpha \rightarrow \beta$	Δ	0.27	−6	[5]
Sb_2O_3	Senarmontite → valentinite	O	−0.44	0	[5]
TaN	CoSn-type → WC-type	O	−0.35	−4	[1]
TiO	Monoclinic → cubic	O	−1.0	−8	[7]
TiO_2	Anatase → rutile	•	−0.57	0	[8]
TiO_2	TiO_2 II → rutile	•	0.14	−6	[8]
Tm_2S_3	Monoclinic → cubic	Δ	−1.1	−16.7	[4]
Y_2O_3	Cubic → monoclinic	O	−0.72	−25	[1]
Yb_2S_3	Rhombohedral → cubic	O	−1.8	3.3	[4]
ZnS	Wurtzite → sphalerite	•	−0.019	−2	[5]

The meanings of V_{atom} and Z_{atom} are mentioned below. It is known that a phase with smaller volume is stable at high pressures. Since compressive pressure is intermittently applied to the powder particles during MM, the local high pressure possibly contributes to the formation of a phase with the smaller V_{atom} value. Such contribution of the pressure has been discussed in some literatures, some of which attribute the metastable phase formations to the effects of the pressure [3, 9]. However, it has not been confirmed yet whether it is applied generally to the milling induced phase transformations.

On the other hand, Z_{atom} represents the complexity of the crystal structure, and is possibly related to the kinetics of the transformation. It has been suggested that recovery and recrystallization as well as defect creation occur during repeated deformation by MM [10, 11]. Recrystallization generally proceeds via nucleation and growth, which may involve the formation of another phase. In the report on the formation of the non-equilibrium solid solution by the MM of β-Al_3Mg_2, it is discussed that long range diffusion is needed to nucleate the equilibrium phase with a complex crystal structure, while the non-equilibrium solid solution appears kinetically easier to form due to its simple structure [12]. In addition, it has been clarified, in solidification of undercooled melts, that the growth rate of a solid phase is related to the complexity level of the crystal structure [13]. Similar mechanism may work during MM.

3 Evaluation of Indicators

The changes of V_{atom} (ΔV_{atom}) and Z_{atom} (ΔZ_{atom}) through the studied polymorphic transformations are plotted in Fig. 1. In this figure, circles and triangles correspond to those in Table 1. From Fig. 1, the ΔV_{atom} values were negative for 20 out of 23 polymorphic transformations studied (87.0%). On the other hand, the ΔZ_{atom} values were zero or negative for 17 out of 23 (73.9%). It is also worth noting that there are no transformations where both ΔV_{atom} and ΔZ_{atom} are positive. This result shows that a phase with the smaller V_{atom} or Z_{atom} values is preferred during MM.

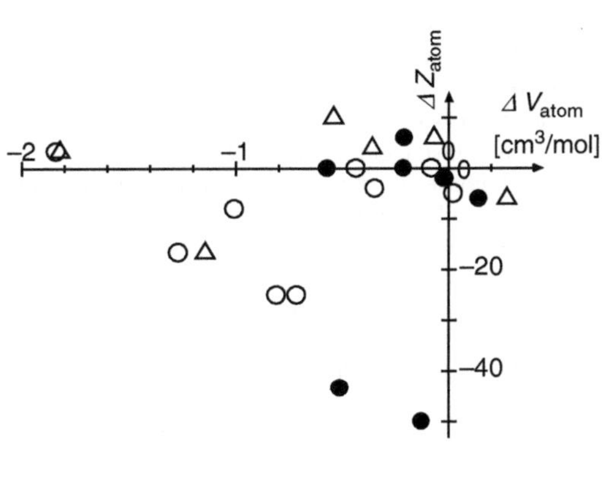

Fig. 1 Polymorphic transformations plotted for the variation of V_{atom} (ΔV_{atom}) and that of Z_{atom} (ΔZ_{atom}). *Circles and triangles* have the same meaning as the marks in Table 1

In addition, the sign of ΔV_{atom} was opposite to that of ΔZ_{atom} for five out of six transformations which proceed partially (Δ, 83.3%), suggesting that, in such transformations, two factors related to V_{atom} and Z_{atom} are competing.

However, there is an exception in Fig. 1. It is Tm_2S_3 (monoclinic → cubic, $\Delta V_{atom} = -1.1$ and $\Delta Z_{atom} = -16.7$, the mark is Δ). Therefore, other factors should be considered to explain this transformation. The values of ΔG were neglected in this study, but it must be one of factors. Furthermore, during MM, the ΔG value is changed intricately due to local high temperatures generated by friction, increased contribution of the interfacial energy and the grain boundary energy, and the structural defects or the strain caused by the defects. Furthermore, from the viewpoint of the driven system, the ballistic jump of atoms caused by external forcing is one of important factors deciding the produced phases [2]. In addition, the activation energy for the transformation is probably one of important factors because if the activation energy of a transformation is low, the non-equilibrium phase easily transforms into the equilibrium one. By taking some of these factors neglected in this study into consideration, more reliable criteria of the phase stability during MM may be developed.

4 Summary

Twenty-three different kinds of reported polymorphic phase transformations during MM were examined in this study. The molar atomic volume (V_{atom}) and the number of atoms in the reduced unit cell (Z_{atom}) were proposed as indicators for evaluating phase stability during MM. To investigate the validity of the indicators, milling-induced polymorphic transformations reported in literatures were classified for the variation of V_{atom} and Z_{atom}. As a result, it was shown that phases with the smaller V_{atom} or Z_{atom} values were preferred during MM.

References

1. Suryanarayana C (2001) Mechanical alloying and milling. Prog Mater Sci 46:1–184
2. Martin G, Bellon P (1997) Driven alloys. Sol State Phys 50:189–331
3. Kwon YS (2007) Decomposition of intermetallics during high-energy ball-milling. Mater Sci Eng A 449–451:1083–1086
4. Han SH, Gschneidner KA Jr, Beaudry BJ (1992) Preparation of the metastable high pressure γ-R_2S_3 phase (R≡Er, Tm, Yb and Lu) by mechanical milling. J Alloys Compd 181:463–468
5. Lin IJ, Nadiv S (1979) Review of the phase transformation and synthesis of inorganic solids obtained by mechanical treatment (mechanochemical reactions). Mater Sci Eng 39:193–209
6. Bokhonov BB, Konstanchuk IG, Boldyrev VV (1995) Sequence of phase formation during mechanical alloying in the Mo–Si system. J Alloys Compd 218:190–196
7. Hara KO, Yamasue E, Okumura H et al (2009) Formation of metastable phases by high-energy ball milling in the Ti–O system. J Phys Conf Ser 144:012021

8. Bégin-Colin S, Girot T, Le Caër G et al (2000) Kinetics and mechanisms of phase transformations induced by ball-milling in anatase TiO_2. J Sol State Chem 149:41–48
9. Liu L, Lun S, Liu S-E et al (2002) Thermodynamic mechanisms of mechanical crystallization of amorphous Fe–N alloy. J Alloys Compd 333:202–206
10. Koch CC (1993) The synthesis and structure of nanocrystalline materials produced by mechanical attrition: a review. Nanostruct Mater 2:109–129
11. Zhang X, Wang H, Koch CC (2004) Mechanical behavior of bulk ultrafine-grained and nanocrystalline Zn. Rev Adv Mater Sci 6:53–93
12. Scudino S, Sakaliyska M, Surreddi KB et al (2009) Solid-state processing of Al–Mg alloys. J Phys Conf Ser 144:012019
13. Li M, Kuribayashi K (2004) Growth kinetics of highly undercooled Al_2O_3 melts. J Appl Phys 95:2342–2347

The Study of CO_2 Fixation in Spent Oil Sand Under the Different Temperature and Pressure

Dong-Ha Jang, Hyun-Min Shim, and Hyung-Taek Kim

Abstract Fossil fuel is one of the energy sources which are used by human beings. Therefore most countries are influenced by the fluctuations of the current oil price. This study deals with oil sand. This study has been focused on the fixation of CO_2 in the spent oil sand after extraction bitumen for oil sand. The physical properties of spent oil sand (components, structures) were analyzed through proximate and ultimate analysis, XRF. Moreover, it was studied about carbon reaction with pretreatment processing. In this paper, carbon fixing in spent oil sand which was conducted as alkaline-earth metal (using CaO). During the experiment, the condition of pretreatment and temperature were changed. And also the CO_2 pressure conditions were changed (1, 25, 50 atm). Mass reduction of TGA analysis was indicated 8.68% in temperature 400°C, 28.74% in temperature 500°C, 14.14% in temperature 600°C, and 17.43% in temperature 700°C, respectively. So, the optimal condition of CO_2 fixation in the spent oil sand is considered near 500°C according to the TGA analysis.

Keywords Carbonation • CCS • CO_2 • Spent oil sand

1 Introduction

In this study it is mainly dealt with oil sand buried in Canada and the United States. As the main resources of fossil fuels are dried up, the human is looking for new energy sources. One of the resources is oil sand. CO_2 generation is occurred in oil sand extraction at the oil refining and upgrading process. CO_2 is generated for this process three times more than typical CO_2 generation process in the oil extraction [1].

The purpose of this study is to fix CO_2 in the spent oil sand, as residue sand from the oil sand refining bitumen. Therefore this study will be able to help to reduce

D.-H. Jang, H.-M. Shim, and H.-T. Kim (✉)
Division of Energy Systems Research, Ajou University, Suwon, South Korea
e-mail: htkim@ajou.ac.kr

T. Yao (ed.), *Zero-Carbon Energy Kyoto 2009*, Green Energy and Technology,
DOI 10.1007/978-4-431-99779-5_34,

the waste and CO_2 generation. To obtain this purpose, we have analyzed the basic properties of spent oil sand and studied effective pretreatment method for fixed CO_2 in the spent oil sand with temperature and pressure conditions. The analyze results were obtained from TGA (thermogravimetric analysis) and GS-MS (gas chromatography-mass spectrometry).

2 Experiment Methods

2.1 Investigation on Physical Properties and Pretreatment in Spent Oil Sand

Before performing CO_2 fixation experiment, it is required to analyze the basic properties of spent oil sand. Because this result can influence performance of fixed CO_2 in spent oil sand and decides whether or not to consider pretreatment in the process. At first, to investigate properties of spent oil sand is basically performed of method of pretreatment for fixed CO_2 in spent oil sand. It was performed ultimate analysis and proximate analysis. This analysis informs the component parts of spent oil sand. In addition, to make sure the configuration of the structure through XRF (X-ray fluorescence) analysis was conducted. This analysis was helpful decide to pretreatment process about alkaline-earth metal. The experiment is performed with 1 M CaO aqueous solutions preprocessed in spent oil sand. The reason is that spent oil sand almost doesn't include alkaline-earth metal which reacts with CO_2. Therefore pretreatment of spent oil sand was necessary.

2.2 CO_2 Fixation Experiments Dependent on Temperature and Pressure Change

The experiment on CO_2 fixation in spent oil sand which physical pretreatment in the CaO aqueous solutions depending on temperature and pressure conditions. The experiment is performed by the temperature condition of 400°C, 500°C, 600°C, and 700°C and the pressure condition of 1, 25, and 50 atm, respectively. As the experiment at temperature, using 5 g of spent oil sand which pretreatment in the CaO is reacted on each temperature with CO_2 into 500 cc min^{-1} in furnace and reaction time with 5 h. The experiment at pressure, using 5 g of spent oil sand is reacted on each pressure with CO_2, and syngas on temperature at 200°C and reaction time with 3 h. Processing of pressure experiment was conducted in the following order, the supercritical reactor after fixing a temperature until 200°C, and connected with the gas booster which connected with the compressor in order to increasing gas pressure, and performed experiment each conditions of pressure.

After performing an experiment, to check the reactive carbonate in spent oil sand samples exposed to CO_2 and syngas is analyzed by GS-MS and TGA.

The methods of analysis is used to improve the CO_2 fixation in 700–800°C of decarbonation section of mass reduction from spent oil sand [2].

3 Experimental Result and Consideration

3.1 Basic Properties of Spent Oil Sand

Spent oil sand samples were obtained from the combustion of oil sand in Canada, Alberta through furnace under the temperature 600°C and time 2 h. The spent oil sand is analyzed by ultimate analysis (Automatic Elemental Analyzer, CHNS-932 Leco) and proximate analysis (Thermogravimetric Analyzer, TGA-601 Leco). And that is represented in Table 1. In addition, result showed Table 2.

Through proximate and ultimate analysis of spent oil sand which is consist of 98% as ash. Also XRF analysis of spent oil sand was announced that most of spent oil sand be consist of SiO_2. According to the results, it is proved that spent oil sand cannot react with CO_2 without any preprocessing. For this reason, digesting CaO aqueous solutions which are an alkaline-earth metal in spent oil sand is required.

Table 1 Proximate analysis and ultimate analysis

Method	Wt%	Spent oil sand
Proximate analysis[a] (wt%)	M	1.06
	VM	0.14
	Ash	98.75
	FC[c]	0.05
Ultimate analysis[b] (wt%)	C	0.15
	H	0.006
	N	0.01
	S	0.03
	O[c]	0
	Ash	99.8

[a] As-received
[b] Moisture free base
[c] By difference

Table 2 XRF analysis

	Wt%
SiO_2	94.1
Fe_2O_3	2.915
Al_2O_3	0.76
K_2O	0.372
CuO	0.057

3.2 Carbonation Reactivity of Spent Oil Sand Dependent on Temperature Change

Pretreatment spent oil sand of CaO can be used an experiment. It is reacted to CO_2 in furnace under the change of temperature condition with 400°C, 500°C, 600°C and 700°C. Results of reactivity carbonation are analyzed the TGA. The TGA analysis result for changing of temperature is shown in Fig. 1.

The (A) section of Fig. 1 shows that chemical bonding water fly away from spent oil sand. And (B) section of Fig. 1 is the mass reduction part from the leaves of the CO_2 in spent oil sand. The reasoning for the validity of the experiment can be found in the expression reaction. In the experiment, reaction of CaO and CO_2 response to the expression $CaCO_3$. And decarbonation is generated at the near 700°C. It is the reason for (B) section is reduction part of CO_2 [3].

The (B) section of Fig. 1 is judged with 29.74% of CO_2 reactions quantity at 500°C.

3.3 Carbonation Reactivity of Spent Oil Sand Under the Change of Pressure

Spent oil sand digest of the CaO is reacted at 200°C under the CO_2 and syngas(N_2:CO_2:CO:H_2=3:1:3:3) of the pressure condition of 1, 25, and 50 atm for 3 h. TGA analysis result dependent on pressure change is shown in Figs. 2, 3, respectively. The results of the experiment in pressure condition, in CO_2 gas, (B) section was decreased rather than expected increasing rate of the mass reduction. And in syngas, (B) section 25 atm was increased until pressure condition, 25 atm, but mass decreased just a little in 50 atm. That is to say, in the (A) section of Fig. 2 amount of the chemical bonding water has a different. But in the (A) section of Fig. 3 have a same shape all three kinds.

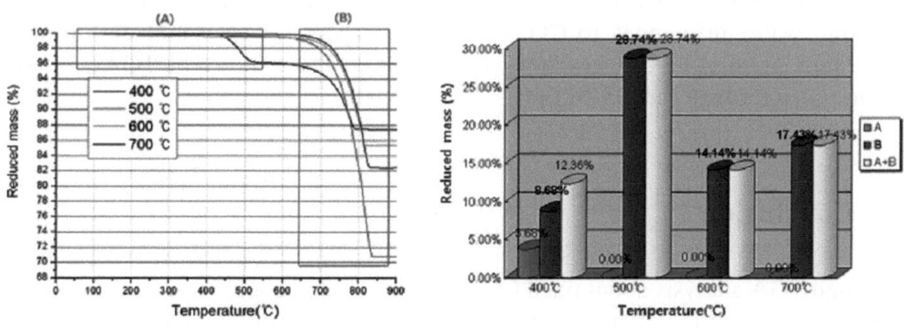

Fig. 1 TGA analysis of temperature change

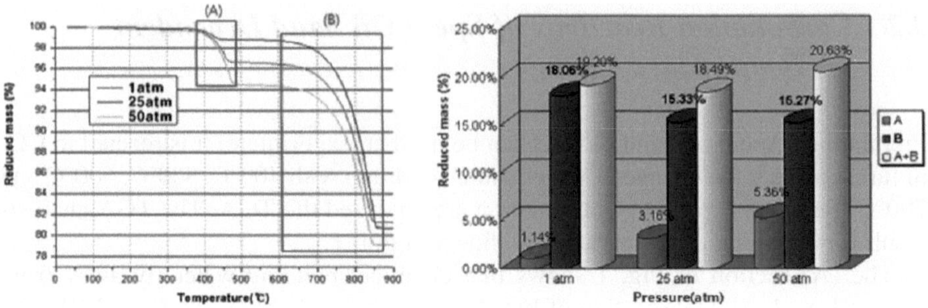

Fig. 2 TGA analysis of pressure change (CO_2)

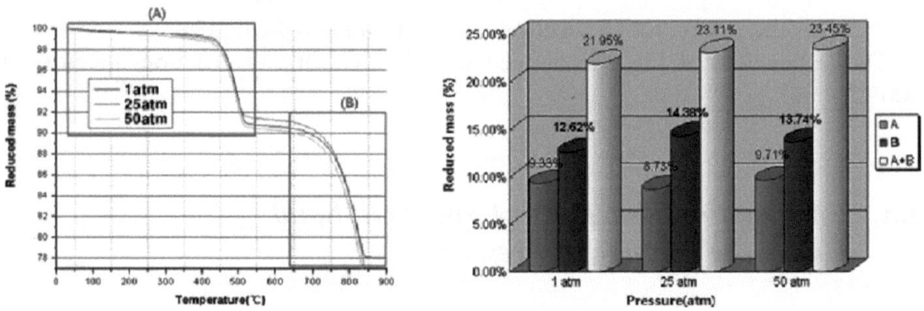

Fig. 3 TGA analysis of pressure change (Syngas)

The reason is that added pretreatment process in experiment which dries spent oil sand in 1 h at 500°C at furnace to remove the chemical bonding water in order to extend reaction of CO_2 in spent oil sand. This case was Fig. 2. In contrast, all of the (A) sections of Fig. 3 were having same reductions. The reason in case Fig. 3 is that spent oil sand reacts rightly without removal of chemical bonding water. So, (A) section of Fig. 3 has same reduction. In the (B) section of Fig. 2 has a result, reduction of mass 18.06% at 1 atm, 15.33% at 25 atm, and 15.27% at 50 atm. And The (B) section of Fig. 3 has a result, 12.62% at 1 atm, 14.3 8% at 25 atm, and 13.74% at 50 atm. As a result of TGA analysis, judged like that, the chemical bonding water influences to CO_2 reactions in spent oil sand and the optimal pressure condition of CO_2 reactions were existed. The validity of TGA analysis could prove through the GS-MS [4].

4 Conclusion

Through TGA analysis about fixed CO_2 in spent oil sand could indicate the optimal temperature condition of about 500°C. Also the pressure condition of optimum will be able to predict between 25 atm. In addition, the removal of chemical bonding water is thought that the experimental process which is the possibility of raising more fixed CO_2 in spent oil sand. In conclusion, results were able to deduce that condition of temperature is more effective than pressure fixed CO_2 in spent oil sand.

Acknowledgment This research is supported by fund of the Energy Resource Technology R&D project from KETEP (Korea Institute of Energy Technology Evaluation & Planning) under the control of the MKE (Ministry of Knowledge Economy).

References

1. Woynillowicz D, Baker CS, Raynolds M (2005) Oil sands fever. Pembina Institute's Publication, Drayton Valley, AB, Canada, pp 16, 22
2. Roo W-H, Kwon T-R, Lee W-M, Lee C-W, Ahn J-Y, Baek I-H (2003) Characteristics of carbonation and decarbonation of carbon dioxide over calcium oxide. Hwahak Konghak 41(5):662
3. Garbev K, Bornefeld M, Beuchle G, Stemmermann P (2008) Cell dimensions and composition of nanocrystalline calcium silicate hydrate solid solutions. Part 2: X-ray and thermogravimetry study. J Am Ceram Soc 91:3015
4. Titelman GI, Gelman V, Bron S, Khalfin RL, Cohen Y, Bianco-Peled H (2005) Characteristics and microstructure of aqueous colloidal dispersions of graphite oxide. J Am Carbon Soc 43(3):647

The Study on Characteristics Upgraded Low Rank Coal (Lignite-IBC) by Changed Temperature and Particle Size

Tae-Jin Kang, Na-Hyung Jang, and Hyung-Taek Kim

Abstract Recently, a coal price is suddenly risen from $53 per ton-coal used in thermal power plant on March 2007 to $129 per ton on March 2008. In present time, it is imported at more than $200 per ton in the steam supply power generation. The low rank coal price is equivalent to one third of present coal, which is difficult to use as generating fuel for two reasons: high moisture, and instability. This study made progress with the aim of using low rank coal by changing it into generating fuel in a dewatering way. As dewatering parameter of this study: temperature and particle size – optimum temperature deduced from TGA results, dividing two part: heating condition and isothermal condition. Along with particle size less than 3 mm, this study was under way on a experiment as divided three sections. 0.3–1 mm, 1–1.18 mm, and 1.18–2.8 mm. According to the study along with different temperatures, moisture content was changed low up to 80°C, but conspicuous up to 150°C. After that, it was nothing noticeable. According to different particle size, in the beginning about five minutes, it seemed a little differences in change of moisture content. After 30 min experiment, it showed no visible differences. Through this progress, it can be found that tendency of variation of moisture content is identical; furthermore, it also indicated that pore structure changed after dewatering. The alteration is shown by means of SEM.

Keywords Dewatering • Low rank coal • Moisture content

1 Introduction

Among the various coals, there are unused low rank coals, which is referred as LRC(Low Rank Coal). The characteristic of LRC has dark-brown, has 4,000–6,000 kcal kg^{-1} of caloric value, and contains 40% of volatile matters.

T.-J. Kang, N.-H. Jang, and H.-T. Kim (✉)
Division of Energy Systems Research, Ajou University, San 5 Woncheon-Dong, Youngtong-Gu, Suwon, Korea
e-mail: htkim@ajou.ac.kr

T. Yao (ed.), *Zero-Carbon Energy Kyoto 2009*, Green Energy and Technology, DOI 10.1007/978-4-431-99779-5_35,

Compared to other coals, LRC has a high moisture content and volatile matter [1]. On the contrary, it is characteristic of low fixed carbon so that it is easy to be wet [2]. When drying, it easily turns to powder. As another attribute of LRC, it also has strong absorption of gas. Out of estimated amount of coal, brown coal accounts for 45%; nevertheless, a great large amount of it remains unexploited. The price of lignite is valued at one third of coal. It is demanded to use because it has high moisture, and in a instable state. Due to such a high moisture content of the coal, moisture removal is the first and essential step in almost any process for upgrading or utilizing them. When the raw lignite is burnt in conventional power stations, up to 20% of the chemical energy of the coal is wasted during the mill-drying process for the evaporation of coal water contained within the lignite structure [3]. For this reason, in recent years efforts have been made to develop efficient dewatering or drying process. The mechanical/thermal dewatering [4–7] has been developed to avoid the disadvantages of the well known thermal [8, 9] and mechanical dewatering [10] processes which are restricted in the technical application due to very high temperatures (>235°C), respectively, pressures (>16 MPa). Hence, in this study, we were carrying out research as to how low rank coal is dried by different size of particles and temperature.

2 Experimental

The lignite used in the experiment is sampled from IBC of Indonesia, those characteristics are shown in Table 1.

The coal is pulverized by using coal crusher as shown in Fig. 1. Particle size is divided by means of sieve shaker. And device used in experiment is EMB(Electronic Moisture Balance) as shown in Fig. 2.

Along with different particle sizes, conducting an experiment on lignite's properties of dewatering, the change of mass is shown during the experiment. The EMB is a device which can measure the quantitative change by setting up temperature from 30°C up to 180°C.

Table 1 Basis properties of the lignite

Item	Result
Total moisture (as received basis)	33.02
Proximate analysis (air dry basis), Wt%	
Inherent moisture	13.63
Ash	3.30
Volatile matter	44.44
Fixed carbon	38.63
Higher heating value (kcal kg^{-1})	
Gross calorific value (air dry basis)	5,370
Gross calorific value (dry basis)	6,220

Fig. 1 Coal crusher and lignite size

Fig. 2 EMB (Electronic Moisture Balance)

3 Result

3.1 Effect on Dewatering Dependent on Temperature

It was examined using the statistical method-EMB in order to find how much water is reduced. It was under way on an experiment using 10 g, of LRC, which has size 1–1.18 mm of particle size for 30 min according to each different temperature. The result is shown in Fig. 3.

Through the result of dewatering dependent on temperature, it shows tendency to decrease rapidly the moisture content from 80°C to 150°C. Above 150°C, it is indicated that dewatering rate has lowered. Based on the result, examined the change moisture content by drying time.

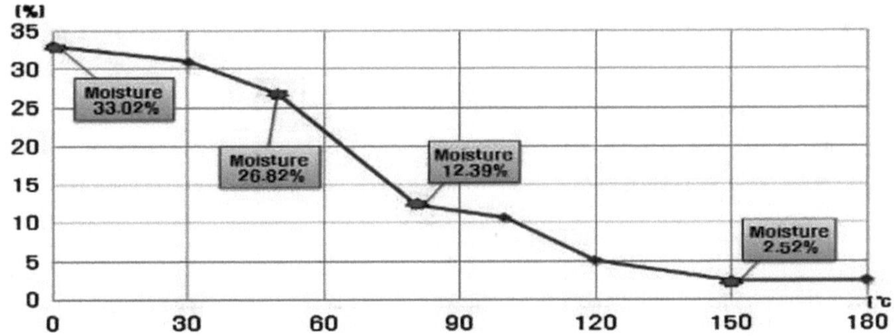

Fig. 3 Moisture content change of lignite by temperature

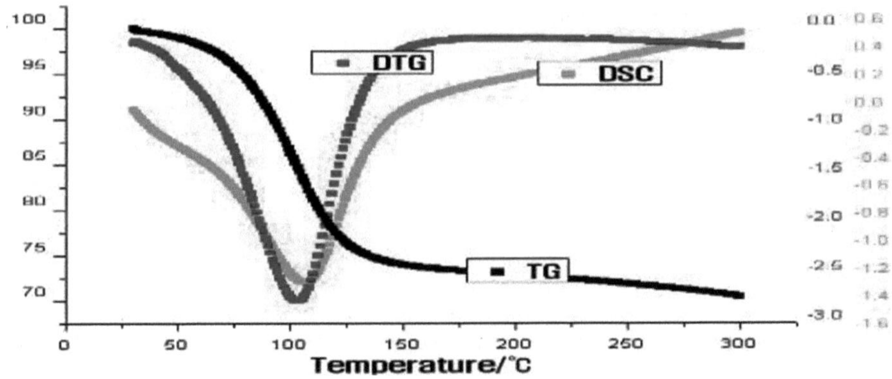

Fig. 4 TGA analysis result

3.2 Effect on Dewatering Dependent on Drying Time

Concerning dewatering parameter, TGA analyze was conducted for getting appropriate temperature. The result of experiment is displayed in Fig. 4.

From the TGA result, it can be deduced that the optimum temperature is 107°C during the dewatering of LRC. The result is shown is Fig. 5.

In the experiment with 10 g, size 1–1.8 mm lignite at 107°C, it found out the fact that decreased water is conspicuous from 15 min. From then on 30 min, it is conducted with optimum 30 min because of little change.

3.3 Effect on Dewatering Dependent on Particle Size

From the isothermal TGA result, it is conducted by different time duration, particles condition from 80°C which is a point changing dramatically; moreover, the experiment

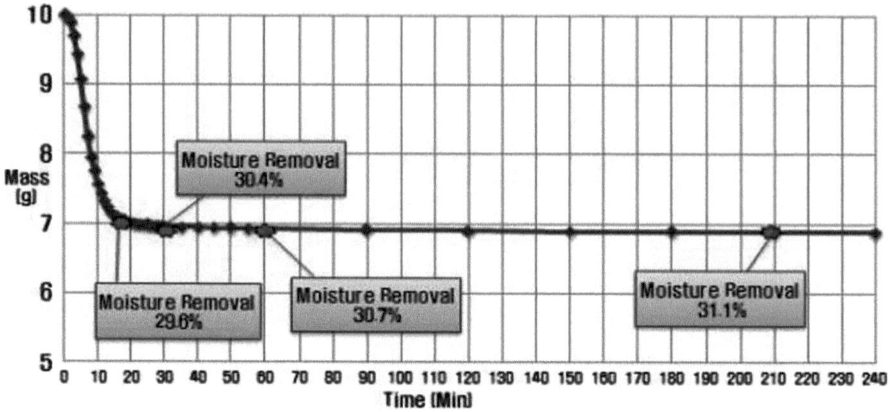

Fig. 5 Moisture content change of lignite at 107°C

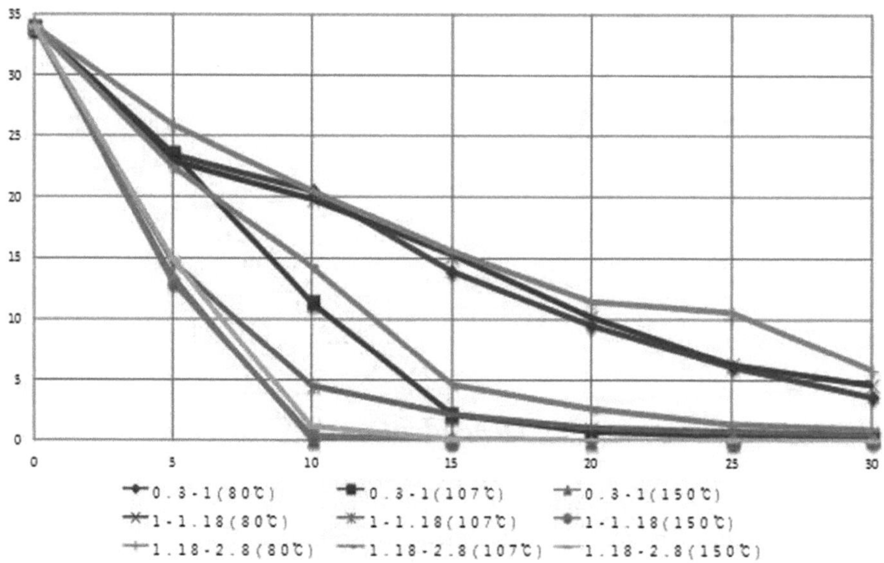

Fig. 6 Moisture content change of lignite by different size at 80°C, 107°C and 150°C

was carried out on appropriate condition – at 107°C, and 150°C that shows little reduction. Figure 6 shows result of dewatering experiment dependent on each temperature and particles size.

As a consequence at 80°C, it is discovered that lignite tend to gradually decrease along with different size and time. In the end of 30 min experiment, it is displayed that the leftover of water remains 5%. As a result of 15 min experiment up to 107°C, there is little change of moisture content; furthermore, at 150°C for 15 min, which shown the most greatest change in the course of experiment. After that, there

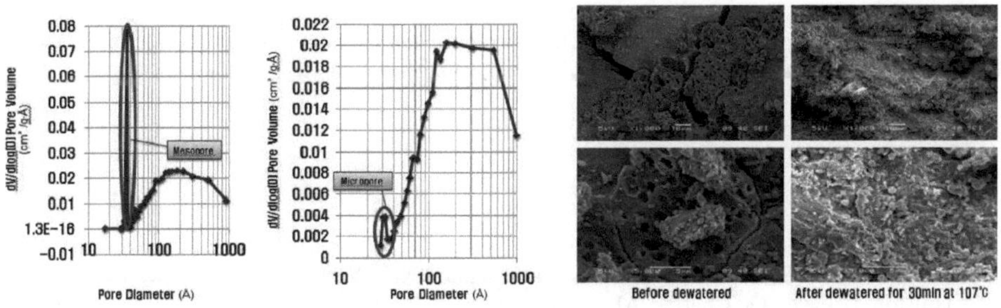

Fig. 7 (**a**) Analysis of BET (**b**) Analysis of SEM

is no change in moisture content. In the last analysis lasting for 30 min dewatering experiment, the moisture content is shown as almost 0%; In addition, as a result of different size, there was no change in moisture content under 3 mm size according to temperature, and time.

3.4 *Effect on Dewatering Pore Structure*

During heating, the coal release parts of the water as a result of the collapse of the colloidal lignite structure [9, 11]. Figure 7a shows the result of BET analysis of lignite dewatered for 30 min at 107°C, particle size 1–1.18 mm. In Fig. 7b, changed pore structure by SEM is shown as follows.

As the result of BET analysis, it can be found that while lignite forms mesopore before dewatering, following that, it produces micropore. Besides, in the result of SEM, in contrast to many pore before dewatering, a number of pore disappeared following that.

4 Conclusion

1. From the TGA result, the point happening dramatic decrease of moisture content is at 107°C.
2. With regard to experiment by particle size, there is no difference in containing moisture.
3. After 30 min at 150°C, initial decreased moisture content is turned out to be high, and the beginning moisture-reduced period of time is short.
4. Lignite progressively transforms pore structure – mesopore into micropore as it is dried. Consequently, it can be found that a total dimension is reduced by a set of dewatering lignite.

Acknowledgment This work was supported by Korea Institute of Energy Research for Development of Drying/Stabilization Mechanism and Establishment of Global Model through the Characteristic Analysis of Low Rank Coal.

References

1. Meyers RA (1981) Coal handbook. Marcel Dekker, New York, pp 1–18
2. Hulston J, Favas G, Chaffee AL (2005) Physico-chemical properties of Loy Yang lignite dewatered by mechanical thermal expression. Fuel 84:1940–1948
3. Bergins C (2003) Kinetics and mechanism during mechanical/thermal dewatering of lignite. Fuel 82:355–364
4. Strauß K (1996) Method and device for reducing the water content of water containing brown coal. Patent EP 0 784 660 B1; WO 96/10064
5. Strauß K, Bergins C, Bohlmann M (2001) Beneficial effects of the integration of a MTE plant into a brown coal fired power station. In: Proceedings of theVGB/EPRI Conference. Lignites and low rank coals: operational and environmental issues in a competitive climate, pp 213–223
6. Bergins C (2001) Mechanismen und Kinetik der Mechanisch/Thermischen Entwasserung von Braunkohle. Dissertation, Universita¨t Dortmund, Shaker, Aachen
7. Umar DF, Usui H, Daulay B (2006) change of combustion characteristic of Indonesian low rank coal due to upgraded brown coal process. Fuel Process Technol 87:1007–1011
8. Fohl J, Lugscheider W, Wallner F, Tessmer G (1987) Entfernung von Wasser aus der Braunkohle: Teil 2: Thermische Entwa¨sserungsverfahren. Braunkohle 39(4):78–87
9. Dunne DJ, Agnew JB (1992) Thermal upgrading of low-grade, low-rank South Australia coal. Energy Sources 14:169–181
10. Banks PJ, Burton DR (1989) Press dewatering of brown coal: part 1: exploratory studies. Drying Technol 7(3):443–475
11. Bak Y-C, Yang U-S, Son J-E (1990) Change in Pore Structure of Chars Obtained under Different Pyrolysis Conditions, Hwahak Konghak 28(6):691–698
12. Duane GL, Richard HS, and Bernard GS, Understanding the chemistry and physics of coal structure (A review), Proc. Natl Acad. Sci. USA , Vol. 79, pp. 3365–3370, May 1982, Review

Energy Efficiency of Combined Heat and Power Systems

Eunju Min and Suduk Kim

Abstract District heat and power (DHP) involving the use of cogeneration facilities is generally regarded to be energy efficient and thus represents an effective way to reduce greenhouse gas emissions. As such, DHP is drawing increasing attention as countries seek ways of honoring commitments to address climate change. This study examines the validity of the assumption that DHP using combined heat and power (CHP) facilities is highly energy efficient in Korea. For this purpose, the definitions of energy efficiency of CHP are examined and the energy efficiency is analyzed based on empirical evidence. It is found that when applying the usual definition of energy efficiency, the index increases as the weight of heat supply increases in CHP. With 95% energy efficiency assumed for single housing boilers, results show that the energy efficiency of separate heat and power (SHP) either supersedes or at most lags behind DHP by 7%, with one exception. Considering the standards of high-efficiency cogeneration defined by Directive 2004/8/EC, a reevaluation seems to be required for the Korean government energy policy to promote DHP.

Keywords District heat and power (DHP) • Combined heat and power (CHP) • Separate heat and power (SHP) • Energy efficiency

1 Introduction

The Korean government has prioritized the promotion of DHP for energy conservation since it is generally regarded to be energy efficient and thus is an effective way to reduce greenhouse gas emissions. But the climate conditions in Korea are not suitable for the full operation of CHP for all seasons. Thus, energy efficiency may

E. Min and S. Kim(✉)
Department of Energy Studies, Division of Energy Systems Research, Ajou University, San 5, Woncheon-Dong, Youngtong-Gu, Suwon, Korea
e-mail: Suduk@ajou.ac.kr

T. Yao (ed.), *Zero-Carbon Energy Kyoto 2009*, Green Energy and Technology, DOI 10.1007/978-4-431-99779-5_36, © Springer 2010

not be as high as other countries, such as those in Northern Europe. Also, with the increasing energy efficiency of combined-cycle gas power plants and commercialized condensing boilers for individual housing, empirical evidence seems to show that CHP systems are less efficient than individual or separate heat and power systems. Park and Kim [6] compared the efficiency of DHP utilizing CHP and heat only boiler (HOB) to SHP with heat from condensing boilers and power from the grid generated by combined-cycle gas power plants. They show that SHP is higher in its energy efficiency than that of DHP. Martens [5] estimated the efficiency of SHP and DHP. He showed that SHP is more efficient than DHP when the energy efficiency of combined-cycle gas power plants and that of single housing boilers are more than 50% and 90%, respectively. Both Kaarsberg et al. [4] and the Austrian Energy Agency [1] used a model to compare the efficiency of two systems by examining the amount of fossil fuel input to provide the same amount of heat and electric power for a typical customer. These studies show that the efficiency of each system depends on the efficiency of SHP, the heat loss of DHP, the power loss of transmission and distribution for SHP, etc.

In this study, these different opinions are examined in the light of empirical data for Korea. This paper is organized as follows. Section 2 explains the data used for the analysis of energy efficiency and examines its definition. Results are discussed in Sect. 3, while conclusions are provided in Sect. 4.

2 Analytical Framework

Energy efficiency is basically defined by comparing energy output to input energy. To define the energy efficiency of CHP of DHP, first we need to define its production efficiency. The production efficiency of CHP can be denoted as $\eta^{P}_{CHP,i} = \frac{P_i^C}{I_i^C}$ which measures the ratio of energy production (P_i^C) to energy input (I_i^C) for CHP, where i denotes either heat (h) or electric power (e). For the energy produced to reach consumers, there will be loss incurred in the process of energy sales. Such loss of energy can be defined as

$$\frac{S_i^D}{P_i^D} = \frac{S_i^C + S_i^H + S_i^K}{P_i^C + P_i^H + P_i^K} = 1 - e_i^L .$$

Since energy sales (heat or electric power) and energy production of DHP are through CHP, HOB, and power import from Korea Electric Power Corp. (KEPCO), each of these can be denoted as S_i^C, S_i^H, S_i^K for energy sales and P_i^C, P_i^H, P_i^K for energy production, respectively. Also, because the loss of energy e_i^L is defined as $e_i^L = \frac{P_i^D - S_i^D}{P_i^D} = 1 - \frac{S_i^D}{P_i^D}$, the above equation can be easily understood.

Final efficiency of CHP, therefore, can be defined as the sum of its efficiency from heat and electric power. If we denote this final efficiency as E_F, E_F will be defined such as following:

$$E_F = \sum_i E_i, \text{ where } E_i = \eta^P_{CHP,i}(1 - e^L_i).$$

For the efficiency of CHP in DHP to be compared with that of SHP, the required energy input for SHP is needed to be calculated under the condition that the same amount of heat and electric power should be supplied to consumers as is in the case of DHP. From the heat sales S^C_h of CHP with e^B the efficiency of separate boiler, the energy input required for the same heat supply to SHP can be calculated such as $I^S_h = \frac{S^C_h}{e^B}$. Similarly, the required energy input for the supply of the same electric power can be calculated such as $I^S_e = \frac{E_e \times (1 + e^L_{S,e})}{e^P_e} S$, where $e^L_{S,e}$ is power loss from T&D, and e^P_e is average efficiency of electricity generation of combined-cycle LNG power plants [8, 9]. The results of input energy calculated for SHP are summarized in Table 1 and compared with final energy efficiency of CHP. It is noted that with two cases, 85% and 95%, of the efficiency of separate boiler, e^B are assumed [2].

3 Results of the Energy Efficiency Analysis

This section summarizes the results of our efficiency analysis of six DHP companies in Korea [7]. Especially noticeable is that those companies with relatively higher volume of heat sales show higher final energy efficiency. The summary result is normalized with the total input energy for electric power and heat as 100. Using the 6 cases' results, relative input energy required for SHP has been calculated and reported for the supply of the same amount of electric power and heat.

4 Conclusion

The current support plan for the promotion of DHP by government entered into force in 1991 and it has been regarded to be a way to actively respond to the climatic change convention, and to contribute to saving energy, thus advancing the national interest. The result of energy efficiency calculation shows that the input energy required for both DHP and SHP does not exhibit a significant difference. The EU council abandoned its plan to increase the promotion of DHP up to 18% by 2017. One of the reasons was because it was found to be difficult to expand DHP when

Table 1 Results of input energy in SHP

	CHP final efficiency			SHP input energy (85%)			SHP input energy (95%)		
	E_h	E_e	E_T	I_h^S	I_e^h	I_T	I_h^S	I_e^h	I_T
Korea District Heating	41.4	26.4	67.8	48.71	58.16	106.86	43.58	58.16	101.74
The Seoul Metropolis	61.2	18.1	79.3	72.00	39.87	111.87	64.42	39.87	104.29
Ansan city Development	57.1	20.7	77.8	67.18	45.60	112.78	60.11	45.60	105.71
GS Power	24.9	37.1	62.0	29.29	81.73	111.02	26.21	81.73	107.94
Incheon Airport Energy	21.1	31.1	52.2	24.82	68.51	93.33	22.21	68.51	90.72
Kenertec	41.8	36.4	78.2	49.18	80.19	129.36	44.00	80.19	124.19
Total	25.6	27.8	53.3	30.12	61.24	91.36	26.95	61.24	88.19

governments took the lead. Also, the EU defined high efficiency of CHP only when there is at least a 10% efficiency difference of DHP from SHP [3].

Considering that the Korean government supports the promotion of CHP/DHP on the basis of its purported high efficiency, it would appear necessary to conduct a further in-depth study of the energy efficiency of DHP based on detailed empirical data.

References

1. Austrian Energy Agency (2002) Cogeneration (CHP) technology portrait. Institute of thermal turbo machinery and Machine dynamics
2. Eco-design of Boilers (2007) Preparatory study on eco-design of boilers
3. European Parliament and of the Council (2004) The promotion of cogeneration based on a useful heat demand in the internal energy market and amending Directive 92/42/EEC, Directive 2004/8/EC of the European Parliament and of the Council of 11 February
4. Kaarsberg T, Bluestein J, Romm J, Rosenfeld A (1998) The outlook for small-scale combined heat and power in the U.S. CADDET Energy Efficiency Newsletter, http://caddet-ee.org/
5. Martens A (1998) The energetic feasibility of CHP compared to separate production of heat and power. Appl Thermal Eng 18:935–946
6. Park HC, Kim HS (2008) Heat supply system using natural gas in the residential sector: the case of the agglomeration Seoul. Energy Policy 36:3843–3853
7. Korea District Heating Corporation (2008) Data book of District heating business
8. Korea power exchange (2009) Electric Power Statistics Information System, http://epsis.kpx.or.kr/
9. Korea power statistic (2007) Business statistics

Behavior of a Boron-Doped Diamond Electrode in Molten Chlorides Containing Oxide Ion

Yuya Kado, Takuya Goto, and Rika Hagiwara

Abstract Behavior of a boron-doped diamond electrode as an oxygen evolution electrode material was investigated at 773 K in molten LiCl–KCl (58.5:41.5 mol%), LiCl–KCl (75:25 mol%), LiCl–$CaCl_2$ (64:36 mol%), LiCl–NaCl–$CaCl_2$ (52.3:13.5:34.2 mol%) containing oxide ion. In molten LiCl–KCl systems, the BDD electrode is stable and its stability does not depend on the concentration of oxide ion and the melt composition. In molten LiCl–$CaCl_2$ and LiCl–NaCl–$CaCl_2$, the BDD electrode is less stable than in molten LiCl–KCl systems.

Keywords Oxygen gas evolution • Inert anode • Molten salts • Metal oxides

1 Introduction

Molten alkali and alkaline earth chlorides are attractive electrolytes having a lot of excellent features such as thermal and/or chemical stability, wide electrochemical windows, high conductivities and high solubilities of other substances. This is why electrochemical processes with molten alkali chlorides are used for the production of alkali, alkaline earth and rare earth metals which are impossible or difficult to obtain by electrochemical processes in aqueous solutions [1].

In recent years, new reduction processes of metal oxides with molten chlorides are proposed to obtain those metals [2–5]. In these reduction processes, oxide ion is generated as a by-product in the electrolytes and smooth removal of it is essential for the improvement of the processes. Carbon has been used for a consumable anode in order to remove oxide ion in the electrolytes, however, carbon anodes evolves carbon monoxide and/or dioxide leading to the dispersion of carbon

Y. Kado, T. Goto (✉), and R. Hagiwara
Graduate School of Energy Science, Kyoto University, Kyoto, Japan
e-mail: tgoto@mail.doshisha.ac.jp

T. Yao (ed.), *Zero-Carbon Energy Kyoto 2009*, Green Energy and Technology,
DOI 10.1007/978-4-431-99779-5_37, © Springer 2010

particles into the melts in the electrochemical reduction of metal oxides in high-temperature molten salts. In our previous studies, a boron-doped diamond (BDD) electrode was found to act as an oxygen evolution electrode in molten LiCl–KCl–Li_2O system at 723 K [6] and molten LiCl–NaCl–$CaCl_2$–Li_2O system at 773 K [7]. In addition, high solubilities of oxides in the electrolytes are also important for efficient electrochemical reduction of metal oxides. The solubilities of oxides have been measured in various molten chlorides for the survey of the system which can dissolve a large amount of oxides [7, 8]. It has been found that the solubility significantly depends on the composition of the melts.

In the present study, electrochemical stability of the BDD electrode was evaluated in eutectic LiCl–KCl (58.5:41.5 mol%), LiCl–KCl (75:25 mol%), eutectic LiCl–$CaCl_2$ (64:36 mol%), eutectic LiCl–NaCl–$CaCl_2$ (52.3:13.5:34.2 mol%) at 773 K. The eutectic compositions were selected based on the reported phase diagrams [9–11].

2 Experimental

Reagent-grade LiCl (Aldrich-APL 99.99%), KCl (Aldrich-APL 99.99%), NaCl (Aldrich-APL 99.99%) and $CaCl_2$ (Aldrich-APL 99.98%) were used for the melt after vacuum drying at 573 K for 24 h. Li_2O (Aldrich. 97%) was used as an oxide ion source which was directly added into the melt after vacuum drying at 453 K for 24 h. All the experiments were conducted in a glove box filled with argon with a gas-refining instrument (MIWA, MS3-H60SN) under dried and deoxygenated atmosphere. The concentration of water and oxygen gas in the atmosphere were always monitored and kept less than 1 ppm.

A boron-doped diamond (BDD) electrode (Diahem^R, Permelec Electrode Ltd., thickness: 2–3 μm, substrate: Si) was used as a working electrode. An aluminum plate (Nilaco Corp., 99.2%) was employed for a counter electrode. The Ag^+/Ag reference electrode was prepared by immersing a silver wire (Japan Metal Service, 99.99%) in each melt containing 0.5 mol% AgCl (Wako Pure Chemical Co. Ltd., 99.5%) in a Pyrex glass tube with the thin bottom. The potential of the reference electrode was standardized against the Cl_2/Cl^- redox couple.

Electrochemical measurements were performed using an electrochemical measurement system (Hokuto Denko Corp., HZ-3000). The sampled gases during and after electrolysis were analyzed by an oxygen gasometer (Panametric Japan Co. Ltd., OX-2) and infrared spectrometer (BIORAD Ltd., FTS-155). The BDD electrode before and after the electrolysis was analyzed by scanning electron microscopy (Hitachi, Ltd., S-2600H), X-ray diffraction (Rigaku, MultiFlex), and Micro-Raman spectroscopy (Jobin-Yvon, Labram spectrometer).

3 Results and Discussion

3.1 Oxygen Gas Evolution on the BDD Electrode in Molten Chlorides

Figure 1 shows cyclic voltammograms of the BDD electrode after the addition of Li_2O into eutectic LiCl–KCl (58.5:41.5), LiCl–KCl (75:25), eutectic LiCl–$CaCl_2$ and eutectic LiCl–NaCl–$CaCl_2$ at 773 K. Anodic currents are observed at around −1.2 V vs. Cl_2/Cl^- in eutectic LiCl–KCl (58.5:41.5), LiCl–KCl (75:25) and −0.7 V vs. Cl_2/Cl^- in molten LiCl–$CaCl_2$ and LiCl–NaCl–$CaCl_2$. These currents are attributed to the oxidation of oxide ion to form oxygen gas by gas analyses, which are similar to the results in our previous study [6, 7].

$$O^{2-} \rightarrow \frac{1}{2}O_2 \uparrow + 2e^- \tag{1}$$

Here, the potentials for the oxygen gas evolution are more positive by 0.5 V in LiCl–$CaCl_2$ and LiCl–NaCl–$CaCl_2$ than that in LiCl–KCl systems. This result suggests that the coordinating environment of oxide ion in the melts is different among them. Conceivable reason is the difference of the interaction between oxide ion and its coordinating cations. The interaction of O^{2-} ion with dipositive Ca^{2+} ion is considered to be stronger than that with Li^+ and K^+ ions. The stabilities of solid oxides improve in the following order: $K_2O<Li_2O<CaO$ [12, 13] and this tendency corresponds to the obtained result in the present study.

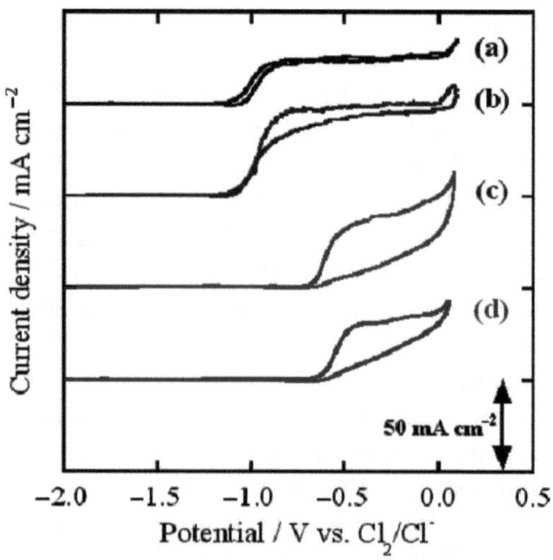

Fig. 1 Cyclic voltammograms on the BDD electrode in molten chlorides containing 1.0 mol% Li_2O at 773 K. Scan rate is 0.1 V s^{-1}. [(**a**) eutectic LiCl–KCl, (**b**) LiCl–KCl (75:25 mol%), (**c**) eutectic LiCl–$CaCl_2$, (**d**) eutectic LiCl–NaCl–$CaCl_2$]

3.2 Electrochemical Stability of the BDD Electrode in a LiCl–KCl Eutectic Melt

Galvanostatic electrolysis at 12 mA cm^{-2} for 20 h was performed in order to evaluate the electrochemical stability of the BDD electrode in LiCl–KCl eutectic melts containing 1.0 and 2.0 mol% of Li_2O at 773 K. The quantity of electricity was 800 C cm^{-2}. The potential of the BDD electrode was kept roughly around −1.0 V (vs. Cl_2/Cl^-) during the galvanostatic electrolysis accompanied by oxygen evolution. Oxygen gas evolution was considered to continuously occur during the electrolysis since CO and CO_2 were not observed by IR spectroscopy in the gaseous sample from the electrolysis cell. Figure 2 shows SEM images of the BDD electrode before and after the galvanostatic electrolysis at 773 K, where the morphologies are almost identical. XRD patterns and micro-Raman spectra of the electrode before and after electrolysis also reveal that the diamond structure is preserved after the electrolysis. In addition, these results suggest that the stability of the BDD electrode does not depend on concentration of oxide ion in the melt. However, apparent consumption of the electrode was observed after electrolysis at temperatures higher than 823 K. Thus the BDD electrode is concluded to be electrochemically stable and act as an oxygen evolution electrode at temperatures lower than 773 K in molten LiCl–KCl–Li_2O systems.

3.3 Dependence of the Stability on the Melt Composition

The stability of the BDD electrode as an oxygen evolution electrode in molten LiCl–KCl (75:25), eutectic LiCl–$CaCl_2$ and eutectic LiCl–NaCl–$CaCl_2$ was examined in the same manner as described for a LiCl–KCl eutectic melt. The potentials during the galvanostatic electrolysis were around −1.0 V (vs. Cl_2/Cl^-) in LiCl–KCl (75:25), −0.05 V (vs. Cl_2/Cl^-) in eutectic LiCl–$CaCl_2$ and eutectic LiCl–NaCl–$CaCl_2$. Figure 3 shows the SEM images before (a) and after galvanostatic electrolysis

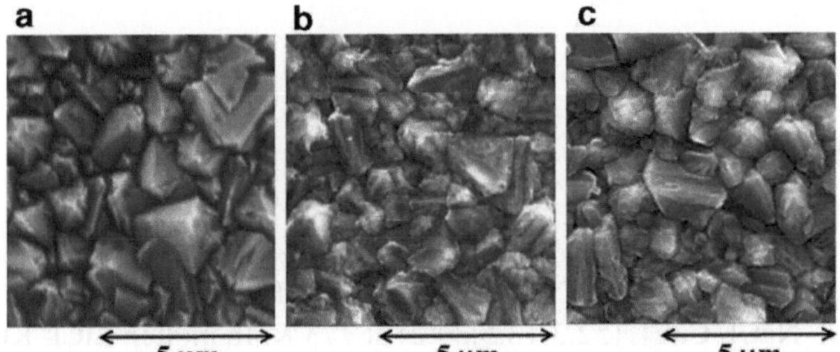

Fig. 2 SEM images of the BDD electrode before and after galvanostatic electrolysis in a LiCl–KCl eutectic melt containing Li_2O [(**a**) as-received, (**b**) 1.0 mol% Li_2O, (**c**) 2.0 mol% Li_2O]

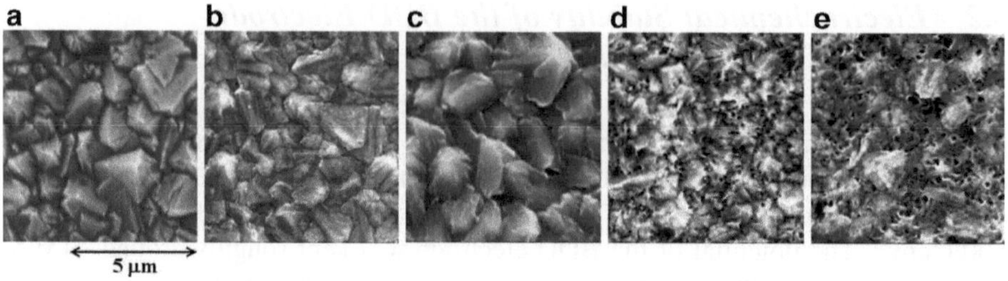

Fig. 3 SEM images of the BDD electrodes before and after galvanostatic electrolysis in molten LiCl–KCl systems containing 1.0 mol% Li_2O [(**a**) as-received, (**b**) eutectic LiCl–KCl, (**c**) LiCl–KCl (75:25 mol%), (**d**) eutectic LiCl–$CaCl_2$, (**e**) eutectic LiCl–NaCl–$CaCl_2$]

at 12 mA cm^{-2} for 20 h in eutectic LiCl–KCl (58.5:41.5) (b), LiCl–KCl (75:25) (c), eutectic LiCl–$CaCl_2$ (d) and eutectic LiCl–NaCl–$CaCl_2$ (e) containing 1.0 mol% Li_2O. The BDD electrode is as stable in LiCl–KCl (75:25) as in eutectic LiCl–KCl (58.5:41.5). This result suggests that the stability of the BDD electrode is not dependent on the melt composition in molten LiCl–KCl systems and the BDD electrode is considered to act as an oxygen evolution electrode in any compositions of LiCl–KCl at 773 K.

In molten LiCl–$CaCl_2$ and LiCl–NaCl–$CaCl_2$, although no notable change was observed by XRD and Raman spectroscopy, SEM images in Fig. 3 apparently show the BDD electrode is less stable as an oxygen evolution electrode than in LiCl–KCl systems. However, the anodic currents attributed to CO and/or CO_2 evolution are not observed in Fig. 1. Conceivable explanation is chemical consumption of the BDD electrode by oxygen gas electrochemically generated on the electrode. This result might be attributed to the low wettability of the melts containing $CaCl_2$, leading stronger adsorption of oxygen gas on the electrode surface and higher overpotential for oxygen gas evolution. Therefore high activity of oxygen gas is considered to chemically facilitate the consumption of the BDD electrode. Another conceivable reason is the catalytic effect of calcium in the electrolyte. It is possible that calcium deposited at the cathode diffuses into the electrolyte and chemically interacts with carbon atom of the diamond surface.

4 Conclusion

The behavior of a boron-doped diamond electrode as an oxygen evolution electrode was investigated in molten LiCl–KCl (58.5:41.5), LiCl–KCl (75:25), LiCl–$CaCl_2$ (64:36), LiCl–NaCl–$CaCl_2$ (52.3:13.5:34.2) at 773 K. In molten LiCl–KCl systems, the BDD electrode is stable and its stability does not depend on the concentration of oxide ion and the melt composition. Thus the BDD electrode has a potential to be employed for the inert anode in molten LiCl–KCl with any compositions at 773 K.

In molten LiCl–$CaCl_2$ and LiCl–NaCl–$CaCl_2$, however, the BDD electrode is less stable than in molten LiCl–KCl systems, which is conceivably due to the low wettability of the melts containing $CaCl_2$. As a result in the present study, it is suggested that molten LiCl–KCl (75:25) is one of the best candidate electrolytes for the reduction processes of metal oxides when combined with the BDD counter electrode taking account of the factors such as high stability of the electrode and the large solubility of oxides at relatively low temperatures.

References

1. Fray DJ (2004) In: Proc. 7th Int. Conf. on Molten Slags, Fluxes and Salts, South African Institute of Mining and Metallurgy, Cape Town, p 7
2. Chen GZ, Flay DJ, Farthing TW (2000) Nature 407:361
3. Sakamura Y, Kurata M, Inoue T (2006) J Electrochem Soc 153:D31
4. Suzuki RO, Ono K (2002) In: Proc. 13th Int. Symp. on Molten Salt, 2002, Electrochemical Society, Pennington, p 810
5. Usami T, Kurata M, Inoue T, Sims HE, Beetham SA, Jenkins JA (2002) J Nucl Mater 300:15
6. Goto T, Araki Y, Hagiwara R (2006) Electrochem Solid-State Lett 9:D5
7. Kado Y, Goto T, Hagiwara R (2008) J Electrochem Soc 155:E85
8. Kado Y, Goto T, Hagiwara R (2008) J Chem Eng Data 53:2816
9. Elchardus E, Laffitte P (1932) Bull Chim. France 51:1572
10. Mahendran KH, Nagaraj S, Sridharan R, Gnanasekaran T (2001) J Alloy Comp 325:78
11. Bukhalove GA, Arabadzhen AS (1962) Zh Neorgan Khim 7:2230
12. Landolt-Börnstein (1999) SGTE. Springer, Berlin
13. Landolt-Börnstein (2001) SGTE. Springer, Berlin

(iii) Advanced Nuclear Energy Research

An Algorithm for Automatic Generation of Fault Tree from MFM Model

Jie Liu, Ming Yang, and Xu Zhang

Abstract Fault Tree Analysis (FTA) is a powerful technique and most widely applied in the domain of reliability engineering and Probabilistic Safety Analysis (PSA). However, the conventional construction of fault trees for a large and complex system is usually hard and time-consuming, and susceptible to human errors, and the validation and modification of fault trees are also difficult. GTST-MFM (Goal Tree Success Tree–Multilevel Flow Modeling) method has been proposed in author's previous work, based on the method, an algorithm for automatic generation of fault tree from the system's GTST-MFM model is presented in this paper. An example is also employed as an application of this translation algorithm.

Keywords Multilevel flow models • Fault tree analysis • Translation algorithm • Reliability analysis

1 Introduction

The FTA is a widely used method for evaluation of reliability and safety [1], however, the conventional method to construct fault tree has the following problems: (1) the construction of fault tree for a large and complex system is usually hard, time-consuming and susceptible to human errors. (2) Due to the different understandings to the system, validating or modifying fault trees by other analysts is usually difficult.

A new reliability analysis method was proposed based on Multilevel Flow Modeling (MFM) in authors' previous work [2], by which the analysts could construct a system's model quickly and efficiently. And then, fault tree can be mapped from GTST-MFM manually, so it's susceptible to human errors. In order to make the work of system reliability analysis easier and more convenient, an algorithm for automatic synthesized fault tree is presented in this paper.

J. Liu, M. Yang (✉), and X. Zhang
College of Nuclear Science and Technology, Harbin Engineering University, Harbin, China
e-mail: yangmingheu@hrbeu.edu.cn

T. Yao (ed.), *Zero-Carbon Energy Kyoto 2009*, Green Energy and Technology.
DOI 10.1007/978-4-431-99779-5_38, © Springer 2010

2 Translation Algorithm

The translation algorithm is briefly shown in Fig. 1. There are three main parts in this algorithm, translation of goal, translation of gate, translation of function respectively. As shown in Fig. 2, each goal is translated into the corresponding event, and all the *AND* gate are translated into *OR* gate and *OR* gate into AND gate. Figure 3 is the translate template of function, function in MFM is corresponding to function fault in fault tree, and this fault may be aroused by system fault *OR* physical component fault which realize the function. And the system fault can be caused by condition fault *OR* the upstream function fault.

An example (Fig. 4) is used here to explain the translate process, and the main goal of this system is raising the water lever in container when the level is low. GTST-MFM model of this system is shown in Fig. 5.

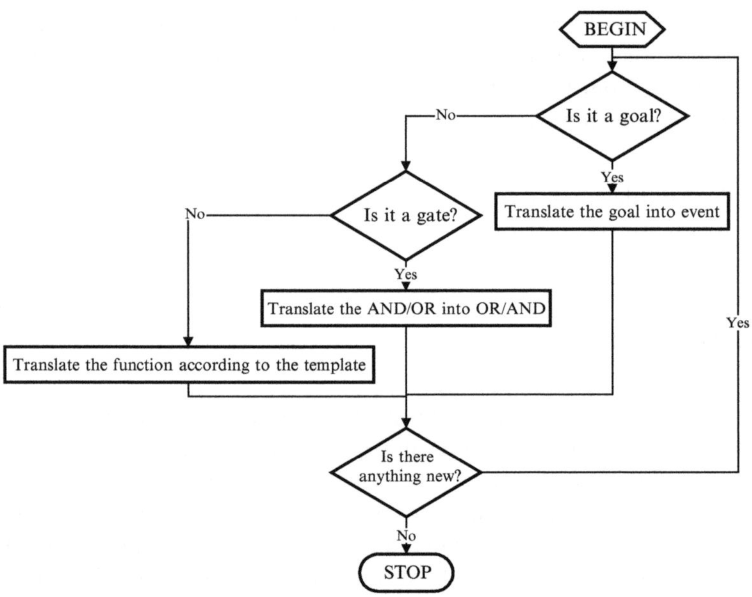

Fig. 1 The chart of translation algorithm

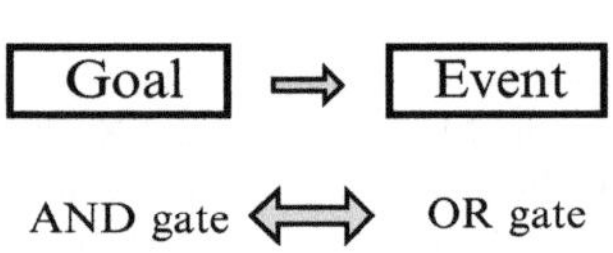

Fig. 2 Translation of goal and gate

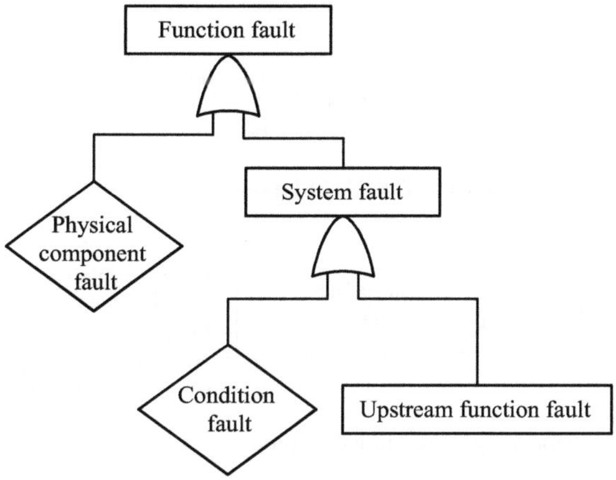

Fig. 3 Translation template of function

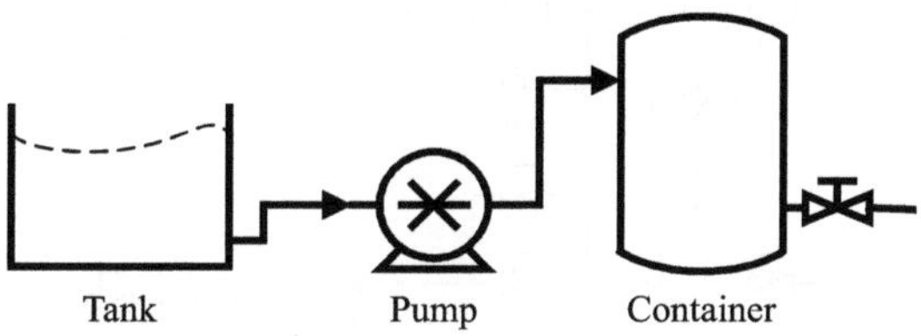

Fig. 4 Diagram of the example system

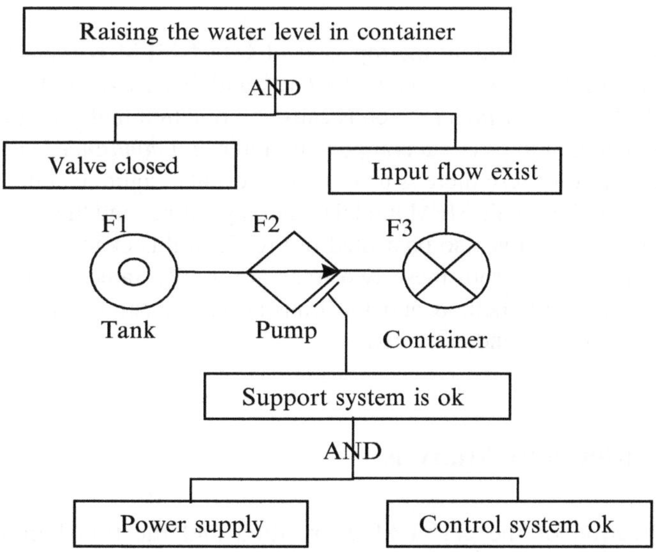

Fig. 5 GTST-MFM model of the system

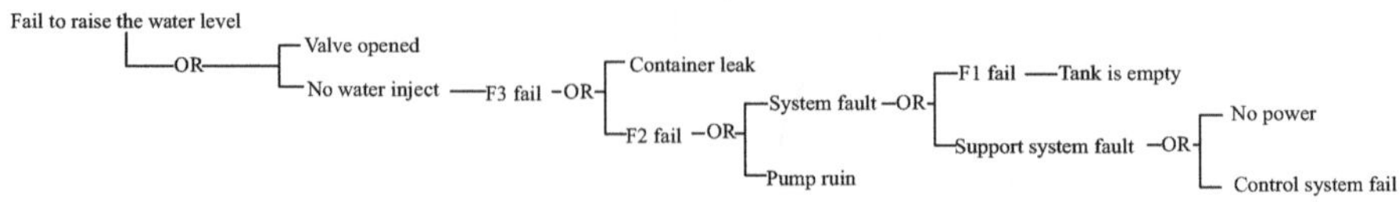

Fig. 6 The fault tree translated from GTST-MFM model

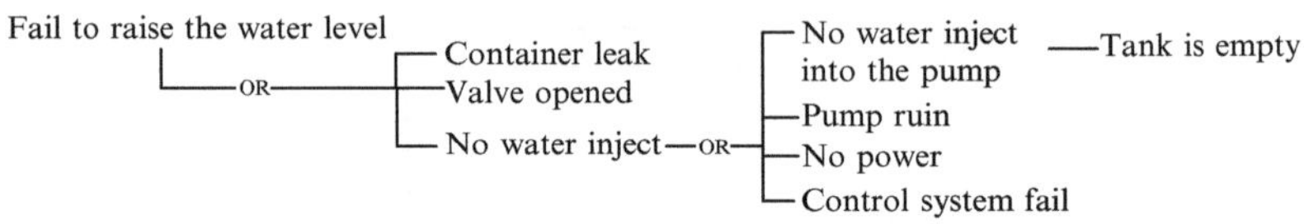

Fig. 7 The fault tree mapped by conventional method

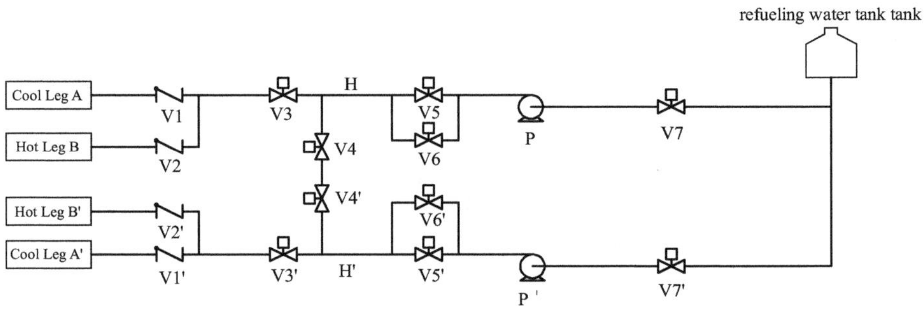

Fig. 8 The LHSIS of a PWR

According to the algorithm, the top goal of GTST-MFM is translated into the corresponding event *fail to raise the water level*, and this event would be caused by *value closed* OR *no water inject* which is cause by function fault event *F3 fail*, then it can be subdivided into physical component fault event *container leak* and system fault event *F2 fail* (because there is no condition event). Figure 6 is the whole fault tree mapped from the GTST-MFM model by our algorithm, and the tree constructed by conventional method is also presented in Fig. 7. In this example, the minimum cut sets (mcs) obtained from these two trees are totally same. Consequently, by using the GTST-MFM method and the translation algorithm, the analyst could model a system quickly and correctly.

3 Application and Analysis

As an application of the GTST-MFM method and the translation algorithm proposed in this paper, the Low Head Safety Inject System (LHSIS) of a PWR (in Fig. 8) is also presented.

A fault tree of this system has been proposed in ZHU's literature [3] by conventional method, and in this study, the algorithm is used to translate the GTST-MFM model which has been presented in literature [2] into fault tree model, then Fault Tree+11.0 (a reliability analysis software)is employed to analysis the fault tree and the mcs of the translated tree are compared with ZHU's results. It is easy to find that our results include all the mcs in ZHU's, and there are 18 sets peculiarly in our result because some failure modes are ignored in ZHU's model, for example, the failure of V7 and V7′. Thereby, the GTST-MFM method can model a system roundly and conveniently, and the algorithm can be used to translate GTST-MFM into fault tree correctly and quickly.

4 Conclusions

In previous study, the GTST-MFM method is proposed to construct system model, base on the GTST-MFM method, analyst could construct the system model quickly and roundly, because the concept of flow is embedded and the conservation law is also complied in the GTST-MFM method, which makes it is possible to model a system without neglect. Further, fault tree can be synthesized from system's GTST-MFM model by the translation algorithm, which greatly reduces the impact of human error in the model translation process. And the results of translated fault tree can be used to validate the ones constructed by conventional method, so this study also provides a new means to resolve the validation problem of fault tree.

Acknowledgments This study is supported by National Natural Science Foundation (NFSC) of China (Grant No. 60604036), Heilongjiang Provincial Foundation for Returned Scholars (Grant No. LC06C06) and the 111 project (Grant No. b08047).

References

1. Roberts NH, Vesely WE, Haasl DF, Goldberg FF (1981) Fualt tree handbook. NUREG-0492. US NRC, Washington, DC
2. Jie L, Yang M, Zhijian Z (2009) A qualitative method of reliability analysis based on multilevel flow models. In: NPIC&HMIT 2009, Knoxville, TN
3. Zhu jizhou (1989) Principle and application of fault tree. Xi'an Jiaotong University Press, Xi'an, pp 97–100

A Method of Generating GO-Flow Models from MFM Models

Xu Zhang, Ming Yang, and Jie Liu

Abstract Multilevel Flow Modeling (MFM) is a graphical modeling method with means-end and whole-part conceptions. GO-Flow is a system modeling technology which could be used for the quantitative calculation of reliability analysis of a large system with multiple operational sequences. The conversion of MFM models into GO-Flow models can not only extend the application of MFM to system reliability analysis, but also makes the construction of GO-Flow models easier and more convenient. In this paper, a method of generating GO-Flow models from MFM Models is proposed, and an example is also employed as an application of this conversion method.

Keywords Multilevel flow modeling • GO-Flow • Reliability analysis

1 Introduction

MFM is a graphical modeling method proposed by Lind in 1990. In recent years, many research efforts have been devoted into applying MFM into the field of reliability analysis [1–3], however, it is still a new research area.

GO-Flow is a system modeling and reliability analysis technology developed by Matsuoka in 1988 based on GO methodology. However, a GO-Flow model is difficult to be understood as well as to apply qualitatively to find out the weaknesses of the system, which limits the wide application of GO-Flow methodology in complex and safety-related systems.

X. Zhang, M. Yang (✉), and J. Liu
College of Nuclear Science and Technology, Harbin Engineering University, Harbin, China
e-mail: yangmingheu@hrbeu.edu.cn

T. Yao (ed.), *Zero-Carbon Energy Kyoto 2009*, Green Energy and Technology,
DOI 10.1007/978-4-431-99779-5_39, © Springer 2010

In this paper, a method of converting MFM models into GO-Flow models is proposed. By using this method, a GO-Flow model could be generated in accordance with the pre-prepared MFM model. Therefore, the precise quantitative calculations of the system reliability at any given time points can be analyzed by GO-Flow methodology.

2 Conversion Method

The corresponding relations between the concepts of MFM and GO-Flow elements are briefly shown in Table 1, the concepts and characteristic parameters mentioned are referring to the Multilevel Flow Models based Reliability Analysis (MFMRA) proposed by Jan [4]. Besides of the above relations, a Type 35 operator (cold standby) or Type 37 operator (hot standby) of GO-Flow should be employed in series when the aging effect of the equipment is considered, and a Type 40 operator should be employed when the signal delays. The flow structures of the generated GO-Flow models are also treated as three types: mass flow, energy flow and information flow. And because the concept of network is preserved during the conversion, the generated GO-Flow models are also of high understandability with multiple levels of means-end and whole-part decompositions.

3 Application and Analysis

The Safety Injection System (SIS) of Pressurized Water Reactor (PWR) under Lose Of Coolant Accident (LOCA) is presented as an case study of the conversion method. The MFM model (Fig. 2) is built in accordance of the schematic diagram of the system (Fig. 1), and a GO-Flow model is generated (Fig. 3) through the GO-Flow program GFED [5]. The operation sequence of SIS and the actions of the equipments are shown in Table 2 and Fig. 4. And based on the generated GO-Flow model, quantitative calculation of the system reliability is done as shown in Fig. 5.

4 Conclusions

In this paper, a method of Generating GO-Flow Models from MFM Models is proposed, and the quantitative calculations of the system reliability can be analyzed accurately by employing the generated GO-Flow models. The significant points of this study are as follows: Firstly, it provides a way of applying MFM to deal with the problems in system reliability analysis and Probabilistic safety analysis (PSA). Secondly, it not only provides an insight into the internal logical relations of the

Table 1 The conversion method

Concept in MFM	Remarks	GO-FLOW element	Formula
Source/observer/manager		Type 25 Operator	Pobj(t)=Poutf(t)=Pfun(t)×Pcond(t)
Sink/actor/direction/barrier		Type 21 Operator	Pfun=Pg,Pobj(t)=Pinf(t)×Pfun×Pcond(t)
Balance	Balance in MFMRA has only one input	Type 21 Operator	Pfun=Pg,Pobj(t)=Pouf1(t)=Poutf2(t)=…=Poutfn(t)= Pinf(t)×Pfun×Pcond(t)
Storage	Two failure modes	Two Type 21 Operators in series	Pfun=Pg1×Pg2,Pobj(t)=Poutf(t)= Pinf(t)×Pfun×Pcond(t)
Transport	Transfer element	Type 21 Operator	Pfun=Pg,Pobj(t)=Poutf(t)=Pinf(t)×Pfun×Pcond(t)
	Switching element	Type 26 Operator	Pobj(t)=Poutf(t)=Pinf(t)×O(t)×Pcond(t), O(t1)=Pp,O(t)= O(t')+[1-O(t')]×P(t)×Pg
Goal		Signal Line	
Component		Type 25 Operator	
Achieve Relation	*G[X]*: the function operation of the Gates.		Pgoal(t)=*G[*Pobj(t)*]*
Condition Relation		An And Gate in series	Pcond(t)=*G[*Pgoal(t)*]*, Pobj(t)=Pinf(t)×Pfun(t)× Pcond(t)
Realize Relation		An And Gate in series	Pobj(t)=Pinf(t)×Pfun(t)×Pcond(t)
And Gate with *N* inputs	In Condition or Realize Relations	*N* And Gates in series	
	In Achieve Relations	An And Gate with Type 40 Operators before every input	
Or Gate with *N* inputs		An Or Gate with Type 40 Operators before each inputs	
Vote Gate(*M*-out-of-*K*)	Assembly of And Gates and Or Gates		

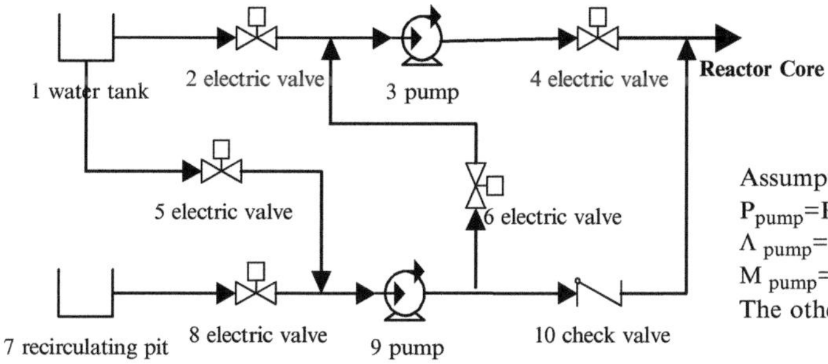

Assumptive reliability parameters:
P_{pump}=Pvalve=0.9
Λ_{pump}=4E-2
M_{pump}=0.1
The others are supposed reliable.

Fig. 1 The SIS of a PWR

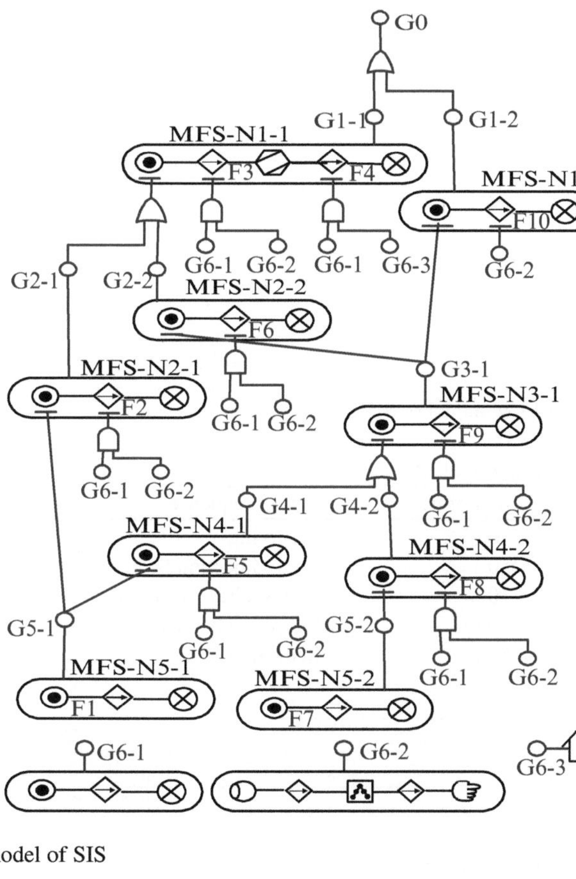

Fig. 2 MFM model of SIS

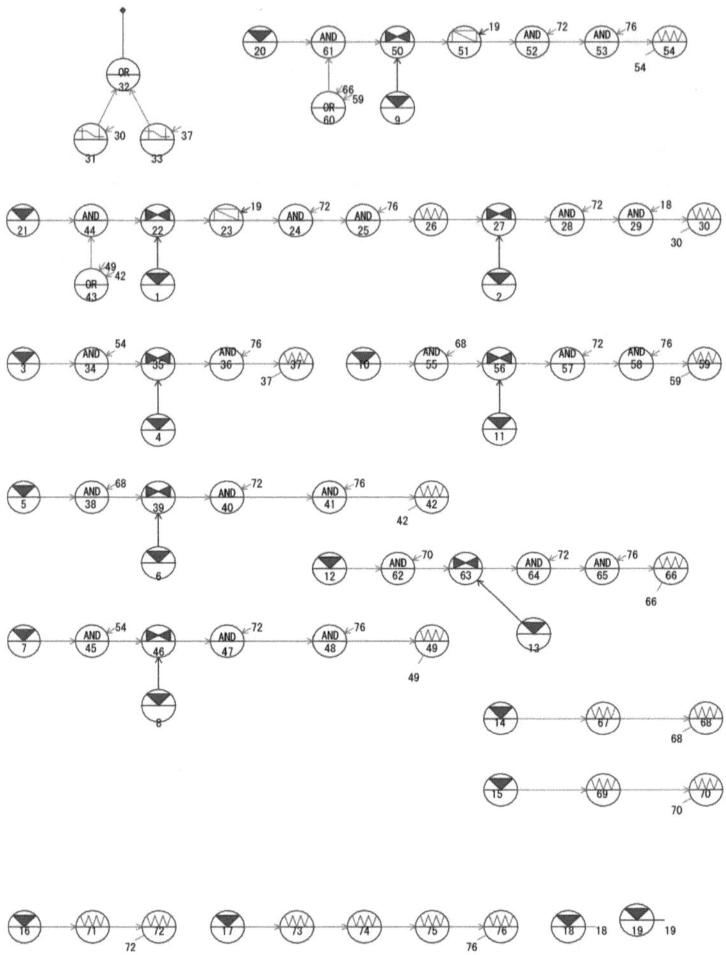

Fig. 3 Generated GO-Flow model of SIS

Table 2 Time sequence of SIS

Time point	Real time	Action sequence
1	0	Reserving
2	50 s	Direct injection begins at 50 s after LOCA
3	10 m	End of direct injection
4	10 m	Recycling begins
5	50 h	Recycling

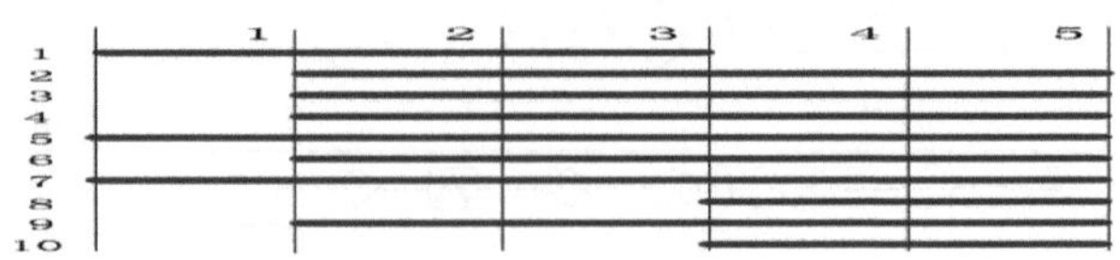

Fig. 4 Actions of system components

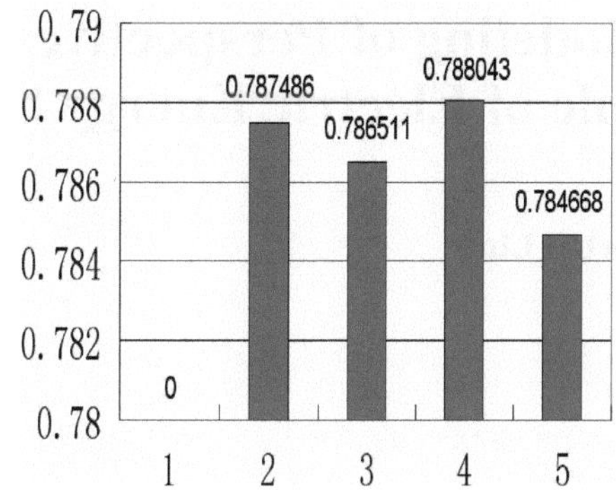

Fig. 5 Result of reliability analysis

system behaviors and functions, but also makes the construction of GO-Flow models easier and more convenient. At last, this study provides a way for the qualitative analysis by GO-Flow since MFM can also be translated into Fault Tree [6], which lays a foundation for the wide applications of GO-Flow in the reliability analysis and PSA for modern power systems such as nuclear power plant.

Acknowledgments The authors thank NFSC (Grant No. 60604036) and the 111 Project (Grant No. b08047). And during this study, through master Hidekazu Yoshikawa of Harbin Engineering University, Prof. Takeshi Matsuoka of Utsunomiya University and Mr. Kazuo Tamura of Itochu Technosolution Co. provided the GO-FLOW software and manual information. The authors express special thanks for their kind helps and collaborations also.

References

1. Jan Eric Larsson (2002) A pilot project on alarm reduction and presentation based on Multilevel Flow Models. In: Proceedings of the Enlarged Halden Program Group Meeting, HPR-358, Storefjell, Gol, Norway
2. M. Yang (2008) Development of an alarm analysis system based on Multi-level Flow Models for Nuclear Power Plant. Chinese Journal of Nuclear Science and Engineering, Vol.28 No.1, China
3. A. Gofuku, A. Ohara (2008) Fault tree analysis of chemical plants based on multi-level flow modeling. In: Proceedings of ISSNP 2008, Harbin, China
4. Zhan J (2007) Application of reliability analysis based on multilevel flow models in nuclear power plant. Dissertation for the Degree of M. Eng, Harbin
5. Matsuoka T, Kobayashi M (1991) Development of the GO-FLOW reliability analysis support system. In: Proceedings of an International Symposium Probabilistic Safety Assessment, IAEA-SM-321/61, pp677–688
6. Liu J, Yang M (2009) A qualitative method of reliability analysis based on multilevel flow models, In: Proceedings of NPIC&HMIT 2009, Knoxville, TN

Functional Modeling of Perspectives on the Example of Electric Energy Systems

Kai Heussen and Morten Lind

Abstract The integration of energy systems is a proven approach to gain higher overall energy efficiency. Invariably, this integration will come with increasing technical complexity through the diversification of energy resources and their functionality. With the integration of more fluctuating renewable energies higher system flexibility will also be necessary. One of the challenges ahead is the design of control architecture to enable the flexibility and to handle the diversity. This paper presents an approach to model heterogeneous energy systems and their control on the basis of purpose and functions which enables a reflection on system integration requirements independent of particular technologies. The results are illustrated on examples related to electric energy systems.

Keywords Functional modeling • Multilevel flow modeling • Intelligent energy systems • Power systems • Frequency control

1 Introduction

We anticipate that sustainable energy systems are more *intelligent* energy systems. The integration of energy systems is a proven approach to gain higher overall energy efficiency. Invariably, this integration will come with increasing technical complexity through the diversification of energy resources and their functionality. With the integration of more fluctuating renewable energies higher system flexibility will also be necessary. All this results in a demand for ever more advanced control of electric power system to handle the mix of resources with increased flexibility, while the system robustness ought to be maintained.

One approach to improve efficiency of the electricity sector is its integration with the heat sector. As heat can easily be stored, this integration also gives way for a cheaper and more effective type of energy storage: flexible demand. For example, the

K. Heussen (✉) and M. Lind
Department of Electrical Engineering, Technical University of Denmark,
2800 Kongens Lyngby, Denmark
e-mails: kh@elektro.dtu.dk; mli@elektro.dtu.dk

T. Yao (ed.), *Zero-Carbon Energy Kyoto 2009*, Green Energy and Technology,
DOI 10.1007/978-4-431-99779-5_40, © Springer 2010

Danish electricity supply relies mainly on combined-heat-and-power (CHP) plants. All larger CHP plants have been equipped with significant heat storage to offset electricity production from the district heating demand. Studies suggest further an addition of heat pumps to the district heating system to enable the integration of wind power into the electricity supply, e.g. [8].

In recent years, many visions of future integrated energy systems have been proposed, some are based on a particular technology domain such as Microgrids or Zero-energy Buildings, others are based on an abstract planning and optimization process that does not involve the technical details of an implementation (they often assume some type of global coordination). Such integrated energy systems depend on separate domains of engineering which have their own way of representing design problems and requirements.

Integration of energy systems means the combination of systems that were previously independent and therefore have partly incompatible conceptualizations. Common system analysis is behavioural is therefore dependent on assumptions about the technical realization. The functional modeling approach applied in this paper instead allows the study of interrelations on a more general level by formalizing the semantic relations between different perspectives. The functional models are presented by Multilevel Flow Modeling (MFM). In this paper the method is outlined with a focus on the underlying semantics. The concept of perspectives is introduced and illustrated on an example related to electric energy systems.

2 Functional Modeling with MFM

Multilevel Flow Modeling (MFM) is an approach to modeling goals and interconnected functions of complex processes involving interactions between flows of mass, energy and information [6, 7].[1] It provides means for a *purpose-centered* (as opposed to *component-centered*) description of a system's functions. MFM enables modeling

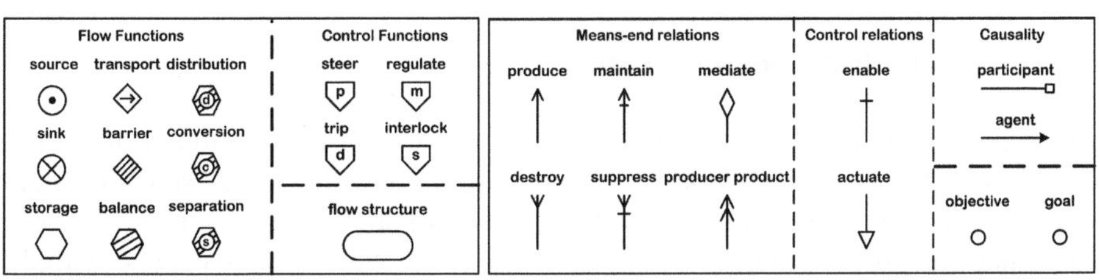

Fig. 1 (**a**) The *box on the left* lists the MFM-symbols, elementary flow-and control-functions as well as the flow structure, which combines an interconnection of functions; (**b**) the *right box* presents all MFM relations and the symbols for objectives and goals

[1]Please contact one of the authors for more information on MFM.

at different levels of abstraction using well-defined means-ends relations and whole-part compositions (Fig. 1b). Process functions are represented by elementary flow functions interconnected to form flow structures which represent a particular goal oriented view of the system (Fig. 1a). The views represented by the flow structures, functions, objectives and their interrelations together comprise a comprehensive model of the functional organization of the system represented as a hypergraph. MFM is founded on fundamental concepts of action and each of the elementary flow and control functions can be seen as instances of more generic action types.

Models created in MFM are a formalized conceptual representation of the system, which support qualitative reasoning about control situations. MFM is supported by knowledge based tools for model building and reasoning.

MFM models can be and have been employed for the purposes of state identification (and representation) and action generation. State identification applications include: model based situation assessment and decision support for control room operators; hazop analysis; alarm design and alarm filtering. Further possible applications include operator support systems or integrated HMI and process-design.

MFM has been used to represent a variety of complex dynamic processes, i.e. in fossil and nuclear power generation and chemical engineering (e.g. oil refineries) and biochemical processes. The method was originally conceived in the context of cognitive systems engineering as an intermediary model for work domain analysis, but has its own path of development now. Its strong semantic concepts and existing software tools make it suitable for integration with modern methods of intelligent control [10]. For IT applications it is useful to formalize all aspects of the modeling technique. An outline of this formalization is given below.

2.1 Underlying MFM Concepts

In this section we discuss the underlying concepts that establish the functional structures of MFM. The goal is to identify the basic operations on a functional description of a system.

2.1.1 Actions, Roles and Functions

MFM is strongly related to the semantics of action, and it is possible to formalize MFM entities in a framework of actions and action-roles. The "semantic deep structure of an action" [1] has been analyzed in relation to MFM in [9]. What is important for MFM is the concept of semantic roles, which are associated with the semantic deep structure of an action. It can be illustrated like this:

This illustration provides an *action* in the centre with *semantic roles* like “slots” to be filled. The kind and number of slots depend on the specific action, but *agent*, *object* and *instrument* are the most generic:

The apple is cut with a knife by John.
OR: *John uses a knife to cut an apple.*

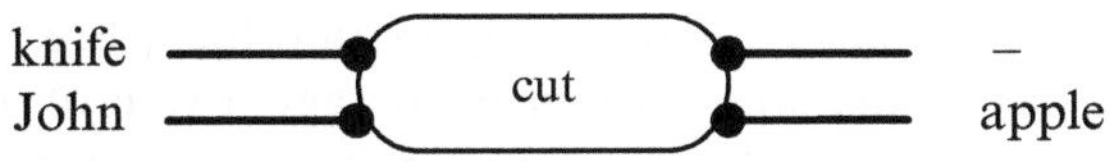

Given this understanding of an action, functional modeling can be described as a modeling approach that formalizes meaningful combinations of actions and roles in the context of a means-ends framework. MFM provides templates for the interconnection of a number of specific actions. These templates are functions, particularly *flow-functions* and *control functions*.

Definition of *function* [9]:

A function of a concrete entity **E**, which is part of a system **S**, is specified in terms of the role **R** of **E** in relation to an action describing an intended state-change in **S**.

According to von Wright [11, 12], elementary actions can be derived from the concept of elementary change. Given a proposition p about the state of the world the four elementary changes are { “p disappears”= $p\mathrm{T}\neg p$; “p happens” = $\neg p\mathrm{T}p$; $p\mathrm{T}p$; $\neg p\mathrm{T}\neg p$ },[2] where “$\neg p$” is “not p” and “T” stands for a transition. An *intentional action* must now be distinguished from a change that does not involve an agent A: Instead of “p happens”, we say “A makes p happen, otherwise $\neg p$ happens”, in short: $\{\neg p\mathrm{T}[p\mathrm{I}\neg p]\}$.[3] Particularly control functions in MFM are directly derived from elementary actions.

In summary, *propositions* about the state of the system define the effect of a *function* (action), and the *semantic roles* of the action capture the relations between entities in a system. Action *phases* structure temporal information aspects of a function.

2.1.2 Flow Structures and Control Structures

There are energy flow structures, mass flow structures and control structures. Most commonly energy- and mass-flow structures are used to represent a particular goal-oriented view of a system.

A *flow* structure allows modeling of a process without direct reference to the agents associated with realizing the process. However, the agent role is associated with each function and can be assumed by an external agent.

---

[2] The latter two are non-changes, $p\mathrm{T}p$; $\neg p\mathrm{T}\neg p$, which lead to the concept of *elementary omissions*, as discussed in [4].

[3] Please refer to [4], [5] and [7] for a thorough introduction.

A *control* structure is meant to represent the purpose of a control action. Von Wright's theory of intentional action sets a framework for the modeling of control actions. The four elementary interventions define the four possible control functions *steer*, *regulate*, *trip* and *interlock*, respectively (Fig. 1a).

A simple control structure is composed of one *process objective*, which is usually an objective associated with another energy or mass flow structure, and a *control function* (steer, regulate, trip or interlock, Fig. 1a) [6, 7]. The control function has an *actuate-relation* to the agent-role of a flow-function in a lower-level means-ends level (example in Fig. 2, p. 288). A control-structure has an external *objective* that describes performance requirements of the control.

2.1.3 Perspectives and Views

The simplest and elementary form of an MFM model is an energy- or mass- flow structure connected with an objective via an achieve-relation (produce, maintain, destroy or suppress). The objective or goal is an expression of the intention (the "Why") that is associated with the functional structure and the system it represents. A flow structure contains a conceptualization of the functions the system utilizes to achieve its purpose (the "HOW"). MFM provides templates or conceptual schemes for the *representation of functions*, as well as for goals, objectives and means-end relations which form the statement of *intention*. A *perspective*, or elementary functional description, consists therefore of a set of two elements:

1. Intention (Objective + means–end relation)
2. The representation of functions in a functional view

Usually, an MFM-model consists of several such *perspectives* that are connected through a number of possible relations (*mediate, producer/product, enable, actuate* [all in Fig. 1b]).

3 MFM Model of Energy System Balancing

The concepts introduced above are illustrated in the following on a number of examples from a modeling application to power systems. The examples have been previously published in [2, 3].

The abstract model in Fig. 2 relates the overall goal $\mathbf{g}_1$ to the intended functional organization of the system. The passive role of the generation side reflects system goal, but an analysis of the realization of Generation shows that this role needs to be enabled by the objective $\mathbf{o}_1$. The enabling objective describes a condition to be fulfilled at a lower level of abstraction.

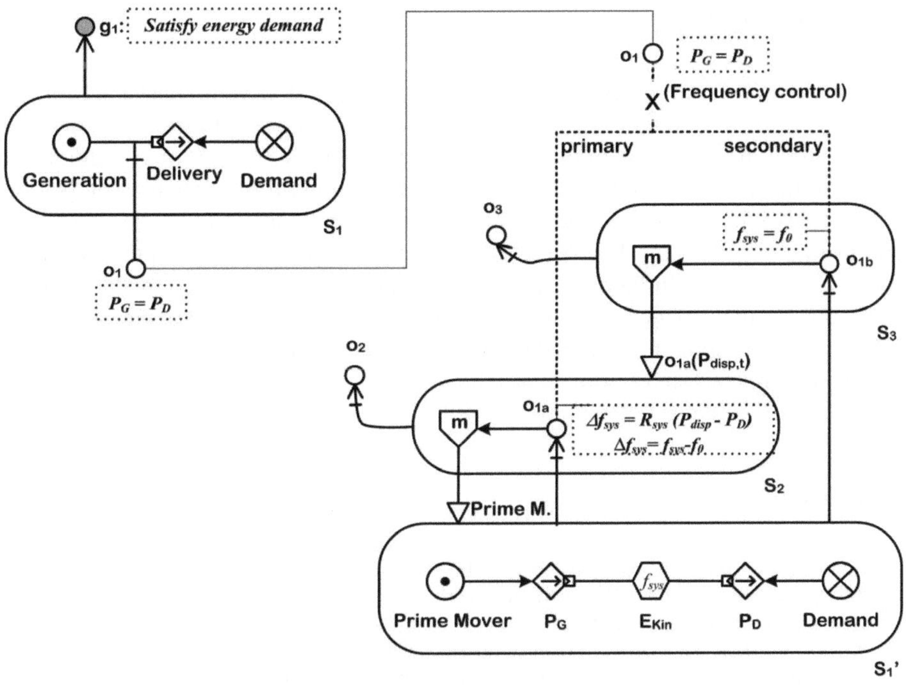

Fig. 2 Abstract (*left*) and more detailed (*right*) representations of system balancing functions

The descriptions followed abstract considerations about the system design, showing a connection between the statement of design intentions ("goals"), functional abstraction and more concrete process objectives.

The objectives are structured into an objective hierarchy, where the original objective is reformulated $\mathbf{o}_1$.= ($\mathbf{o}_{1a}$ *and* $\mathbf{o}_{1b}$) with consideration of the flow-structure of the lower-level functional view, from a (mathematical) decomposition of the original frequency control objective $\mathbf{o}_{1.}$

This decomposition is based on AC power systems with synchronous generators. In AC power systems the common frequency reflects the energy stored in the rotating mass of the generators and therefore is a measure of the energy balance. Restoring the frequency therefore is eventually restoring the energy balance.

The objectives of the objective hierarchy are achieved by a combination of a flow structure $\mathbf{S}_1$, representing the energy system, and two control structures representing primary ("droop") and secondary ("integral") frequency control ($\mathbf{S}_2$ and $\mathbf{S}_3$). The objectives are maintained by a cascade of control structures $\mathbf{S}_2$ and $\mathbf{S}_3$, which employ the system frequency measure and actuate the generators to *maintain* their respective control objectives – which means to balance the system. Note that there are three strongly connected perspectives in this MFM-model.

4 Conclusion

This paper presented an overview of semantic and action theoretical concepts in Multilevel flow Modeling. The concept of *perspective* as a set of *intention* and functional *representation* was introduced. This concept of perspective forms a framework for the formal representation of the role-shifts that occur in MFM-relations – integrating action-roles with the means-ends levels of MFM. An example from the domain of energy systems illustrates how these "shifts in perspective". The work presented here forms a platform for further research. Future work branches out into two directions: (a) Computer-implementation of the formalizations and development of new reasoning rules; (b) The modeling approach can be applied to analyze possible integrated energy systems or "smart grid" control concepts.

References

1. Fillmore CJ (1968) The case for the case. In: Batch EA, Harms RT (eds) Universal linguistic theory. Holt, Heinhart and Winston, New York
2. Heussen K (2009) Decomposing objectives and functions in power system operation and control. In: IEEE PES/IAS Conference on Sustainable Alternative Energy, Valencia, 2009
3. Heussen K, Saleem A, Lind M (2009) Control architecture of power systems: modeling of purpose and function. In: Proceedings of the IEEE PES General Meeting, 2009
4. Lind M (2002) Promoting and opposing. Technical report, Tech. Univ. of Denmark, 2002
5. Lind M (2004) Description of composite actions. Technical report, Tech. Univ. of Denmark, 2004
6. Lind M (2005a) Modeling goals and functions of control and safety systems in MFM. In: Proc. of the International Workshop on Functional Modeling of Engineering Systems, Kyoto
7. Lind M (2005b) Modeling goals and functions of control and safety systems in MFM. Technical report, Tech. Univ. of Denmark, 2005
8. Lund H, Münster E (2006) Integrated energy systems and local energy markets. Energy Policy 34(10):1152–1160
9. Petersen J (2000) Knowledge based support for situation assessment in human supervisory control. Ph.D. thesis, Tech. Univ. of Denmark
10. Saleem A, Heussen K, Lind M (2009) Agent services for situation aware control of power systems with distributed generation. In: Proceedings of the IEEE PES General Meeting, 2009
11. Von Wright GH (1963) Norm and action. Routledge & Kegan Paul, London
12. Von Wright GH (1968) An essay in deontic logic and the general theory of action. Acta Philosophica Fennica 21:1–55

Mechanical Properties and Microstructure of SiC/SiC Composites Fabricated for Erosion Component

Min-Soo Suh, Akira Kohyama, and Tatsuya Hinoki

Abstract To accomplish "Zero CO_2 Emission society" development of alternative energy and reduction of energy dissipation are both necessary. SiC_f/SiC composites are considered as enabling technology for both new advanced energy system and high efficiency system. A novel challenge, employing *in situ* fiber crystallization method, has been made in order to develop a new concept of SiC/SiC composites. The optimization of SiC/SiC fabrication process has been carried out through the observed microstructural defects. Worn behaviors and erosion resistance will be also discussed. The depending issues of prototype process were rather serious in this fabricating concept therefore the process optimization has been done on following issues. (1) Non-uniform fiber deformation caused by non-uniform PyC coating. (2) Excess pressure and stress concentration influencing fiber deformation ratio around 2.44. (3) Easy fiber detachment due to the weakened bonding energy and generation of aperture between fiber and interface. (4) Serious fiber volume contraction around 24.5% during the crystallization process in high temperature. (5) Generated pores due to interface cleavage.

Keywords SiC/SiC composite • Erosion • In situ fiber crystallization • Fabrication optimization

1 Introduction

To accomplish "Zero CO_2 Emission society" by solving current issues of global environment and energy crisis, the development of alternative energy and reduction of energy dissipation have to be carried out together. Continuous SiC fiber-reinforced

M.-S. Suh (✉)
Graduate School of Energy Science, Kyoto University, Gokasho, Uji, Kyoto, Japan
e-mail: Min.S.Suh@gmail.com; mssuh@iae.kyoto-u.ac.jp

A. Kohyama
Department of Materials Science and Engineering, Muroran Institute of Technology, Muroran 050-8585, Japan

T. Hinoki
Institute of Advanced Energy, Kyoto University, Gokasho, Uji, Kyoto 611-0011, Japan

T. Yao (ed.), *Zero-Carbon Energy Kyoto 2009*, Green Energy and Technology,
DOI 10.1007/978-4-431-99779-5_41, © Springer 2010

SiC matrix (SiC_f/SiC) composites are considered as prime candidates for structural materials where demands both high temperature and high wear resistance environments such as advanced aero and space propulsion [1–4]. CO_2 emission can be reduced by developing a new advanced SiC/SiC composite; to serve a long-term high temperature component for high efficient energy systems and controlling wear characteristics of materials to reduce not only energy dissipation but also material squander [2–4]. Significant advances have been made in recent years regarding all constitutes, fibers, interface and matrix of SiC/SiC, thereby resulting in pure near-stoichiometric small-diameter fibers, exact-demanded interface thickness and highly crystallized near-full dense matrix that provide most of the composite requirements [1,3,4]. Consequently, *state-of-art in advanced SiC/SiC composites* has solved most of technical issues [3,4]. In spite of all the superiorities the extreme high cost of SiC/SiC composites are still an issue.

This study focused on two basic parts; development of a new advanced SiC/SiC composite and evaluation of mechanical properties for structural reliability to answer the purpose. Microstructure of newly fabricated materials is also observed to find the crucial parameter of fabrication, and examining the wear behaviors to clarify the appropriateness for erosion resistance application.

2 Experimental Procedures

A novel challenge of fabrication has been proposed by employing Pre-SiC fiber and PyC (Pyrolytic Carbon) as a fiber-reinforcement and a fiber-coating. Erosion wear test has carried out to evaluate issues of durability and reliability for FOD resistance. Densitometry test was carried out by the Archimedean method to examine the density and content of porosity after fabrication. Nano indentation was held by applying 16 gf load on each matrix of materials. Erosion wear test which coincide with ASTM G76 was carried out by impinging SiC particles on the composite surface to evaluate the erosion resistance (Fig. 1). Microstructural analysis validate the optimized fabrication condition of LOT17 comparing with prototype LOT11.

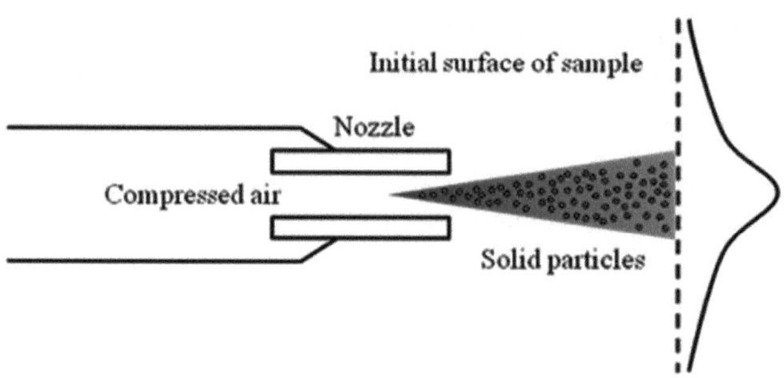

Fig. 1 Schematic of the solid particle impingement using gas jet

Image analysis was used to confirm microstructural defects mainly focused on crystallized fiber deformation and porosity generation.

3 Result and Discussion

Table 1 shows the mechanical properties of newly fabricated specimens by hot-press. *Porosity* was mainly generated in two forms, one is pore in matrix and another is aperture around the fiber. Dominant reasons for porosity were excess pressure and stress concentration caused by non-uniform PyC coating around the fibers and also volume contradiction of fiber itself (see Figs. 2 and 3). These porosities influenced on easy detachment of fiber and matrix by particle impact. Nano indentation results show that the typical hardness of matrix is over 16 GPa for three of all fabricated composites except some of defected areas on LOT11 and 12 where inadequate densification of matrix and excess oxide zone.

Fiber deformation in LOT11 was comparatively serious, result of image analysis shows that the volume contraction ratio was around 24.5% and deformation ratio was around 2.44 in case of oval shape; these mean that there will be quite a lot

Table 1 Mechanical property of fabricated specimens by hot-press

	LOT17	LOT12	LOT11
Apparent density (g cm^{-3})	3.08	2.89	2.77
Bulk density (g cm^{-3})	3.06	2.74	2.66
Open porosity (%)	0.50±0.037	5.08±0.04	4.2±0.04
Hardness of matrix (GPa), 16g	16.8±0.6	15.9±1.3	16.3±0.5
Young's modulus (GPa)	336±8	285±15	299±6
Wear volume (mm^3)	114.9	216.2	195.9

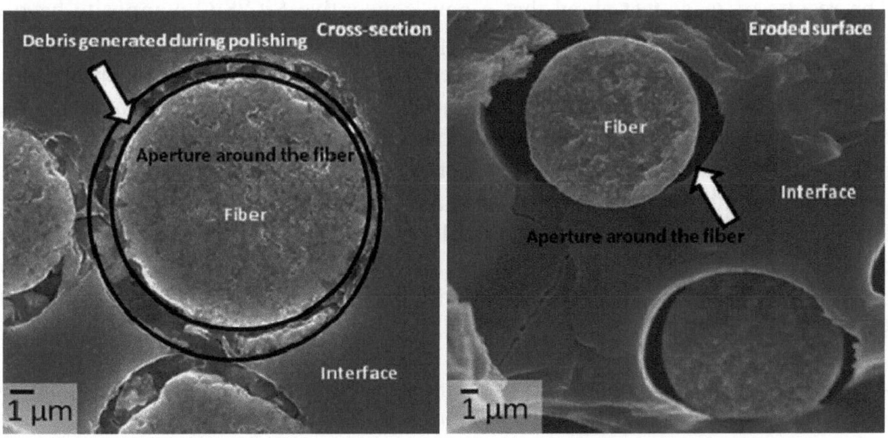

Fig. 2 Aperture generated around the fiber due to reduction of fiber volume

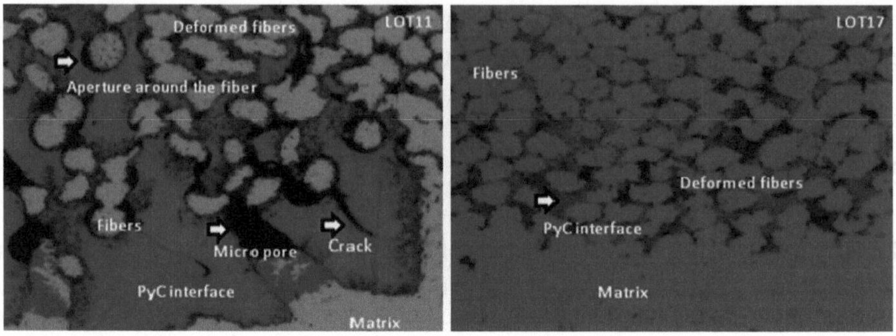

Fig. 3 Cross section of newly fabricated materials. (**a**) Deformed fibers and generated crack during hot-press. (**b**) Near full-densed LOT17

geometrical discordance and generation of pore in the form of aperture around the fiber. *Interface cleavages* were conspicuously observed in prototype materials but LOT17 (see Fig. 3). In case of LOT17, micro pores and interface cleavage were hardly observed due to the adequate amount of employed PyC.

Dominant erosion mechanism was detachment of fiber and interface, and eroding of matrix. Aperture along the fibers and pore in the matrix provoke easy detachment of material constituent. These generated wear, all most in all system, cause issues of durability and reliability due to geometric discordance, irregular performance, and energy dissipation.

4 Concluding Remarks

The results of wear test show that a new conceptive SiC/SiC composite have a successful improvement comparing with prototype materials for erosion resistance. Due to the disproportional pressure delivered to each areas where contains different volume fraction of fiber and PyC interface; huge scale of fiber deformation were observed with 2.4 deformation ratio. Another dominant reason is fiber itself in the process of crystallization during the hot-press in high temperature. *Cracks* also propagated along the interface of matrix and other constituents where the bonding energy has been weakened. Aperture along the fiber was generated by volume contraction of Pre-SiC fiber itself during the crystallization process where pressure delivery hardly occurred. Consequently, it provokes easy fiber detachment, which shows deleterious effect on erosion resistance.

Acknowledgments The Author would like to thank the MEXT of Japan for scholarship and the 21 Global COE program of Kyoto University for partial-financial support.

References

1. Kohyama A et al (2002) Development of SiC/SiC composites by nano-infiltration and transient eutectoid (nite) process. Ceram Eng Sci Proc 23:311–318
2. Suh M-S et al (2008) Friction and wear behavior of structural ceramics sliding against zirconia. Wear 264(9–10):800–806
3. Suh M-S, Kohyama A (2009) Special issues on "in situ" crystallized SiC/SiC composites. In: Proceeding of ISAE2009, pp 439–442
4. Suh M-S, Kohyama A (2009) Effect of porosity on particle erosion wear behavior of lab scale SiC_f/SiC composites. Int J Mod Phys B (this article is under review)

Diffusion Bonding of Tungsten to Reduced Activation Ferritic/Martensitic Steel F82H Using a Titanium Interlayer

Zhihong Zhong, Tatsuya Hinoki, and Akira Kohyama

Abstract Development of materials and related fabrication process is one of the most important technologies for fusion energy development. In fusion reactor, joining of tungsten (W) to reduced activation ferritic/martensitic steel is required. In this work, diffusion bonding between W and ferritic/martensitic steel F82H using a Ti interlayer was investigated. The results indicated that all the joints were successfully obtained. The interfacial microstructure was analyzed by scanning electron microscopy. The chemical composition of these reaction products were determined by energy dispersive spectroscopy. W–Ti solid solution was found at W/Ti interface, while Ti/F82H interface formed complex phases which dependent on the joining temperature. Bond strength was evaluated and the maximum shear strength was obtained for the joint bonded at 900°C. The failure was occurred at Ti/F82H interface during shear testing.

Keywords Tungsten • F82H steel • Diffusion bonding • Interlayer • Titanium

1 Introduction

Tungsten (W) and its alloy has been selected as plasma facing material for fusion nuclear application due mainly to its high melting point, high resistance against sputtering, and low tritium retention [1, 2]. Reduced activation ferritic/martensitic (RAFM) steels, which developed for simplify special waste storage of highly radioactive structures of fusion reactor after service, is one of the candidates to be used as first wall and blanket structural materials in fusion reactors [3], F82H is one such

Z. Zhong
Graduate School of Energy Science, Kyoto University, Gokasho, Uji, Kyoto 611-0011, Japan
e-mail: zh_zhong@iae.kyoto-u.ac.jp

T. Hinoki and A. Kohyama (✉)
Institute of Advanced Energy, Kyoto University, Gokasho, Uji, Kyoto 611-0011, Japan
e-mail: kohyama@iae.kyoto-u.ac.jp

T. Yao (ed.), *Zero-Carbon Energy Kyoto 2009*, Green Energy and Technology, DOI 10.1007/978-4-431-99779-5_42, © Springer 2010

steel [4]. A joint between W and RAFM steels is required for some components such as divertor in fusion reactor according to the design [5]. Joining of W to steel, however, is difficult due to the large differences in their physical properties, particularly the mismatch of their coefficients of thermal expansion (CTE) which leads to a large residual stress in the joints. In addition, the formation of brittle intermetallic compounds (FeW and Fe_7W_6) in the diffusion zone is highly possible which harmful to the joint.

The joining techniques such as active metal brazing [6], plasma spraying [7], and diffusion bonding [8, 9] have been developed for joining of W to RAFM steels. The metallic brazing produced high strength joints and provided well reproducible results. Nevertheless, the brazing temperature of 1,150°C is high and causes grain coarsening in steel which is undesirable [6]. Additionally, it is difficult to obtain a high density W coating on steel for the components produced by plasma spraying. Diffusion bonding seems to be a suitable way for joining of W with RAFM steel due to the acceptable bonding temperature and the joints can be used at high temperatures.

For a diffusion bonded joints, inserting an interlayer between dissimilar substrates is necessary to alleviate the residual stress and to prevent the formation of intermetallic compound in the joints. We have used a nickel interlayer for bonding of W to ferritic steel and found that the results was promising [9]. In the present work, we tried to use titanium (Ti) as interlayer for joining of W to F82H steel by diffusion bonding route.

2 Experimental Procedures

The commercially available W and IEA heated F82H steel used in this work were cut to dimension of $10^L \times 5^W \times 2^T$ mm from the as-received plates. The commercial Ti sheet with 0.6 mm thick was cut to size of $10^L \times 5^W$ mm. The chemical compositions of these materials are shown in Table 1. Prior to diffusion bonding, the joining surfaces of all the materials were polished by an emery paper with 1,500 grit. The materials then were ultrasonically cleaned in acetone for 10 min. The assemblies of W/Ti/F82H were joined in a hot-press furnace at a temperature range of 800–1,000°C for 1 h under a uniaxial load of 10 MPa in vacuum with a heating rate of

Table 1 Chemical compositions (wt%) of the materials used in this work

Alloy	Cr	C	N	P	S	Al	Si	V	Ti	Mn	Ta	W	O	Fe
F82H	7.84	0.09	0.007	0.003	0.001	0.001	0.07	0.19	0.004	0.1	0.04	1.98	0.1	Bal.
W	–	0.02	0.01	–	–	–	–	–	–	–	–	Bal.	0.02	–
Ti	–	0.02	–	–	–	–	–	–	Bal.	–	–	–	0.15	0.01

10°C min^{-1}. Once the bonding process was completed, removed the load, and the joints were cooled slowly in the furnace to room temperature.

The cross-sectional microstructure observation of the joints was conducted with field-emission scanning electron microscopy (FE-SEM). The chemical compositions of the reaction phases were analyzed with energy dispersive X-ray spectrometry (EDS). The elemental intensity profiles of the chemical species across the interfaces were drawn from the electron probe microanalysis (EPMA). The shear strength of the joints was evaluated at room temperature using a tensile-testing machine (Instron 5581) with a crosshead speed of 0.5 mm min^{-1} in a specially designed jig. Five samples were tested for each processing. The fracture surfaces of the samples after shear testing were observed under FE-SEM.

3 Results and Discussion

3.1 *Interfacial Microstructure Analysis*

W was successfully bonded with F82H by using Ti interlayer under all the employed experimental conditions. Figure 1 is a typical general view of the transition joint bonded at 950°C. Both W/Ti and Ti/F82H interfaces are free from discontinuities or cracks. The higher-magnification images showing detailed microstructure of W/Ti interfaces and W concentration profiles in the diffusion zone are given in Fig. 2. It can be seen from these figures that the W/Ti interfaces are planar in nature and a thin diffusion zone whose thickness depends on the joining temperature. The thickness of diffusion zone increases from ~2 μm to ~12 μm when the joining temperature increases from 850°C to 950°C. The variation nature of concentration

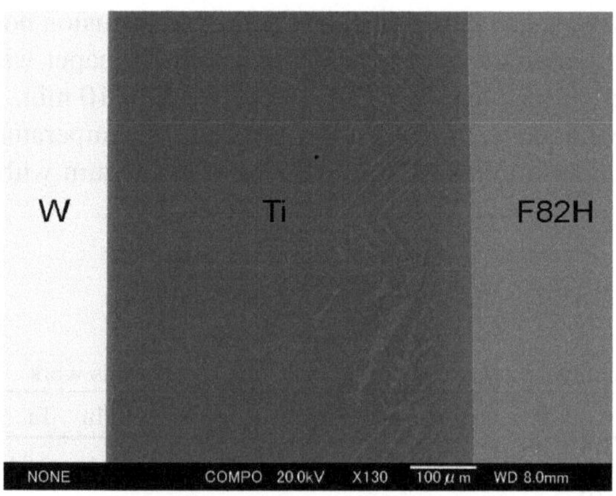

Fig. 1 SEM micrograph of the cross section of the specimen diffusion bonded at 950°C

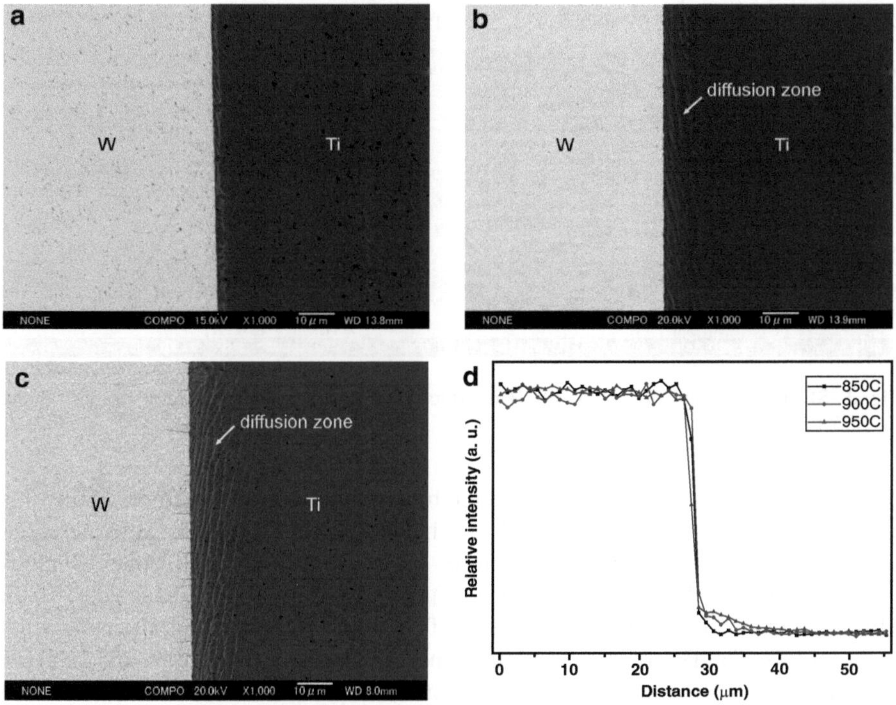

Fig. 2 SEM micrographs of the W/Ti interfaces of the specimens bonded at (**a**) 850°C, (**b**) 900°C, (**c**) 950°C for 1 h, and (**d**) intensity profiles of tungsten across the W/Ti interfaces

profiles indicate that the diffusion layer is a solid solution, as predicted in the W–Ti binary phase diagram [10]. The relatively long tail of W in Ti could be attributed to the higher diffusivity of W in Ti which is three orders of magnitude higher than that of Ti in W [11].

The formation of solid solution between W and Ti is expected based on the W–Ti phase diagram which shows complete solubility in the β-region [10]. Obviously, the diffusion zones shown in Fig. 2 are the consequence of the atomic interdiffusion between W and Ti, and developed as a result of eutectoid transformation. An adequate amount of W in this zone promoted the eutectoid formation of Ti, and thus the observed brighter β-Ti needles containing W were transformed in the darker α-Ti matrix by the decomposition of β-Ti during cooling [12].

At Ti/F82H interface, a very limited interdiffusion of elements has been found for the joints bonded below 850°C. However, when the bonding temperature was raised to above 900°C, a distinct change can be observed as shown in Fig. 3. In Fig. 3b, adjacent to the Ti, similar like that of W/Ti interface, α–β structure was formed. Close to α–β Ti, a shade zone has been observed containing Ti (~91 at. %), Fe (~8 at. %), and Cr (~1 at. %), suggesting that Fe and Cr atoms from F82H penetrated into Ti. Fe and Cr are strong β-structure stabilizers for Ti [13] and

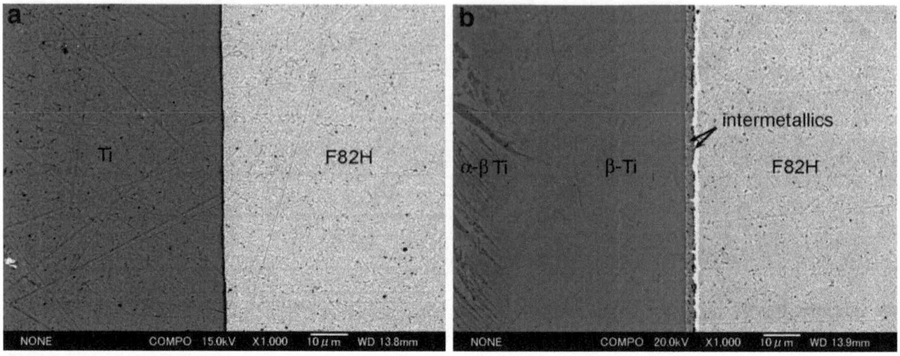

Fig. 3 SEM micrographs of the Ti/F82H interfaces of the specimens bonded at (**a**) 850°C and (**b**) 950°C

substantial quantity of them help to retain high temperature β structure; hence, this area can be designated as stabilized β-Ti. The widths of β-Ti regions were increased with increasing the joining temperature. Next to β-Ti, a small zone which is enriched with Ti (~58 at. %), Fe (~40 at. %) and Cr (~2 at. %) was detected by EDS. This zone is considered to be composed of FeTi and β-Ti based on the isothermal section of Fe–Cr–Ti ternary phase diagram [14]. Between FeTi+β-Ti and F82H, a small bright zone was found, which consists of Fe (~58 at. %), Ti (~30 at. %) and Cr (~9 at. %) and a little W (~3 at. %). Hence, the composition indicates the λ phase containing some W was formed, the λ phase is the solid solution of Fe_2Ti and Cr_2Ti [15]. Ti (<10 at. %) also was found to be penetrated into F82H and presumably this region was α-Fe phase retained from high temperature in association with a small amount of χ phase ($Fe_{17}Cr_7Ti_5$).

3.2 *Shear Strength Evaluation and Fracture Surface Observation*

Shear strength of the W/Ti/F82H joints is presented in Fig. 4. The strength value increased when the joining temperature increased from 800°C to 900°C. For the low bonding temperature like 800°C, the extent of interfacial deformation remains incomplete and interdiffusion of chemical species were limited at both interfaces. Thus, the mating surfaces were lack of contact and produced low strength joint. Conversely, for the 900°C joining temperature, betterment in plastic deformation of the mating surfaces improves bond quality. In particular, the extent of interdiffusion of chemical species was significantly enhanced at higher temperature. Thus the shear strength enhanced and reaches its maximum level (102 ± 11 MPa) for 900°C joint, although a small width intermetallic layer was generated.

A further rise in joining temperature though leads to enhancement of interdiffusion and better plastic deformation, shear strength was decreased and the same downward trend is observed for the joining temperature of 1,000°C. Benefit from

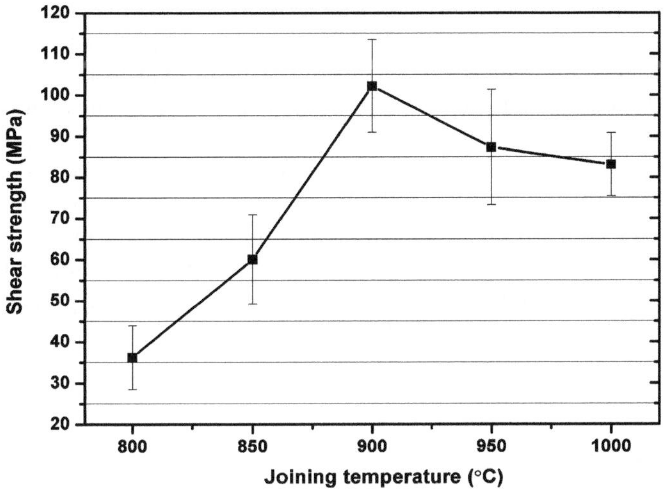

Fig. 4 Shear strength vs. joining temperature of the W/Ti/F82H transition joints

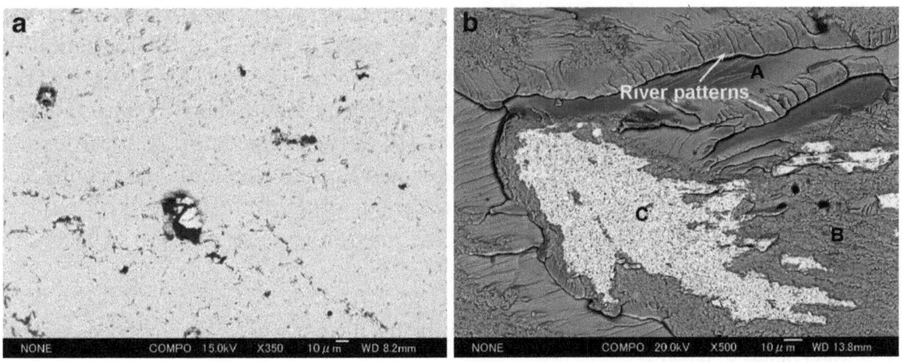

Fig. 5 Fracture surfaces of the joints bonded (**a**) at 850°C, F82H side; (**b**) at 900°C, Ti side

plastic deformation and detriment effect by the growth of brittle intermetallics may balance each other, and it seems that the shear strength is mainly governed by the latter factor. Hence the bond strength decreased. In addition, the Kirkendall voids formed at the interface due to the imbalance in flux transfer during diffusion bonding are also responsible for the reduction in shear strength.

All the joints fractured at Ti/F82H interfaces during shear tests and the fracture surfaces of the joints are given in Fig. 5. The fracture surface of the joint processed at low temperature is basically featureless (Fig. 5a). The discrete black islands attached on the fracture surface are Ti, indicating that some areas are well bonded. When the joining temperature is raised to above 900°C (Fig. 5b), transgranular brittle failure under a shear loading becomes evident by the presence of river patterns on both the fracture surfaces of F82H and Ti. The average composition of the cleavage facets (region A) on Ti side is Fe (~5 at. %), O (~6 at. %), Cr (~1 at. %) and Ti (the balance), suggesting the existence of stabilized β-Ti and may some oxide. While the gray region B is comprised of Fe (~20 at. %), Cr (~2 at. %) and Ti (~80 at. %), FeTi + β-Ti phase mixture can be expected in this region. The white region C is Fe_2Ti, with a composition of Fe (~53 at. %), W (~1 at. %), Cr (~3 at. %) and Ti (the balance). Some W and Cr dissolved in this intermetallic phase. It is well known that the brittle intermetallic compounds in the reaction zone are mainly responsible for strength decreasing for the diffusion bonded joints. The fracture surfaces of the joints in this work indicate that the failure occurred from the intermetallic phases (FeTi and Fe_2Ti) and in some areas the fracture propagated in stabilized β-Ti (close to intermetallics).

4 Conclusions

Successful solid-state diffusion bonding was achieved between W and F82H steel using a Ti as interlayer at 800–1,000°C for 1 h under 10 MPa in vacuum.

Cross-sectional examinations indicate that both the W/Ti and Ti/F82H interfaces were bonded well. Interdiffusion of W, Ti, Fe, and Cr promotes the formation of α–β Ti solid solution, stabilized β-Ti, and intermetallic phases of Fe_2Ti and FeTi in the reaction zones. Elemental diffusions are enhanced greatly above the α–β transformation temperature. The joints failed at Ti/F82H interface during shear testing due to the incomplete interfacial deformation and limited interdiffusion occurred at the interface for the low joining temperature joints or the formation of brittle intermetallics for the high temperature joints. The maximum shear strength of 113 MPa has been obtained for 900°C joint.

Acknowledgment The authors would like to thank the finance support from National Institute for Fusion Science, Japan.

References

1. Dux R, Bobkov V, Fedorczak N et al (2007) Tungsten erosion at the ICRH limiters in ASDEX upgrade. J Nucl Mater 363–365:112–116
2. Zhou ZJ, Du J, Song SX et al (2007) Performance of W/Cu FGM based plasma facing components under high heat load test. J Alloy Compd 428:146–150
3. Klueh RL, Nelson AT (2007) Ferritic/martensitic steels for next-generation reactors. J Nucl Mater 371:37–52
4. Hishinuma A, Kohyama A, Klueh RL et al (1998) Current status and future R&D for reduced-activation ferritic/martensitic steels. J Nucl Mater 258–263:193–204
5. Chehtov T, Aktaa J, Kraft O (2007) Mechanical characterization and modeling of brazed EUROFER-tungsten-joints. J Nucl Mater 367–370:1228–1232
6. Kalin BA, Fedotov TV, Sevrjukov ON et al (2007) Development of brazing foils to join monocrystalline tungsten alloys with ODS-EUROFER steel. J Nucl Mater 367–370:1218–1222

7. Greuner H, Bolt H, Böswirth B et al (2005) Vacuum plasma-sprayed tungsten on EUROFER and 316L: results of characterization and thermal loading tests. Fusion Eng Des 75–79:333–338
8. Hirose T, Shiba K, Ando M et al (2006) Joining technologies of reduced activation ferritic/martensitic steel for blanket fabrication. Fusion Eng Des 81:645–651
9. Zhong ZH, Hinoki T, Kohyama A (2009) Effect of holding time on the microstructure and strength of tungsten/ferritic steel joints diffusion bonded with a nickel interlayer Mater. Sci Eng A. doi:10.1016/j.msea.2009.04.043
10. Massalski TB, Okamoto H, Subramanian PR et al (1990) Binary alloy phase diagrams, 2nd edn. William W Scott, New York
11. Kazakov NF (1985) Diffusion bonding of materials. Mir, Moscow
12. Kundu S, Chatterjee S (2006) Interfacial microstructure and mechanical properties of diffusion-boned titanium-stainless steel joints using a nickel interlayer. Mater Sci Eng A 425:107–113
13. Kurt B, Orhan N, Evin E et al (2007) Diffusion bonding between Ti-6Al-4V alloy and ferritic stainless steel. Mater Lett 61:1747–1750
14. Raghavan V (1987) Phase diagrams of ternary iron alloys, vol 1. ASM International, Metals Park, OH
15. Ghosh M, Kundu S, Chatterjee S et al (2005) Influence of interface microstructure on the strength of the transition joint between Ti-6Al-4V and stainless steel. Metall Mater Trans A 36:1891–1899

The Simulation of Corium Dispersion in Direct Containment Heating Accidents

Wei Wei and Xin-rong Cao

Abstract In this paper, an improvement model for corium dispersion in direct containment heating (DCH) accidents has been proposed according to the mechanisms of the two-phase flow, In addition the momentum and mass equations have been solved through some special numerical approaches. On the basis of the model, the simulated code of the corium dispersion in the (DCH) reactor accident scenario has been developed. The final dispersion fraction and the droplet size in the reactor containment dome can be obtained using this code.

Keywords DCH • Corium • Dispersion

1 Introduction

In the DCH accidents, the degree of molten corium dispersion is one of the most important factors for containment heating and pressurizing. Because the degree of heat transfer and chemical reactions that may lead to containment over-pressurization are basically proportional to the available surface area of the molten corium. If the corium is highly dispersed and the final molten droplet size is very small, the available heating surface can be huge and the probability of containment failure can be very high. Therefore, it is important to investigate the corium dispersion process concerning the mean corium droplet size and the dispersal fraction of the total discharged corium. To understand the corium dispersion phenomenon, intensive experimental studies have been conducted at SNL, ANL, and Purdue University [1] for the Zion reactor geometry, which constructed a large database for analytical modeling. However, none of the present severe accident analysis procedures such as SCDAP/RELAP5, CONTAIN, MELCOR could deal with the problem, therefore, to devise a corium dispersion process analysis procedure is valuable and necessary.

W. Wei (✉) and X. Cao
College of Nuclear Science and Technology, Harbin Engineering University, Harbin, China
e-mail: goodww2004@yahoo.com.cn

T. Yao (ed.), *Zero-Carbon Energy Kyoto 2009*, Green Energy and Technology, DOI 10.1007/978-4-431-99779-5_43, © Springer 2010

2 Model and Calculation Method

2.1 *Description of the Model*

In this paper, the model that describes the corium dispersion in the direct containment heating (DCH) reactor accident scenario was developed by Wu in 1996, and the sources that contribute to the final dispersion fraction F_{disp} in the reactor containment dome are summarized in Fig. 1 [2].

In this model, F_1 is the fraction of corium rushing out of the cavity before gas blowdown, F_2 is the entrainment fraction of corium in the cavity, and F_3 is the re-entrainment fraction on the seal table bottom wall. Therefore, the total dispersion fraction F_{disp} in the containment dome is the summation of F_{21}, F_{22} and F_{31}.

2.2 *The Improvement of the Model*

The cavity is treated as a horizontal channel and the flow regime in the cavity belongs to annular mist flow. According to the mechanisms of the two-phase flow, the momentum and mass balance equations are modified:

Mass balance equation:

$$\left(\pi L D_h \rho_f\right) \frac{dh}{dt} = -(\pi D_h L)\varepsilon_e(t), \tag{1}$$

where, as shown in Fig. 2, L and h denote the film block length remaining in the cavity and the thickness of the liquid film sheet respectively; ε_e is the droplet mass entrainment rate per unit film surface area [3].

Momentum balance equation:

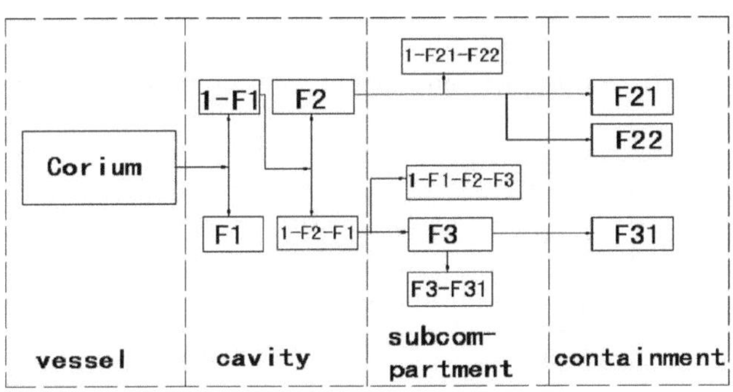

Fig. 1 Corium dispersion process

Fig. 2 Predicted droplet size in the cavity for 1.4 MPa air water tests

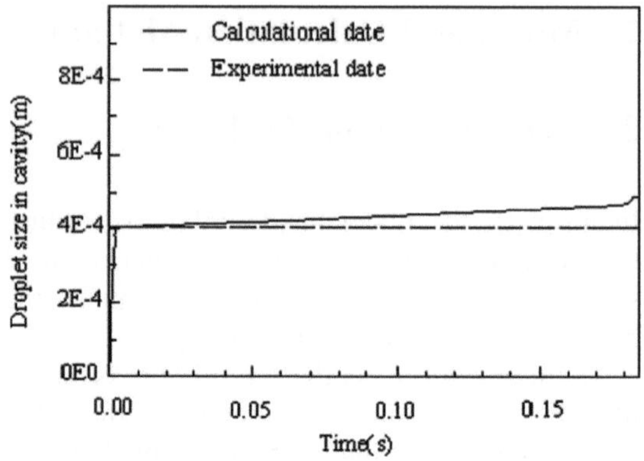

$$-\frac{d^2L}{dt^2}=\frac{\Delta\rho P}{L\rho_f}+\frac{f_i}{2h}\left(\frac{\rho_g}{\rho_f}\right)\left(V_g+\frac{dL}{dt}\right)^2-\frac{\varepsilon_e}{h\rho_f}\frac{dL}{dt}-\frac{f_w}{2h}\left(\frac{dL}{dt}\right)^2, \tag{2}$$

where ΔP is the pressure difference between the film tail and the cavity exit, f_i is the interfacial friction, and f_w is the wall friction factor.

Defining $y=\frac{dL}{dt}$, (1) and (2) can become the following format:

$$\begin{cases} \frac{dy}{dt}=\frac{\varepsilon_e}{h\rho_f}y+\frac{f_w}{2h}y^2-\frac{\Delta P}{L}-\frac{f_i}{2h}\left(\frac{\rho_g}{\rho_f}\right)\left(V_g+y\right)^2, y(0)=-V_f(t=0), \\ \frac{dL}{dt}=y,\ L(0)=L_c, \\ \frac{dh}{dt}=-\frac{\varepsilon_e}{\rho_f},\quad h(0)=\frac{(1-F_1)\vartheta_m}{\pi D_h L_c}. \end{cases} \tag{3}$$

The onset Giteria for droplet entrainment were proposed by Ishii and Grolmes in 1975 [3]:

$$\begin{cases} \frac{\mu_f V_g}{\sigma}\sqrt{\frac{\rho_g}{\rho_f}}\geq 11.78 N_\mu^{\ 0.8}\mathrm{Re}_f^{\ -1/3},\quad N_\mu\geq\frac{1}{15}, \\ \frac{\mu_f V_g}{\sigma}\sqrt{\frac{\rho_g}{\rho_f}}\geq 1.35\mathrm{Re}_f^{\ -1/3},\quad N_\mu>\frac{1}{15}, \end{cases} \tag{4}$$

where N_μ is the viscosity number.

2.3 Calculation Method

Based on the work of above, the simulation code of the corium dispersion in the containment is developed according to the modified model. In the numerical code, (3) is solved by the Runge–Kutta numerical approach, and the Aitken iterated algorithm has been used to calculate the critical size of the droplet that can move through the seal table exit into the containment dome. The integral calculus has been solved by Legendre–Gauss numerical approach.

3 Calculation Results

The models are critically evaluated in this section compared with the results of the 1:10 scale Purdue DCH separate effect experiments [1]. In the experiments, the volume of total discharged water or woods metal was 7 L with a test vessel break size of 3.5 cm.

In Figs. 2 and 3, the predicted mean droplet sizes and entrainment fraction in the cavity for 1.4 MPa have been compared with the experiment data. In addition, Figs. 4 and 5 have showed the vessel pressure effect on both the dispersion fraction and the maximum droplet size.

Fig. 3 Predicted entrainment fraction in the cavity for 1.4 MPa tests

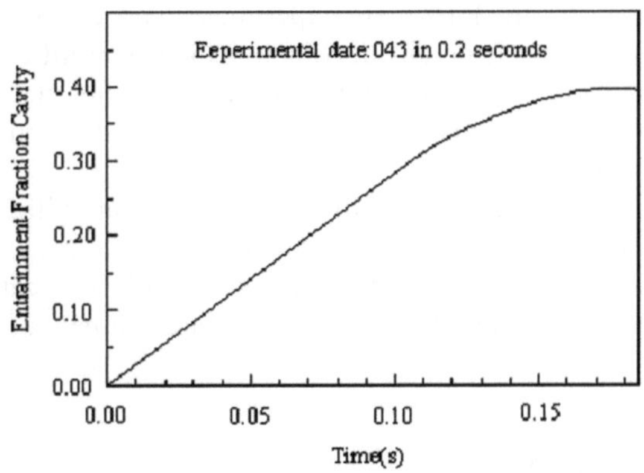

Fig. 4 Vessel pressure effect on the dispersion fraction

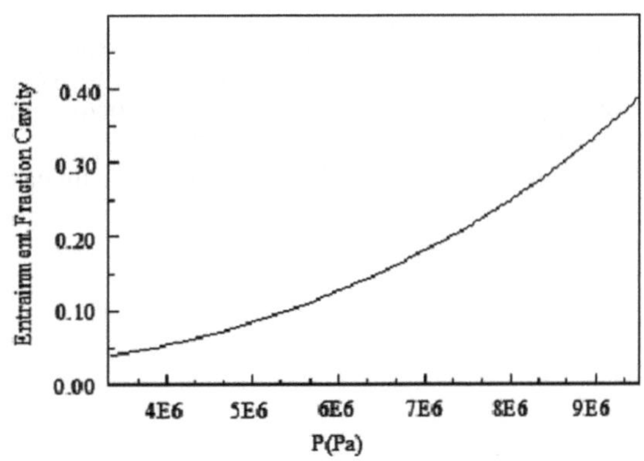

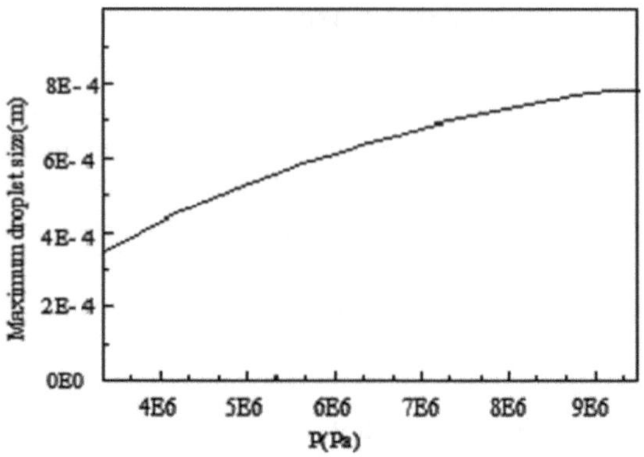

Fig. 5 Vessel pressure effect on the maximum droplet size

4 Conclusion

In the new model the entrainment onset criteria has been replaced by Ishii's onset criteria, and according to the mechanisms of the two-phase flow, the momentum and mass equals of the old model are modified. And some special numerical approaches have been used to solve the equations of the model. Compared with the data of the 1:10 scale experiments carried out at Purdue University, fairly good agreement is obtained. Although many simplifications have been made in the development of the present models, the dominant mechanisms in each step are observed, which make us understand the corium dispersion problem in the DCH accident scenario better.

The future work will use this part as an initial condition to calculate the heating of the containment atmosphere, at last the peak value of the containment pressure and temperature will be obtained. With the peak pressure and temperature value, the frequency of containment failure can be gained by simple analysis.

References

1. Wu Q, Zhang G (1996) Experimental simulation of corium dispersion phenomena in direct containment heating. J Nucl Eng Des 164(1–3):237–255
2. Wu Q, Zhang G, Ishii M, Revankar ST (1996) Modeling of corium dispersion in the DCH accidents. J Nucl Eng Des 164(1–3):211–235
3. Ishii M, Grolmes MA (1975) Inception criteria for droplet entrainment in two-phase concurrent film flow. AIChE J 21:308–318

Study on Three-Dimensional Thermal Hydraulic Simulation of Reactor Core Based on THEATRe Code

Zhaocan Meng and Zhijian Zhang

Abstract The nuclear steam supply system of 300 MWe power plant is simulated in this paper. One fuel assembly is treated as one channel and there are 121 channels in the whole reactor core. And every channel is divided into 12 nodes axially, while the whole NSSS is divided into 1,509 nodes. The simulation of NSSS with 3D reactor core is achieved in high speed calculation method, and the results are acceptable.

Keywords Thermal-hydraulics • 3D simulation

1 Introduction

For the purpose of improving the transient characteristics of nuclear power system simulation, it is very important to obtain both the steady-state and transient calculation of coupled neutronics and thermal-hydraulics behavior of reactor core in three-dimensional geometry. In the existing simulation of nuclear power system, single channel or several channels are still used in the core thermal-hydraulic simulation because of the limitation of computing time. As a result, the system simulation cannot give the 3D thermal-hydraulic distribution of the core, which restricts the simulation accuracy and range of working conditions. To realize 3D thermal-hydraulic simulation of the whole reactor core, with reasonable high-speed calculation method is in the urgent need to be developed.

Therefore, the authors tried to perform such 3D fast running simulation of 300 MWe Nuclear Power Plant with NSSS system included in the model by using THEATRe code (Thermal Hydraulic Engineering Analysis Tools in Real Time developed by the GSE Power Systems, Inc.) [1].

Z. Meng (✉) and Z. Zhang
College of Nuclear Science and Technology, Harbin Engineering University, Harbin, China
e-mail: mengzhaocan@hotmail.com

T. Yao (ed.), *Zero-Carbon Energy Kyoto 2009*, Green Energy and Technology,
DOI 10.1007/978-4-431-99779-5_44, © Springer 2010

2 Mathematical Model and Node Dividing

The THEATRe code is generalized thermal hydraulic code developed for real-time operator training as for best-estimate engineering analysis in both nuclear and non-nuclear plants. It uses 5-equation drift flux model for two-phase flow calculation and Nodal Momentum Nodal Pressure (NMNP) method. The THEATRe code can be used for simulation of thermal non-equilibrium, non-homogeneous two-phase flow systems involving steam–water mixture, noncondensible gaseous species, and non-volatile solute [2].

In this paper, one fuel assembly is treated as one channel and there are 121 channels in the whole reactor. Since every channel is divided into 12 nodes axially, the reactor core is divided into (121 × 12) nodes, while the part of NSSS is divided into 1,509 nodes. Because of the scale limitation of the pressure matrix in THEATRe, the core is divided into five sections in radial direction. These five sections are calculated independently, and then integrated with the NSSS section. In order to cope with these five divisions of the reactor core, the calculation of THEATRe was modified in the part of pressure–velocity solution subroutine to give the pressure boundary of the core regions. And a segment was added into THEATRe code to couple the five core regions and NSSS region. The NSSS system node dividing is shown in Fig. 1. The core regions and node dividing is shown in Fig. 2.

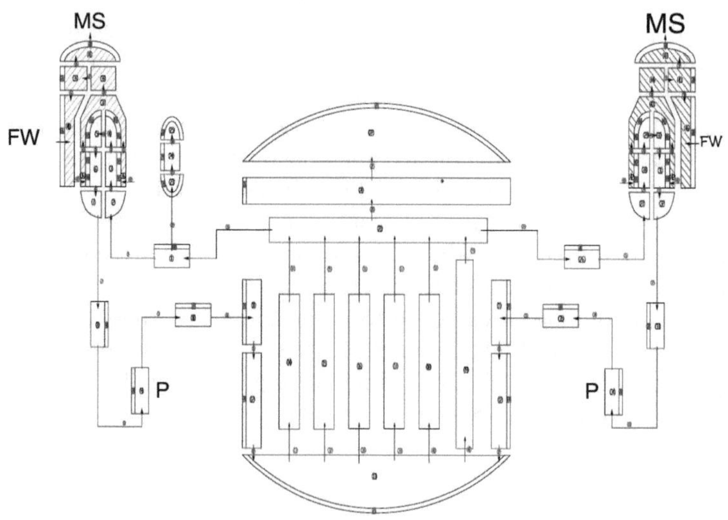

Fig. 1 Node dividing of NSSS

Fig. 2 Node and region dividing of reactor core

3 Simulation Results

3.1 Evaluation of the Coupling Module of Core Regions and NSSS Region

The core was divided into five regions, and it's the key point of the system simulation to assure that the five core regions have the same inlet pressure with the down plenum in NSSS region. The coupling module has been generated, and the goal had been achieved very well. From Fig. 3, it has been calculated that from the beginning to over 300 trend time there is an abruptly change of pressure for the five core regions inlet and the pressure profile is shown, where the pressure is always the same nevertheless of great changing happens.

3.2 Flow Distribution

Take the second core region for example. The inlet volume flow, outlet volume flow and power distribution of the 24 channels are given out in Fig. 4. Outlet volume flow rate of one channel is high with high power of this channel. Inlet volume flow rate is low with high power. The trend of mass flow distribution is the same with

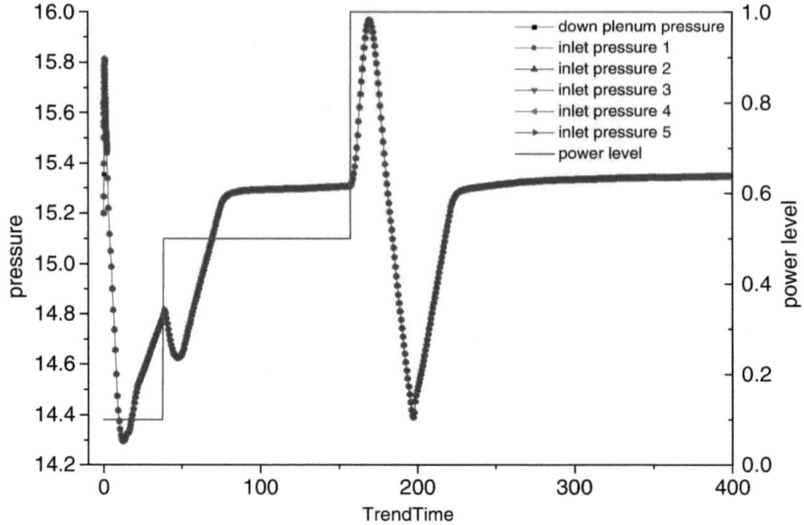

Fig. 3 Pressure of inlet node of core regions

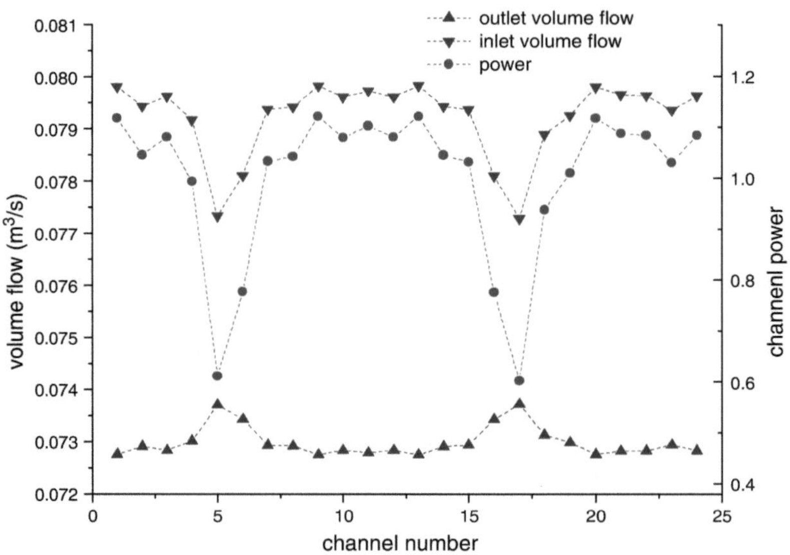

Fig. 4 Core volume flow rate and power

inlet volume flow, because of all the inlet nodes connect with one node. Volume flow rate increase with liquid heated in flow channel, and this result in that the outlet volume flow rate is greater than the inlet volume flow rate of one channel. Drag increases with the volume flow rate increase. Inlet volume flow rate decrease as the result of the combined effects of outlet volume flow rate and drag increasing.

3.3 3D Distributions of Parameters

The core is divided into 121 × 12 nodes. The 3D flow field, temperature field and pressure field of steady-state and transient state have been achieved. Take the second core region with full power steady-state for example, the temperature field is shown in Fig. 5, and the flow field is shown in Fig. 6. There are 12 values in every channel, corresponding to the 12 axial nodes. The power distribution profile in core node is

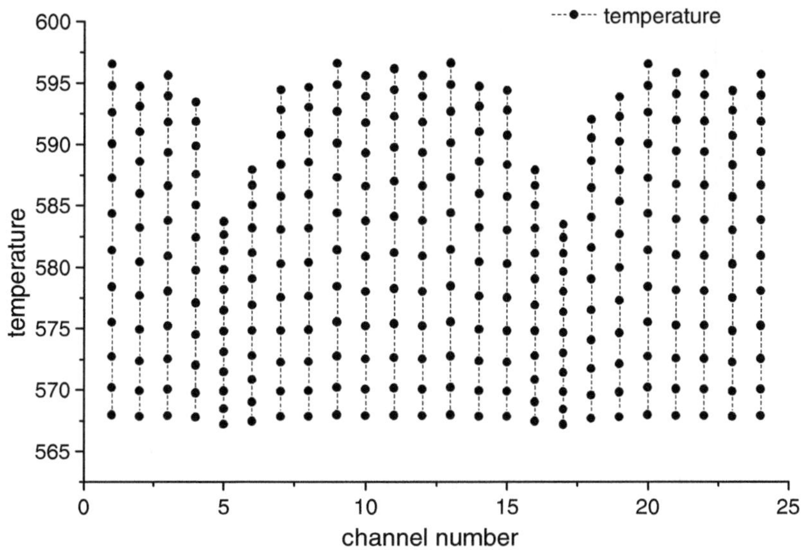

Fig. 5 Temperature field

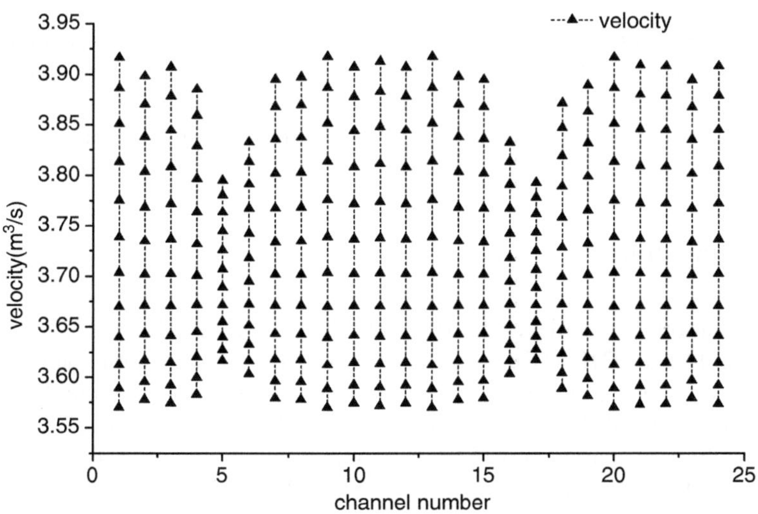

Fig. 6 Velocity field

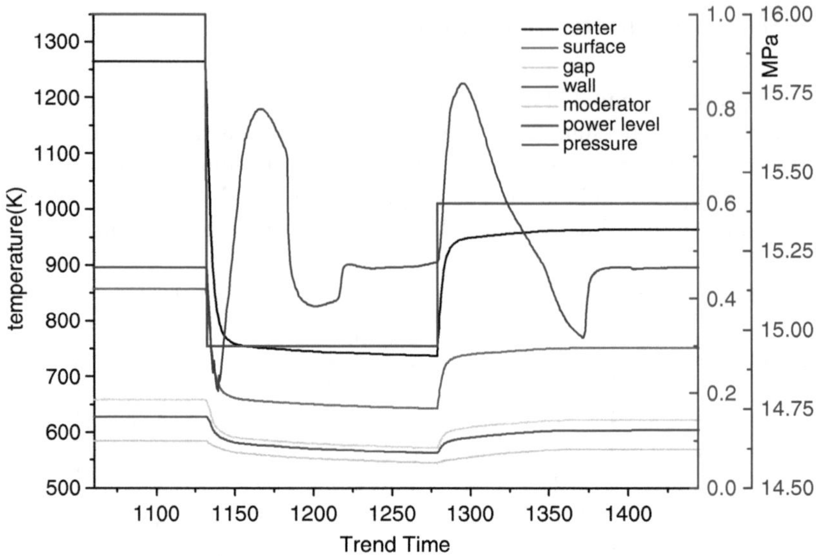

Fig. 7 Transient process

unchanged, and there is no consideration of cross flow between channels. So the temperature and velocity of two adjacent channels may be different greatly.

3.4 Transient Process of the Simulation System

In Fig. 7, the system work at full power at beginning, then power level is set to 30% at 1,132 s, 60% at 1,280 s. Changes with the power level in pressure and temperature of fuel pellet center, fuel pellet surface, fuel rob gap, fuel rob wall are shown. It can be concluded that the transient process with power level change is

4 Conclusions

The simulation was achieved by PC with CPU: AMD Athlon 64 X2 Dual Core Processor 4400+. THEATRe costs about 60 ms for one step computation and this computation speed can meet with the requirements of real-time simulation, with acceptable results of flow distribution of the core 121 channels, and 3D distribution of core thermal parameters both for steady-state and transient conditions. Therefore, 3D fast-running simulation by THEATRe could be achieved successfully.

This module can be used at the simulator development and reactor design. There's no consideration about the cross flow between assemblies. This module will be more suitable for the conditions with little cross flow.

References

1. The RELAP5-3D Code Development Team (2005) Code structure, system models and solution methods. RELAP5-3D Code Manual, Vol I. April 2005
2. GSE Power Systems. THEATRe MTH.15.V1.1D.2003

Study on Turbine System of Nuclear Power Plant Based on RELAP5/MOD3.4 Code

Shao-wu Wang, Min-jun Peng, and Jian-ge Liu

Abstract In this paper, the secondary-loop steam turbine system has been modeled using RELAP5/MOD3.4 code. The operating conditions for the turbine load have been changed from 100% FP (full power) to 80% FP that have been simulated and which will explain transient analysis. These result shows that the transient performance of the steam turbine system components and theoretical analysis are identical, which demonstrated the reliability of the requisite program model.

Keywords Relap5/Mod3.4 • Turbine system • Secondary-loop

1 Introduction

In this paper, the major work has been carried out by considering only the secondary-loop system of nuclear power plant using the renowned safety and simulation code, Relap5/Mod3.4 which confirms the steady-state operation of the turbine system [1]. We have focused on the transient analysis of the main coolant system when the load changes in the secondary-loop happens under normal operation conditions, in fact at the steady-state conditions we always have a simplified modeling method for the safety and operational analysis of the secondary-loop by taken into account the inlet and outlet parameters consistent with temperature, pressure, mass flow rate of feed water and the steam but it cannot give us the full detail when there is load changes will happens so there must be a need of the major system interrelated with equipment analysis models for the secondary-loop in which steam turbine, condenser and pump components may impact hardly on the operating conditions of the main coolant system so we developed the computational model by using the safety and simulation code Relap5/Mod3.4 in order study the characteristics and evaluated the transient analysis of not only the turbine system but also the whole nuclear power plant.

S. Wang(✉), M. Peng, and J. Liu
College of Nuclear Science and Technology, Harbin Engineering University, Harbin, China
e-mail: shaowu_84@yahoo.com.cn

T. Yao (ed.), *Zero-Carbon Energy Kyoto 2009*, Green Energy and Technology, 286
DOI 10.1007/978-4-431-99779-5_45, © Springer 2010

2 Introduction of the Model and Node Dividing

Now we will discuss the procedure of how to make the steam turbine, condenser, and pump models in detail.

2.1 Steam Turbine

Turbine system in the nuclear power plant work on the principle of converting the high-speed steam flow into mechanical energy which then converted into electricity as desired [2].

In this paper, the research has been made in a single steam turbine which is connecting with six stage group, i.e. there is no steam exhausted out of the turbine to any other components as shown in Fig. 1 in which schematic modeling for the steam turbine is in accordance with the characteristics of steam turbine by using lumped parameter model and this model system simulation has a wide range of application.

RELAP5 code use the method of classification groups, as shown in Fig. 1, in which the first and last stage groups are artificial turbine components and its efficiency is zero. The inertia and friction of that turbine components factor should be entered into the input data cards, but smaller than the normal turbine. The two ends of the model are time-dependent control volumes (TDV). There are three procedures to choose the type of turbine designs as discussed in the relap5 code, we prefer the second type model turbine whose efficiency formula of the steam turbine is given in relap5 manual.

The turbine component connected to a control variable shaft component, which in turn connected to a control variable generator component. With this arrangement, the speeds, loads, and inertias of shaft, and generator are determined consistently.

2.2 Condenser

In the RELAP5 code, there is no condenser model, but contain two-phase flow condensation and heat transfer aspects ground on the mathematical model of

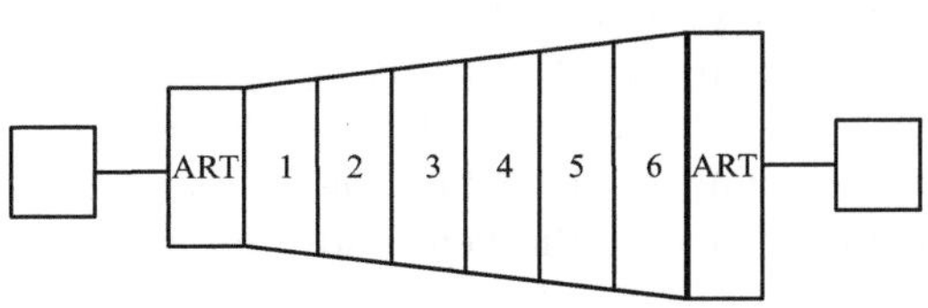

Fig. 1 Steam turbine model

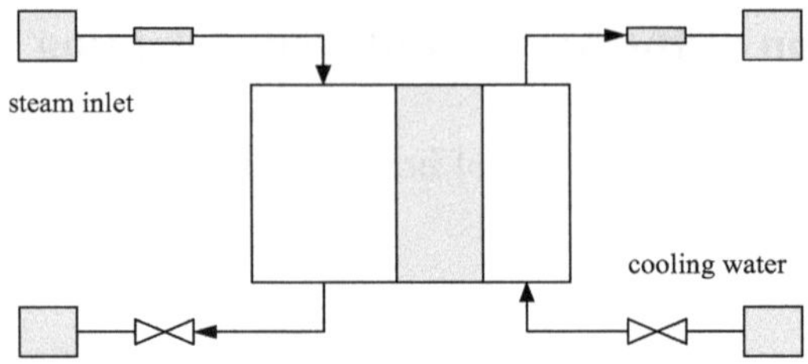

Fig. 2 Condenser model

physics and the material structural thermal conductivity model. So we can model the condenser which is based on both the structural components of the condenser and its physical realization features, combined with RELAP5 code [3].

Figure 2 shows, the turbine exhaust steam flow through the condenser from top to bottom, the circulating cooling water flow inside the condenser tube and exchange this heat and condensate into water. According to the fluid flow characteristics of the condenser, the tube and the shell of the condenser are simulated into the tube-type control volume, which consists of two single control volume, while the wall thickness of the tube are modeled by the heat structure component. The heat is exchanging between the steam and water via the heat structure.

During modeling we found that the number of the control volumes has a relatively large impact on the debugging of the condenser and it is very difficult to maintain the pressure throughout the condenser, moreover the small changes in parameters will affect the pressure of each control volume, this changes in pressure will lead to further fluctuations in steam flow, it is difficult to get the steady-state. So we concluded that, under the low pressure, the calculation of the physical characteristics of the various parts must be consider, but RELAP5 model of condensation heat transfer using lumped model may need further implementation.

2.3 Condensate Pump

A condensate pump is a specific type of pump used to pump the condensate produced in a heating, ventilating and air conditioning, refrigeration, condensing boiler furnace or steam system. In this paper, we used condensate pump model because RELAP5 code itself includes it. We used built-in, homologous date of Westinghouse pump model which has dimensionless quantities. The dimensionless quantities involve the head ratio, torque ratio, volumetric flow ratio, and angular velocity ratio, where the ratios are actual values divided by rated values.

3 Transient Analyses

One of the most important and viable issue in this regards is to explain the transient analysis which is in-coupled with components like turbine system, condenser and condensate pump modeling that we did as explain earlier for the transient performance. Taken into consideration the simulation process using code we have changed the operating condition for the load of steam turbine system from 100%FP (full power) to 80%FP (full power).

In Fig. 3 the transient calculation of steam flow curve is shown and in Fig. 4 we have seen that if the steam flow level is changed the power output of the steam turbine is also changed because the steam flow in the turbine is decreases to 80% as the load changes, corresponding to 100%FP by adjusting the area of the regulating valve.

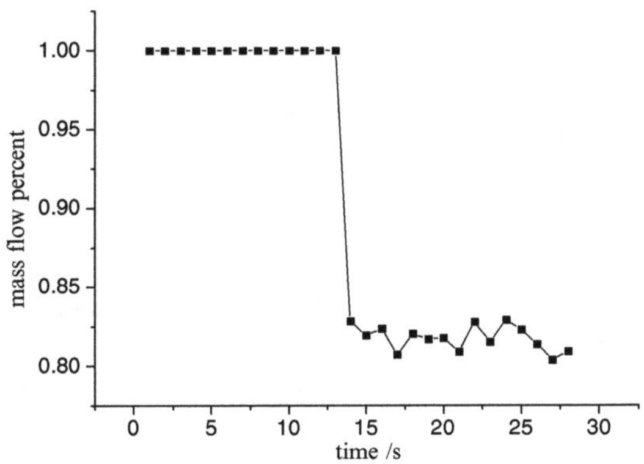

Fig. 3 The inlet steam flow percent of turbine

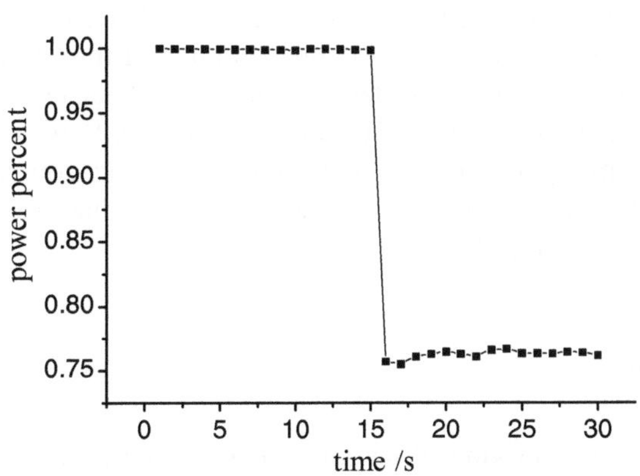

Fig. 4 The power percent of turbine

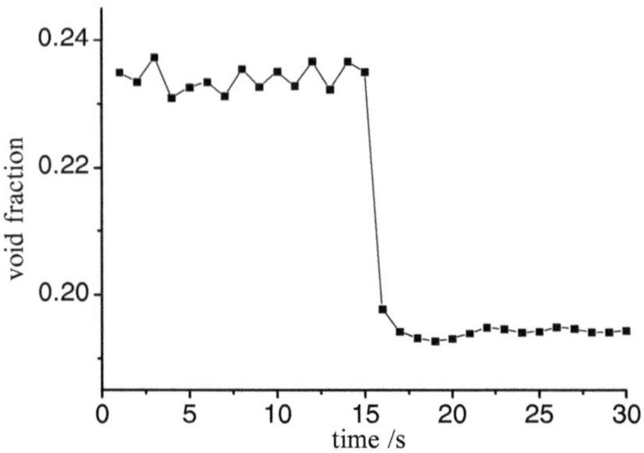

Fig. 5 The void fraction of condenser

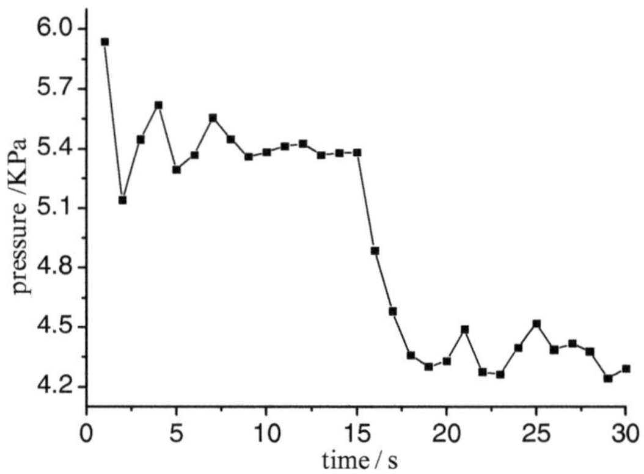

Fig. 6 The inlet pressure of pump

In Fig. 5, it can be seen that the void fraction is present in the middle of the low control volume of the condenser which is at 0.23 and when steam low decreases the void fraction in the middle of the low control volume of the condenser is changed to 0.16 due to circulating cooling water flow that is not been changed. In Fig. 6 the internal pressure of the pump is changed due to the less flow of the steam in the condenser at the same spatial location in the control volume compared with the full power steam flow hence the inlet pressure of the pump has been changed correspondingly as shown in Fig. 6.

4 Conclusion

In this paper, we have introduced the conceptual and detailed procedure to model the steam turbine system, condenser, and pump using the renowned simulation code Relap5/Mod3.4 which can best-estimate the transient simulation with safety features of light water reactor coolant systems. The results showed that Relap5/ Mod3.4 code not only can satisfy the accuracy of the models but also testified the rationality and accuracy of the concern field as we have designed.

References

1. The RELAP5 Code Development Team (1995) SCDAP/RELAP5/MOD3.4 code manual [M]. Idaho National Engineering Laboratory, Idaho Falls, ID, USA
2. Feng-ge P, Min-jun P (2003) Marine nuclear power plant. Harbin Engineering University Press, Harbin, pp 82–84
3. Allison CM, Hohorst JK (2008) Role of RELAP/SCDAPSIM in nuclear safety. In: International topical meeting on safety of nuclear installations, Dubrovnik, Croatia, 3.09–3.10.2008

Analysis of Instability in Narrow Annular Multi-channel System Based on RELAP5 Code

Geng-lei Xia, Min-jun Peng, and Yun Guo

Abstract The flow instability in narrow annular multi-channel system is analyzed in this paper using RELAP5/MOD3.4 code. The sensitivity of the number of the nodes and the number of the channels are studied firstly. It is found that the calculation results are sensitive to the number of control volumes and the number of the parallel channels. Base on optimized numbers, the two-phase flow instability between multi-channels (FIBM) is studied under different system pressures, inlet and outlet resistance coefficients, inlet sub-cooling, and the influence of them on the critical power are obtained.

Keywords Parallel channel • Flow instability • Once-through steam generator

1 Introduction

The phenomenon of two-phase flow instability is of interest in the design and operation of many industrial systems and equipments where two-phase flow are involved. The instability may lead to the mechanical vibration, the disturbance of the electronic control device, the local overheating of the heat transfer surface and the high thermal stress of the solid walls.

Because of the significance of various instabilities, this problem absorb many scientists. Experiments, theories and numerical codes were carried out currently. Two-phase flow instabilities were introduced by Ledinegg [1] first. Boure et al. [2] classified the various types and analysed the different mechanisms of two-phase flow instabilities. In recent years, a great deal of attention has been paid to the study of dynamic two-phase flow boiling instability between multichannels (FIBM), FUKUDA [3], Yun et al. [4].

G. Xia (✉), M. Peng, and Y. Guo
College of Nuclear Science and Technology, Harbin Engineering University, Harbin, China
e-mail: xgl-0907@163.com

T. Yao (ed.), *Zero-Carbon Energy Kyoto 2009*, Green Energy and Technology, DOI 10.1007/978-4-431-99779-5_46, © Springer 2010

2 Influence Factors for the Flow Instability

The nodalization is shown in Fig. 1. Specified boundary conditions are considered in order to study the effects of different parameters. The inlet fluid temperature boundary conditions is fixed using a time dependent volume (208). The fixed outlet pressure boundary condition is imposed by another time dependent volume (214). The inlet flow variation is specified using a time dependent junction (206).

2.1 Research Method

Under constant mass flux condition, the heat flux keeps on increasing with time. If the magnitude of oscillation grows continuously or is a limit cycle oscillation following the perturbation, the corresponding state is considered unstable, and the onset of flow oscillation (OFO) was recorded. The procedure was repeated under different parameters. Figure 2 presents a curve showing the inlet mass flux when unstable flow boiling. The mass flux kept on oscillating under constant heat flux condition. As the heat flux further increased, so did the magnitude of the flow oscillation.

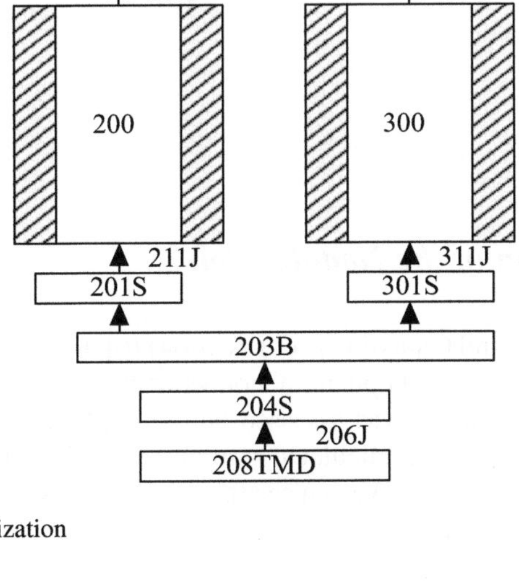

Fig. 1 RELAP5 nodalization

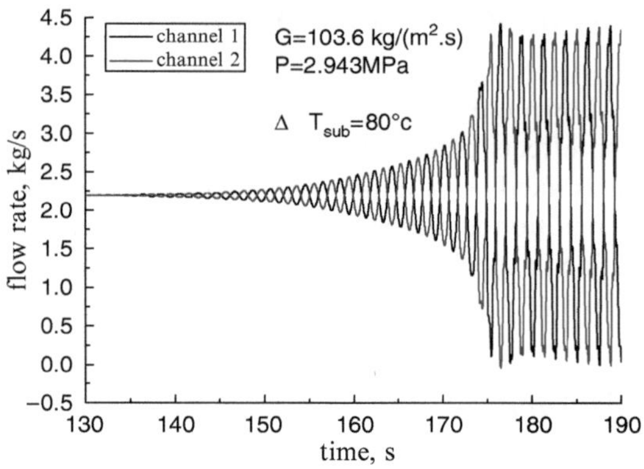

Fig. 2 Flow instability in parallel channels

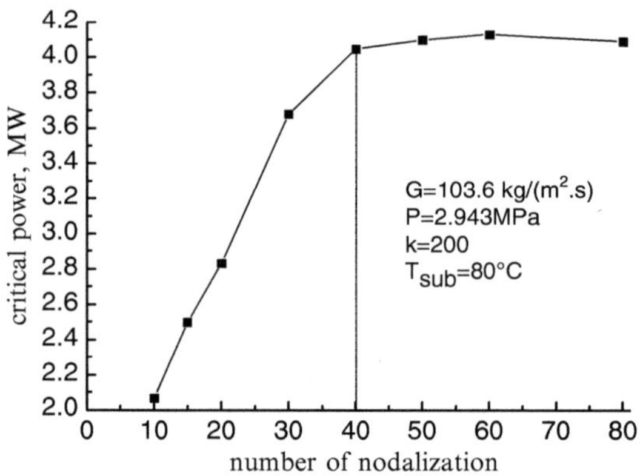

Fig. 3 Effect of nodalization

2.2 *Qualification of the Nodalization*

According to D'Auria and Galassi [5] (IAEA 2004) [6], the complex RELAP5 code may predict unrealistic transient phenomenon when the nodalization is not properly qualified, so the nodalization was assessed before applying the RELAP5 code to the system. It is found that the calculated results are sensitive to the number of CV in the heated channels (Fig. 3). We can see that 40 is the best choice for calculation accuracy and computer time.

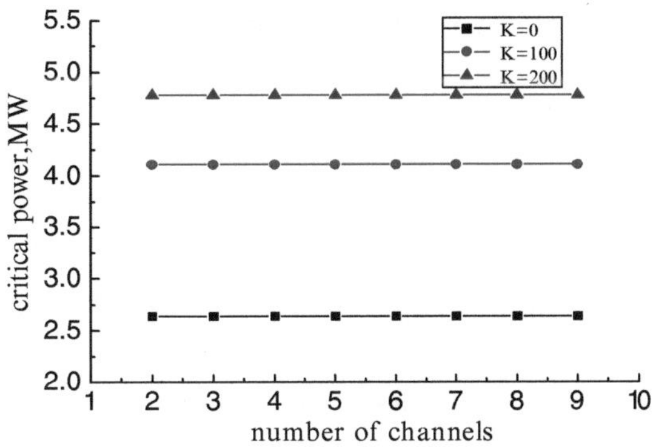

Fig. 4 Effect of channels

2.3 *Influence of Number of Heated Channels*

The critical power for multi-channel systems are compared at different inlet throttling and the onset of flow oscillation are observed (Fig. 4). In a one-channel system oscillations is not observed, even though the power increases in both cases at the same time. It is possible to draw a conclusion that a tow-channel system can represent a multi-channel system for the onset of flow instability.

3 Results for the Twin-Channel System

Two-phase flow instability are more complicated because several effects may occur simultaneously and play a role in a coupled way, so is difficult to study there effect united. The onset of boiling instability of water in a given channel can be studied by single paremeter such as system pressure, inlet and outlet throttling, inlet subcooling, inlet and rising sections.

3.1 *Effect of Inlet and Outlet Throttling*

The results of the present study suggest that the inlet throttling can advance stability and the outlet throttling have the contrary effect. Figure 5 demonstrates that when the inlet resistance coefficient increases, the critical power become bigger, but with the increase of outlet resistance coefficient, the maximum heating power reduced. And the inlet throttling diminishes the amplitude visible along with the increase of resistance coefficient, but the outlet has no prominence effect.

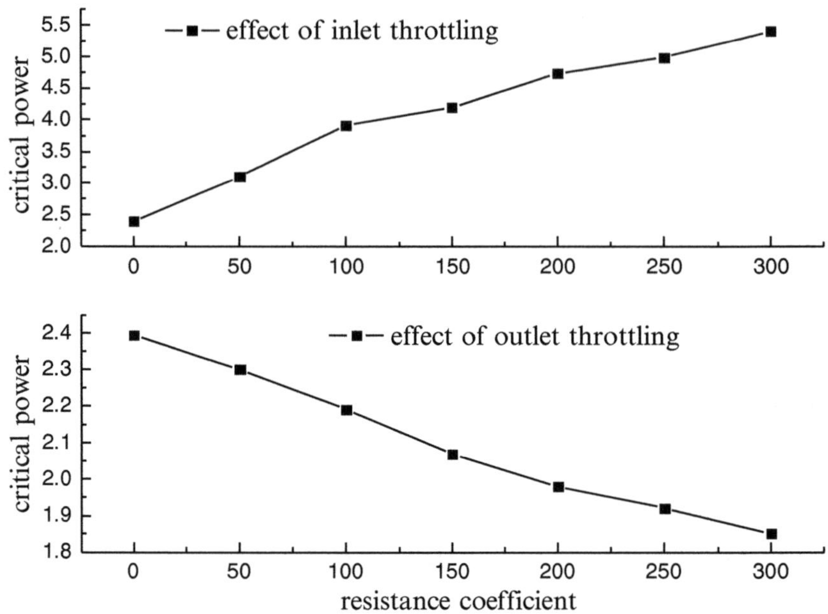

Fig. 5 Effect of restriction on instability

3.2 *The Influence of Inlet and Rising Sections*

The results found that rising section has unstable effect on the system and the system stability is strengthened with the increase of inlet sections. Figure 6 shows calculation results with different inlet and rising sections, the critical power of the parallel channel increased with the increase of inlet sections, but the increase of rising section diminishes the critical power of the parallel channel.

3.3 *Effect of System Pressure*

System pressure was found to significantly affect flow instabilities. According to Fig. 7, the critical power increased with the increase of system pressure, boiling instabilities were significantly delayed. Moreover, Fig. 8 shown that the oscillation amplitudes decrease with the increase of system pressure.

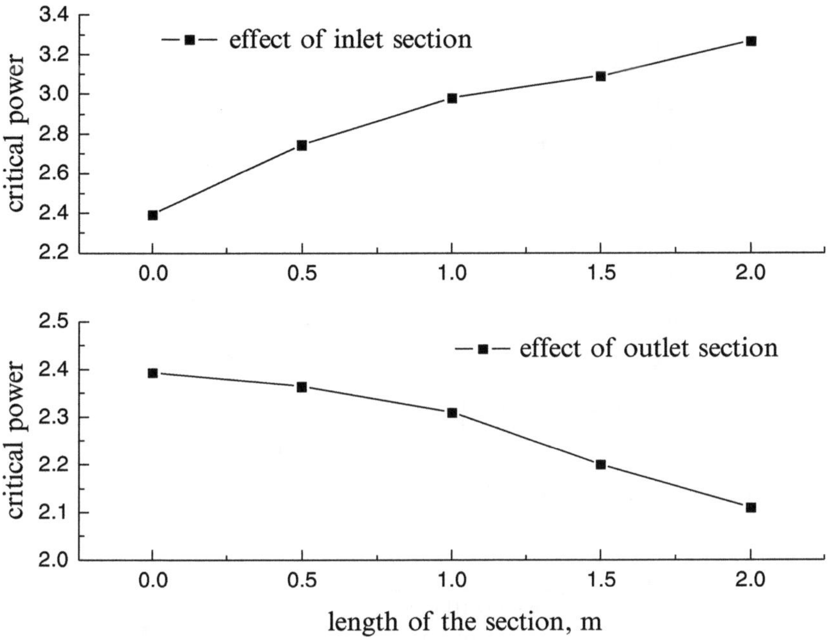

Fig. 6 Effect of the length of inlet and rising section on instability

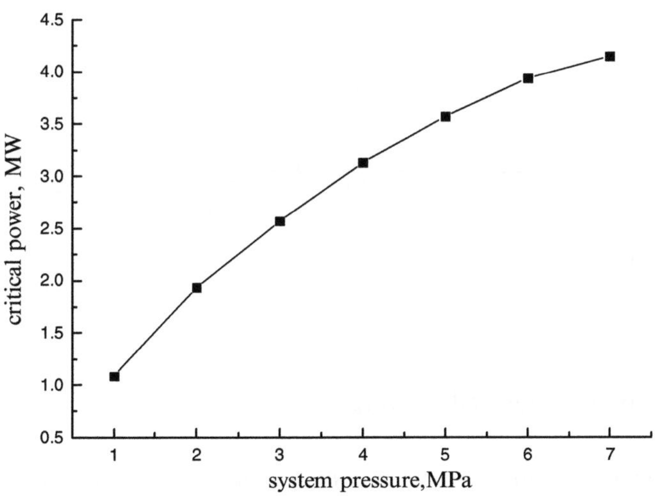

Fig. 7 Effect of pressure on instability

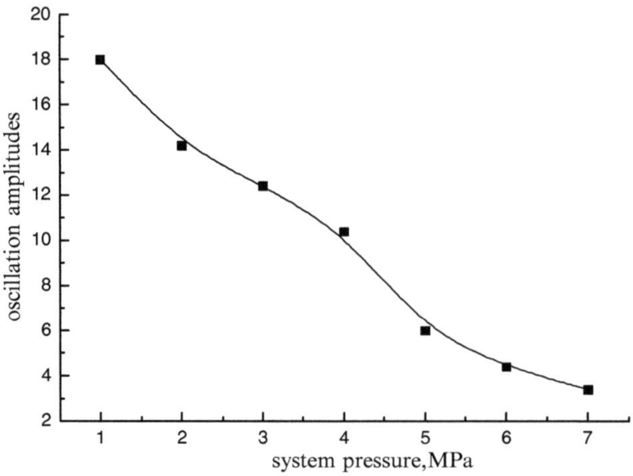

Fig. 8 Oscillation amplitudes of instability

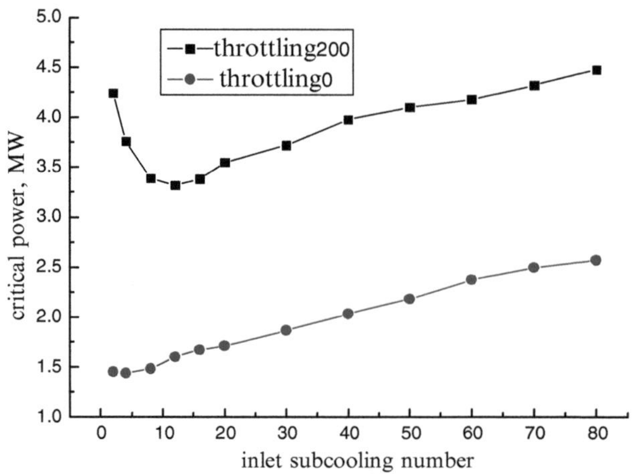

Fig. 9 Effect of inlet subcooling on stability

3.4 Effects of Inlet Subcooling

Figure 9 shows the critical power of a twin-channel system under different inlet subcooling when the system pressure is 3 MPa. We can see that the influence of inlet subcooling number on the instability of system is not single-valued at some conditions and a critical value is confirmed. Exceeding this value the stability of the system will be improved with the increase of inlet subcooling, and lower than this value the phenomena will be opposite. This critical value increases with the system

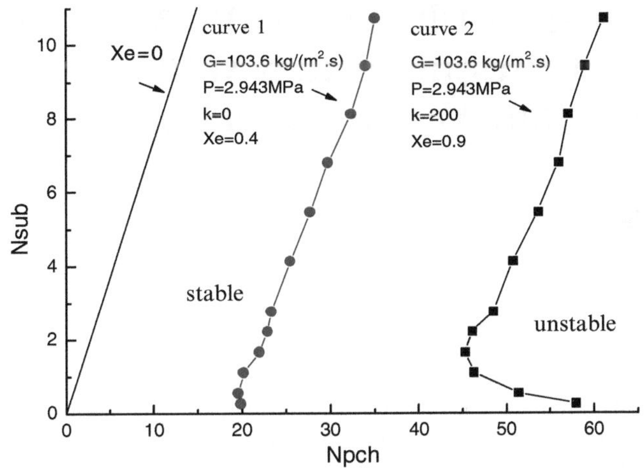

Fig. 10 Effect of inlet subcooling on stability

pressure. The instability boundary are shown in Fig. 10. If setting the equilibrium quality Xe line as the instability boundary, with the aid of throttling coefficient, a higher unstability boundary were observed.

4 Conclusions

In this study, the phenomenon of intertube pulse in parallel two-channel system has been investigated. We attained the conclusions that increasing the system pressure, inlet throttling, inlet sections can intensify the system stability, the influence of inlet subcooling is not single-valued. Further experiments are needed to investigate other possible instability mechanisms.

References

1. Ledinegg M (1938) Instability of flow during natural and forced circulation. Waerme 61(8):891–898
2. Boure JA, Bergles AE, Tong LS (1973) Review of two-phase flow instability. Nucl Eng Des 25:165–192
3. Fukuda K (1979) Classification of two-phase flow instability by density wave oscillation model. J Nucl Sci Technol 16(2):5–108
4. Yun G, Qiu SZ, Su GH, Jia DN (2008) Theoretical investigations on two-phase flow instability in parallel multichannel system. Ann Nucl Energy 35:665–676
5. D'Auria F, Galassi GM (1998) Code validation and uncertainties in system thermal-hydraulics. Prog Nucl Energy 33:175–216
6. IAEA-TECDOC-1387 (2004) Guidelines for the Review of Research Reactor Safety

Development of Ultrafast Pulse X-ray Source in Ambient Pressure with a Millijoule High Repetition Rate Femtosecond Laser

Masaki Hada and Jiro Matsuo

Abstract High intensity Cu Kα X-ray was generated in helium at atmospheric (760 Torr) using a commercial millijoule high-repetition rate Ti: sapphire laser. The characteristic Kα X-ray was generated by focusing the 0.06–1.46 mJ, 100 fs and 1 kHz repetition femtosecond laser onto a solid Cu target to a spot with a 5 μm diameter. We obtained the characteristic Kα X-ray of 5.4×10^9 photos per second into 2π sr at a 1 kHz repetition rate with 5.0×10^{-6} conversion efficiency. The X-ray intensity and conversion efficiency in helium achieved the almost the same level as in vacuum. Such vacuum-free femtosecond X-ray source with a table-top laser can be a promising and easy accessible tool for time-resolved X-ray diffraction and other radiographic applications.

Keywords Laser-induced plasma • X-ray • Femtosecond laser

1 Introduction

Hard X-ray from femtosecond laser-produced plasma has gained much interest, as unique time resolved X-ray diffraction (XRD) experiments demonstrate and reveal the atomic dynamics of chemical reactions which are induced by the interaction between phonons and materials (semiconductors, semimetals organic materials) in solar photovoltaic panels or other photo-electronic devices [1-3]. Laser-produced plasma X-ray sources have consisted of high-power and low-repetition-rate lasers of above 100 mJ and 10 Hz and the Kα-X-ray intensity using this large size laser was reported to be 10^8–10^{11} cps/sr with conversion efficiency of 10^{-4} to 10^{-5} [4-7]. Generally an X-ray intensity of more than several 10^8 cps/sr is required for time-resolved XRD experiments, and a laser-plasma X-ray of 10^9–10^{11} cps/sr is desirable.

M. Hada
Department of Nuclear Engineering, Kyoto University, Kyoto, Japan

J. Matsuo (✉)
Quantum Science and Engineering Center, Kyoto University, Kyoto, Japan

T. Yao (ed.), *Zero-Carbon Energy Kyoto 2009*, Green Energy and Technology, DOI 10.1007/978-4-431-99779-5_47, © Springer 2010

This intensity is sufficiently high for the X-ray radiographic applications. Nevertheless, the utilization of such ultrafast pulsed X-ray has been limited because of the complexity of the huge vacuum system and difficulty in managing a high-power laser. To date, a huge laboratory-top laser and a large size vacuum chamber are requisites to generate ultrafast pulsed X-ray radiation. In this large-scale ultrafast X-ray source, there are also some problems with target forms and debris from the target caused by laser ablation. Regarding the target form, thin tape or wire typed targets have been used because the space in the vacuum chamber is limited. However, the small tape or wire target can be put in a vacuum chamber, and target lifetime is quite short at most a few days. It is also difficult to control the surface of the thin tape or wire within a few micrometers. There is also the problem of debris from the target. When the high-power laser is focused onto the target, target materials are blown off and deposited on the focusing lens and other optics. Thin polymer tape covers have been employed to avoid the debris problem. Thus, such ultrafast pulsed X-ray sources were required to be more compact, easier to access, and have higher conversion efficiency into characteristic X-ray.

Recently, the compact designed table-top millijoule femtosecond laser has been reported to be available for generating hard X-ray with an intensity of about 10^8–10^{10} cps/sr with the Kα X-ray conversion efficiency of 10^{-5} to 10^{-6} [8-10]. Although the experimental scale of a femtosecond laser could be successfully reduced with a table-top femtosecond laser, difficulties remain when using a huge and complex vacuum chamber system, such as target manipulation, target lifetime, and debris emissions. Therefore, a high-intensity X-ray source that can operate in atmospheric pressure with a tabletop laser could be a desirable tool for ultrafast time-resolved measurements. Laser-induced plasma X-ray sources in helium atmospheric condition have been reported, however the X-ray intensity from these sources was quite low [11]. Very recently, J. A. Nees et al. has been developed a high intensity X-ray source (~5×10^9 cps) in helium atmospheric condition at the high plasma intensity of above 1.0×10^{18} W cm^{-2} [12, 13]. Nevertheless, the high intensity plasma above 1.0×10^{18} W cm^{-2} would extend the pulse duration of X-ray up to picosecond order. As for the pulse duration of pulsed X-ray, computer simulated studies have been demonstrated that the pulse duration of X-ray was extended with increase of the laser plasma intensity [14]. At the plasma intensity of 1.0×10^{16} W cm^{-2} and 1.0×10^{18} W cm^{-2}, the extension of pulse duration was ~100 fs and ~1 ps, respectively. Thus, laser-induced plasma source with lower (at most 5.0×10^{16} W cm^{-2}) plasma intensity is required for the time-resolved XRD applications.

In this study, we demonstrated a compact and high-intensity ultrafast pulsed X-ray source constructed in a helium atmosphere at plasma intensity of 0.15–4.0×10^{16} W μm^{2}cm^{-2}. It is possible to reduce the overall size of the X-ray source system without the complexity of a vacuum system. It is also feasible to set a long-lived and large-sized target regardless of the vacuum chamber in air. This vacuum-free X-ray system also allows us to avoid the debris problem. At atmospheric pressure, the debris cannot reach the focusing lens or other optics, which are placed some ten millimeters away from the focusing spot. With helium or other gas jet, the debris can be easily collected with a filter.

2 Experimental

Figure 1 shows the experimental setup for ultrafast pulsed X-ray generation. The mode-rocked Ti: sapphire laser generated femtosecond optical pulses of about 100 fs duration with wavelength at 800 nm, and the optical pulses were amplified at about 3.5 mJ per pulse through a regenerative amplifier (Spectra Physics/Model Spitfire Pro XP) with repetition rate of 1 kHz. The laser pulse profile was TEM_{00} and it was *p*-polarized. The optical pulse generated through the regenerative amplifier was focused into a rotating copper target with an infrared achromatic lens (f=40 mm, N.A. = 0.2). The focusing spot size was 5 μm, which was measured by the crater size of the focusing spot. This spot size was well corresponded to the diffraction limit of this infrared achromatic lens (4.8 μm). The pulse duration of the laser pulse on the surface of copper target was 100 fs. The power of the optical main pulse was 0.06–1.46 mJ. The copper target was a circular cylinder of 100 mm diameter and 300 mm length. The surface position of rotating copper target was controlled within ±1 μm with precision bearings and adequate tension of springs, and was measured with a micrometer during rotation. The copper target rotates at a rate of 1 rpm (>0.96 rpm), which could be varied in the range 0.24–1 rpm and moves in the axial direction at the rate of 10 μm min^{-1} (>5 μm min^{-1}), which gives a fresh copper surface with each

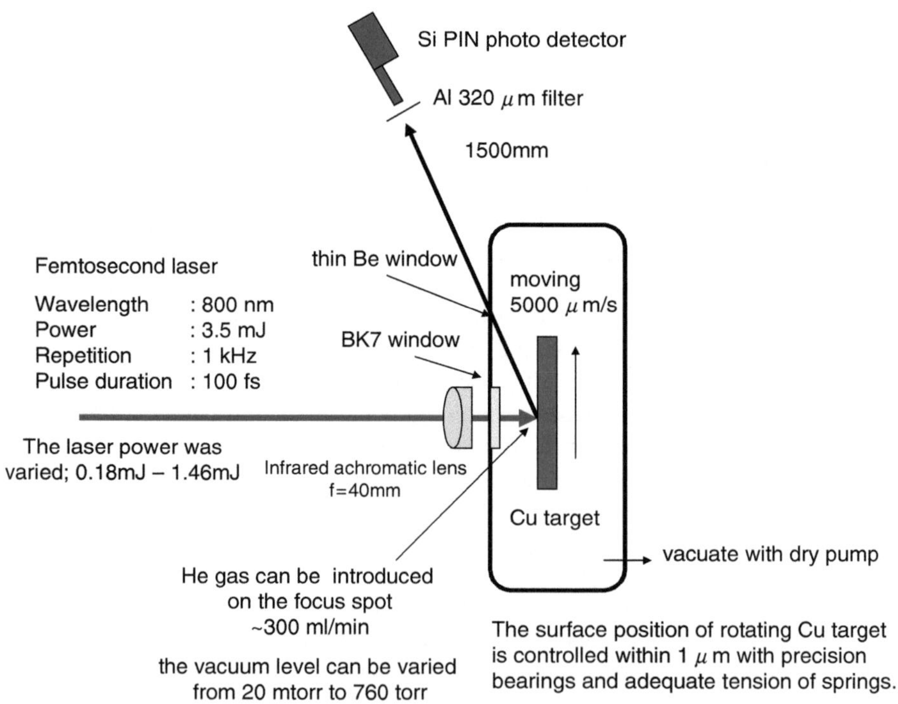

Fig. 1 The experimental setup for ultrafast X-ray generation in various atmospheric conditions

laser shot. Near the focus point atmosphere in the Cu target system can be changed between air or helium. The helium gas was introducing with 1/4 inches gas nozzle and the flow rate of helium gas was 500 ml min^{-1}. In air, the debris from the target was not deposited on the focusing lens or other optics. We also made same X-ray generation system with Cu target in a vacuum chamber to compare with the X-ray intensity at atmospheric pressure in air and under vacuum (20 mTorr).

The X-ray generated from focusing the laser pulse onto a copper target was measured with a PIN-Si photo detector (Amptek/XR-100CR) which has a 300 μm thick, 7 mm^2 square silicon sensor. The detection efficiency of this PIN-Si photo detector is approximately 100% for an 8 keV X-ray. The detector was sealed with 25 μm thickness Be window through which 8 keV X-ray passes without any losses. A spectroscopy amplifier (CANBERRA/2022 Spectroscopy Amplifier) was used to amplify the signal, which was tallied up with a multichannel analyzer (SEIKO EG&G/TRUMP-MCA-2 k). The detector was placed 1,500 mm away from the focusing spot at an angle of 60°. In a vacuum condition, the X-ray went through 3 mm in vacuum and 1,497 mm in air. In helium atmosphere or in vacuum conditions, a thin aluminum filter of 320 μm was placed before the detector. This filter reduces the X-ray intensity to 1.72% for a Cu Kα X-ray. The long distance between the X-ray focusing spot and the detector and the Al filter allowed us to measure the X-ray photons as a Poisson distribution single photon counting. Two different photons cannot be detected at once in the order of 1 ms using a Si photo detector and multichannel analyzer. If more than one photon is in the same detecting time, the total energy of the phonons will be measured with the Si photo detector. In any conditions, it took 40 s to obtain each X-ray spectrum.

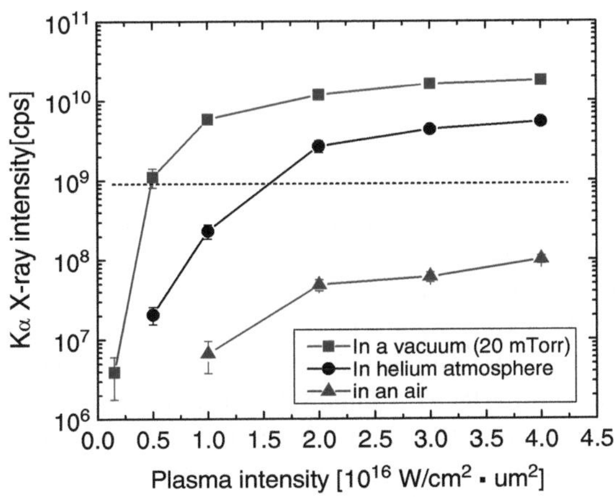

Fig. 2 Kα X-ray intensity in various atmospheres: *solid line with square*, vacuum (20 mTorr); *solid line with circles*, in helium; *solid line with triangle*, in air

3 Results and Discussion

In Fig. 2, the generated Kα X-ray intensity was plotted as a function of the plasma intensity in various atmospheric conditions. We obtained high intensity Cu Kα X-rays with 2.6–5.4$\times 10^9$ photons/sr/s above the plasma intensity of 2.0×10^{16}W μm^{2}cm^{-2} in He atmosphere. This intensity was more than 60 times that in air, and close to the intensity of 1.2–1.8$\times 10^{10}$ photons/sr/s obtained in vacuum. Generally, a value of at least several 10^8cps/sr characteristic X-ray is required for time-resolved XRD or other applications. In Fig. 2, the horizontal dashed line shows the X-ray intensity of 1.0×10^9 cps/sr. Above the plasma intensity of 2.0×10^{16}W μm^{2}cm^{-2}, the Kα X-ray conversion efficiency calculated from the X-ray intensity and incident pulse energy in helium at atmospheric pressure was 5.0 (±0.2)$\times 10^{-6}$, a value 60 times higher than that in air (8.2×10^{-8}) and close to the value in a vacuum (1.8–2.1$\times 10^{-5}$). A Kα conversion efficiency of 10^{-6} to 10^{-5} is also need to obtain X-ray intensity of more than several 10^8cps required for time-resolved XRD. However, in air the Kα X-ray intensity or conversion efficiency was quite low; in helium at atmospheric pressure, the Kα X-ray intensity and conversion efficiency achieved the value of 5.4×10^9cps and 5.0×10^{-6}, which are sufficiently high for time-resolved XRD or other X-ray applications. This vacuum-free, compact and easy-to-access X-ray generation system can be operated in helium at atmospheric pressure. The conversion efficiency obtained by other groups using high repetition millijoule laser has been reported to be about 4.6×10^{-6} to 3.2×10^{-5} [8-12] in vacuum conditions and ~5.0×10^{-6} in helium atmospheric conditions [13, 14]. The Kα X-ray intensity obtained in helium at atmospheric pressure was also of the same order as the other works using high repetition millijoule laser in vacuum conditions. Thus this compactly designed high-intensity Kα X-ray at lower plasma intensity source that can operate at atmospheric pressure without using a large and complex vacuum chamber can be a useful and promising tool for the time-resolved XRD or other radiographic applications.

4 Conclusion

High intensity Kα X-ray generation at a high repetition was demonstrated with table-top commercial ultrafast laser system in helium at atmospheric pressure. Millijoule, 100 fs laser pulses were focused onto a well-controlled Cu surface at intensities of 0.15–4.0$\times 10^{16}$W μm^{2}cm^{-2}. The intensity of the generated Kα X-ray in helium at atmospheric pressure was 5.4×10^9cps/sr with 1 kHz repetition rate and conversion efficiency of 5.0×10^{-6}. This X-ray intensity is close to that obtained in vacuum condition and is high enough for the time-resolved XRD or other X-ray applications. Such a high-intensity, vacuum-free femtosecond X-ray source with a table-top laser could be a promising tool for the time-resolved XRD or other radiographic applications.

Acknowledgment This work is partially supported by the JST, CREST.

References

1. Rose-Petruck C et al (1999) Nature (London) 398:310
2. Sokolowski-Tinten K et al (2003) Nature (London) 422:287
3. Lindenberg AM et al (2005) Science 308:392
4. Yoshida M et al (1998) Appl Phys Lett 73:2393
5. Eder EC, Pretzler G, Fill E, Eidmann K, Saemann A (2000) Appl Phys B 70:211
6. Fill E, Bayerl J, Tommasini R (2002) Rev Sci Instrum 73:2190
7. Siders CW et al (1999) SPIE Proc 3776:302
8. Rettig CL, Roquemore WM, Gord JR (2008) Appl Phys B 93:365
9. Hagedorn M, Kutzner J, Tsilimis G, Zacharias H (2003) Appl Phys B 77:49
10. Serbanescu CG et al (2007) Rev Sci Instrum 78:103502
11. Jiang Y et al (2003) J Opt Soc Am B 20:229
12. Hou B, Easter J, Krushelnick K, Nees JA (2008) Appl Phys Lett 92:161501
13. Hou B et al (2008) Opt Exp 16:17695
14. Reich Ch, Gibbon P, Uschmann I, Förster E (2000) Phys Rev Lett 84:4846

Development of Small Specimen Technique to Evaluate Ductile–Brittle Transition Behavior of a Welded Reactor Pressure Vessel Steel

Byung Jun Kim, Ryuta Kasada, and Akihiko Kimura

Abstract Small specimen test techniques (SSTT) for the evaluation of irradiation embrittlement of reactor pressure vessel steel (RPVS) is considered to be essential to operate light water reactors over 40 years old. In this research, specimen size effects were investigated for RPVS to apply small specimen test technique to surveillance test method. Specimens used in this study were machined from a welded A533B steel plate for RPV. Different size of specimens, Standard, CVN-1/2, CVN-1/3, and CVN-1.5 mm were fabricated from the weld bond. It was found that the ductile-to-brittle transition temperature (DBTT) and upper shelf energy (USE) were reduced by decreasing specimen size. The effect of notch position on DBTT was independent of specimen size.

Keywords Small specimen test techniques (SSTT) • A533B steel • Impact properties • Weld

1 Introduction

Small specimen test techniques (SSTT) for the evaluation of irradiation embrittlement of reactor pressure vessel steel (RPVS) is considered to be essential to operate light water reactors over 40 years old, because the number of surveillance impact test pieces has shortened especially for those welded portion. The reduction of the specimen size can provide enough number of surveillance test specimens for extended operation period. However, it is well known that the impact properties depend on the specimen size and the notch location of the weld bond [1–4]. The heat affected zone (HAZ) closed to the weld fusion line has been known to have lower fracture toughness values than the other regions due to the coarse grained microstructures in this region [5, 6]. In the most of the test regulation of surveillance weld specimens, HAZ is defined as

B.J. Kim (✉), R. Kasada, and A. Kimura
Graduate School of Energy Science, Kyoto University, Kyoto, Japan
e-mail: byungjun_kim@iae.kyoto-u.ac.jp

T. Yao (ed.), *Zero-Carbon Energy Kyoto 2009*, Green Energy and Technology,
DOI 10.1007/978-4-431-99779-5_48, © Springer 2010

Table 1 Nominal compositions (wt %) of A533B steel

Material	Fe	C	Si	Mn	P	S	Ni	Cr	Cu
A533B	Balance	0.18	0.16	1.55	<0.003	0.0014	0.64	0.16	0.03

the specimen whose notch is located at the position 1 mm away from the fusion line. However, resent studies have shown that the fracture toughness of the HAZ(1 mm) is better than that of base metal [4, 7].

In this study, specimen size effects on the DBTT and USE were investigated for a RPVS of A533B to apply SSTT to the surveillance test method.

2 Experiments

The chemical compositions of A533B steel used in this study are shown in Table 1. The weld of the reactor pressure vessel steel was produced by means of a submerged arc welding method. Post-weld heat treatment for 15 h at 625 °C was carried out and followed by furnace cooling.

Charpy impact tests were performed according to the standard test method for notched specimens. Different sizes of specimens, Standard, CVN-1/2, CVN-1/3 and CVN-1.5 mm were fabricated from the weld bond. The specimens are prepared such that the notch is located at different positions. The notch position of the Charpy impact specimen is in base metal, at HAZ 1 mm, at HAZ 4 mm and in weld metal. Charpy impact tests were carried out at temperature from 73 to 473 K. A specimen was set up in a cold bath filled with isopentane, which was cooled by liquid nitrogen for low temperature test, and a silica oil bath was used for high temperature test. The bath was held at a desired constant temperature for 15 min before testing. The data were fit to a curve of hyperbolic tangent function:

$$E = a + b \tanh [c (T - d)]$$

where E is the absorbed energy (J), a, b, c and d are constants where d = DBTT and a + b = USE, and T is testing temperature (K).

3 Results and Discussion

3.1 Effects of Specimen Size on the Impact Properties

The ductile to brittle transition curves of the A533B steel in different specimen sizes are shown in Figs. 1a, b, c and d for standard, 1/2, 1/3 and 1.5 mm, respectively. The shifts in DBTT for half-, third-, 1.5 mm size were −5°C, −45°C, and −100°C, respectively compare to that of standard size specimen.

The USE is reduced with decreasing specimen size primarily because of a decrease in the cross section of the specimens.

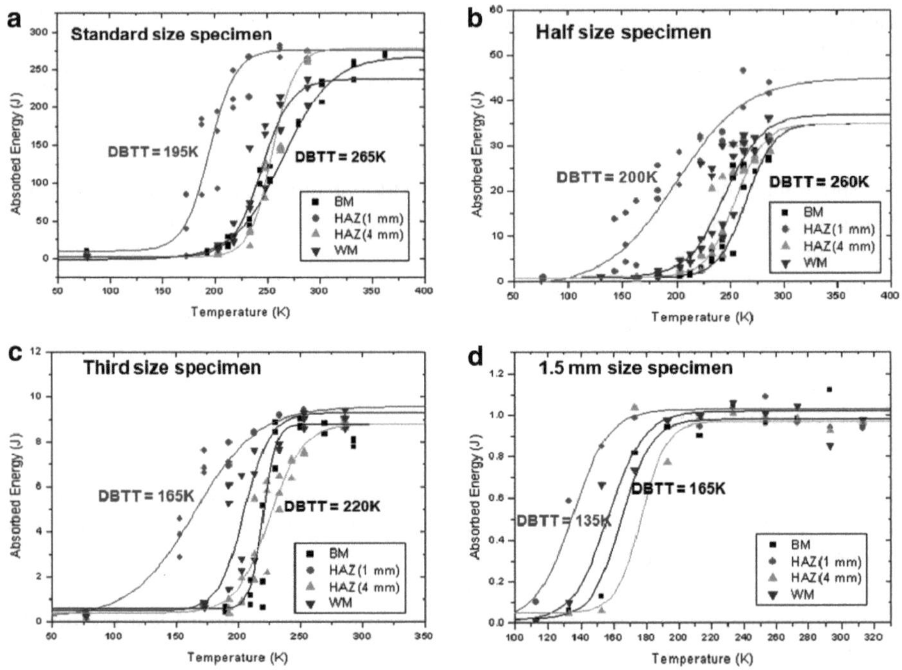

Fig. 1 Effects of specimen size and notch position on the impact properties of A533B steel. (**a**) Standard size specimen (**b**) half size specimen (**c**) third size specimen (**d**) 1.5 mm size specimen

3.2 Effects of Notch Location on the Impact Properties

It is well known that the impact properties depend on the notch location of the weld bond. The various microstructures of HAZ affect the mechanical properties and toughness. Figure 1a shows the effect of notch position on impact properties in standard size specimen. The DBTT of HAZ 1 mm showed lowest DBTT (195 K) while that of base metal is highest (265 K). However, it shows the larger scatter than that of other region (base metal and weld metal). This phenomenon is usually observed in HAZ of which the microstructure is changed by the distance from fusion line through the difference in welding heat and cooling rate during the welding process.

4 Conclusions

Effects of specimen size on the impact properties, DBTT and USE, were investigated for a RPVS A533B. The obtained results are summarized as follows:

(1) The DBTT and USE is reduced with decreasing specimen size.
(2) The DBTT of welded A533B steel is affected by the notch location of the weld bond. HAZ 1 mm is the best in the impact properties for the steel.
(3) The effects of notch position on DBTT are independent of the specimen size.

References

1. Schubert LE, Kumar AS, Rosinski ST, Hamilton ML (1995) Effect of specimen size on the impact properties of neutron irradiated A533B steel. J Nucl Mater 225:231–237
2. Liu C, Bhole SD (2002) Fracture behavior in a pressure vessel steel weld. Mater Des 23:371–376
3. Jang YC, Hong JK, Park JH, Kim DW, Lee Y (2008) Effects of notch position of the Charpy impact specimen on the failure behavior in heat affected zone. J Mater Process Technol 201:419–424
4. Kim JH, Yoon EP (1998) Notch position in the HAZ specimen of reactor pressure vessel steel. J Nucl Mater 257:303–308
5. Kameda J, Takahashi H, Suzuki M (1978) Residual stress relief and local embrittlement in a A533B reactor pressure vessel weldment. Int J Press Vessels Piping 6:245–274
6. Parthasarathy R, Wood WE, Atteridge DG, Devletian JH (1992) International trends in welding science and techonology. ASM International, pp 551
7. Barbaro FJ, Krauklis P, Easterling KE (1989) Mater Sci Technol 5:1057–1068

Research on Distributed Monitoring and Prediction System for Nuclear Power Plant

Yingjie Sun, Min-jun Peng, and Ming Yang

Abstract The study addresses the use of distributed strategy to develop the nuclear power plant (NPP) Primary Loop and Auxiliary System condition monitoring and prediction system (PSCMPS) by analyzing the structure and fault characteristics of the Primary System of Qinshan nuclear power plant. The PSCMPS system can identify abnormality and faults in anticipation and display the result on the human–machine interface. The results proved that the export system with distributed strategy is feasible on multiple faults and the combination of the export system and ANN can forecast the trends of the important parameters of the NPP Primary Loop and Auxiliary System condition well. The program are developed by Visual C++ 6.

Keywords Nuclear power plant (NPP) • PSCMPS system • Artificial neural networks (ANN) • ELMAN network

1 Introduction

Nuclear safety is the main goal of nuclear power plant which should be ensured in the whole span of nuclear lifetime from system design, operation until decommission, where human errors are reported to be the main contribution to the An important means for reducing human errors is to develop advanced and practical condition monitoring and prediction systems for nuclear power plants [1].

The study addresses the distributed strategy and its application in the development of a condition monitoring and prediction expert system (PACMPES) of the Primary Loop and Auxiliary System for Chinese Qinshan nuclear power plant (NPP). The prediction model takes advantage of the expert system and the prediction capability of artificial neural networks (ANN).

Y. Sun (✉), M. Peng, and M. Yang
College of Nuclear Science and Technology, Harbin Engineering University, Harbin, China
e-mail: reboothit@hotmail.com

T. Yao (ed.), *Zero-Carbon Energy Kyoto 2009*, Green Energy and Technology, DOI 10.1007/978-4-431-99779-5_49,

2 Methodology

2.1 Distributed Strategy

The main idea of distributed monitoring & prediction is to decompose the tasks of condition monitoring & prediction according to the characteristics that nuclear power plant has distribution and hierarchy in the structure and function, That is decomposing the whole system into some subsystems to monitor and predict. Figure 1 shows the main idea of Distributed system Design. Based on the primary system and auxiliary system, the structure can be divided into a number of fault categories each corresponding to a subsystem, so that the entire primary system monitoring task can be subdivided into multiple subtasks.

2.2 Elman Recurrent Networks

The ELMAN network is recognized as a two-layer network with feedback from the first-layer output to the first-layer input, see Fig. 2. This recurrent connection allows the ELMAN network to both detect and generate time-varying patterns.

The ELMAN network has tansig neurons in its hidden (recurrent) layer and purelin neurons in its output layer. This combination is special in that two-layer networks with these transfer functions can approximate any function (with a finite number of discontinuities) with arbitrary accuracy. The only requirement is that the hidden layer must have enough neurons. More hidden neurons are needed as the function being fitted increases in complexity [2].

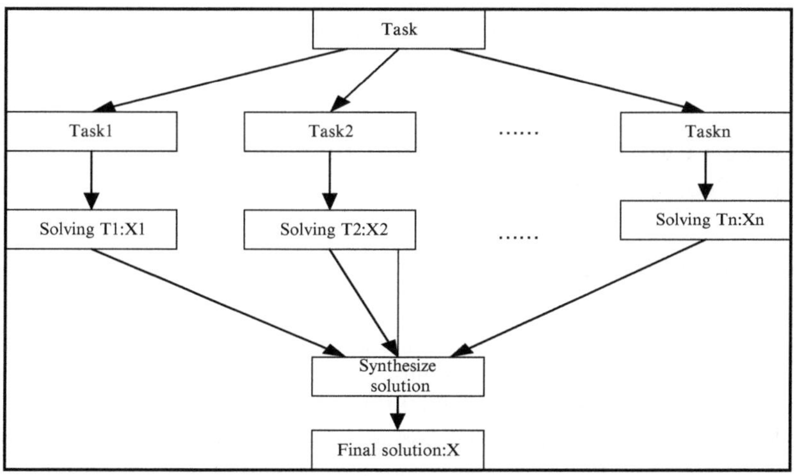

Fig. 1 The main idea of Distributed system Design

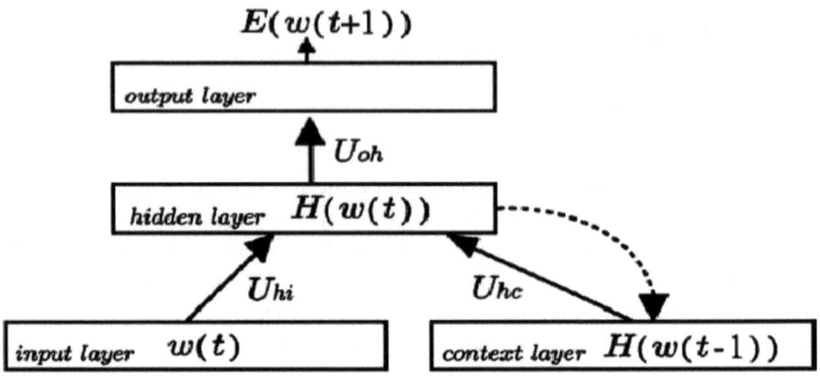

Fig. 2 The Elman network

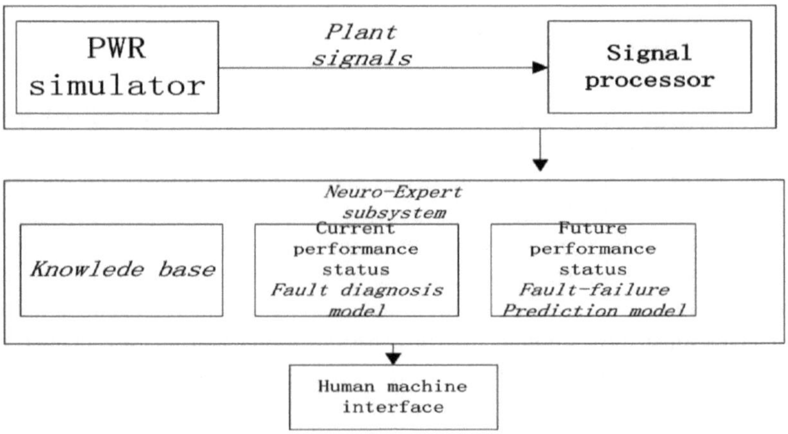

Fig. 3 The structure of monitoring and prediction system

3 Design Functions and Contents of Monitoring and Prediction System

3.1 System Structure

The structure of this monitoring system is shown in Fig. 3. Main purpose of nuclear reactor monitoring is to identify the current status of the operational plant using process signals. We can see that PSCMPS has the following functional units: Knowledge Database, Operation Monitoring and Fault Diagnosis Unit, Fault-failure Prediction Unit. The human machine interface show the monitoring information and the status of NPP diagnosed by neural networks and an expert system.

3.2 Neural Network for Prediction

The Elman neural network designed here has three layers: input, one hidden and output layer. The three layers are composed of 21 input nodes, 23 hidden nodes and 21 output nodes. The output signals supposed to be the same as the input signals at the next time step, so that the neural networks can predict the next outputs of the system. This is direct analogy with the concept of one-step-ahead prediction and can be effectively implemented by neural networks [3].

4 Performance Testing

We emulate a SGTR accident on the simulator. According to the simulation data, we make a 30 s prediction on primary system pressure. Figure 3 shows the result of the simulator and the result we predict by PSCMPS. From the Fig. 4, we can know that the curve we predict coincides with the result of the simulator acceptably.

5 Conclusion

The PSCMPS system has been tested with the PWR simulator. It was shown that the monitoring&prediction system successfully detected the symptoms of anomalies in the early stage.

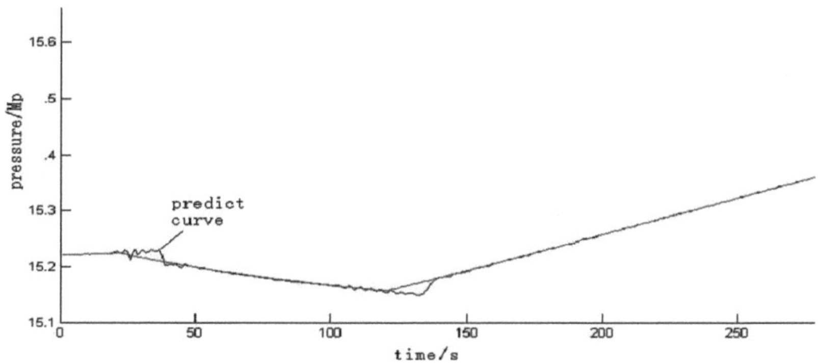

Fig. 4 Comparison the 30 s prediction of pressure of primary system

References

1. Seker S, Ayaz E, Turkcan E (2003) Elman's recurrent neural network applications to condition monitoring in nuclear power plant and rotating machinery. Eng Appl Artif Intell 16:647–656
2. Salazar-Ruiz E, Ordieres JB, Vergara EP, Capuz-Rizo SF (2008) Development and comparative analysis of tropospheric ozone prediction models using linear and artificial intelligence-based models in Mexicali, Baja California (Mexico) and Calexico, California (US). Environ Model Softw 23:1056–1069
3. Nabeshima K, Suzudo T, Ohno T, Kudo K (2002) Nuclear reactor monitoring with the combination of neural network and expert system. Math Comput Simul 60:233–244

Multiple Scale Nonlinear Phenomena in Nature: From High Confinement in Fusion Plasma to Climate Anomalies

Miho Janvier, Yasuaki Kishimoto, and Jiquan Li

Abstract Linear phenomena represent only a very small part of physical processes happening in nature. Climate is among the uttermost nonlinear behaviors found around us. As recent studies showed the possible effect of cosmic rays on the Earth's climate, we investigate how complex interactions between the planet and its environment can be responsible for climate anomalies. A parallel is drawn with the nonlinear behavior of high temperature plasma (needed for the creation of a new energy, fusion power), studied within the frame of Magneto-hydrodynamics.

Keywords Climate anomalies • Interaction sun-earth • Plasma simulation • Nonlinear behaviors

1 Introduction

Instrumental observations of the Earth climate all agree on one thing: elevations in temperatures around the globe are such that we can talk of a global warming. One phase, from the 1910s to the 1940s with an elevation of 0.35°C followed by an increase of 0.55°C from the 1970s to the present have rung urgent bells, warning human on the possible impact of their activities. The latter is known to release high amounts of greenhouse gases that act as a blanket for radiation coming from the surface of the Earth and therefore that warm the atmosphere. There is little doubt that increases in the atmospheric concentration of greenhouse gases are due to intense human activity (the beginning of the increase no surprisingly matches the beginning of the industrial era), and that industrial revolution has changed the interaction between human beings and their environment. However, the Earth's climate acts as a complex machinery where non-trivial dynamics at various scales

---

M. Janvier(✉), Y. Kishimoto, and J. Li
Graduate School of Energy Science, Kyoto University, Gokasho, Uji, Kyoto, 601-0011, Japan
e-mails: janvier.miho@ay7.ecs.kyoto-u.ac.jp; kishimoto@energy.kyoto-u.ac.jp; lijq@energy.kyoto-u.ac.jp

T. Yao (ed.), *Zero-Carbon Energy Kyoto 2009*, Green Energy and Technology, DOI 10.1007/978-4-431-99779-5_50, © Springer 2010

are taking place. As anthropomorphism is not objectivity, we have to, as scientists, reconsider complexity, stand on critical positions, in order to propose explanations and solutions for the global warming. Here, we first investigate the role of the Sun's magnetic activity as a possible criterion to explain some of the changes in the Earth's climate. As the ecological impact of human activities is still a number one priority, we also present recent results in simulation of plasma that are essential understandings for the achievement of fusion power.

2 Nonlinear Anomalies of the Climate Change

Talking about nonlinear systems for the climate makes sense as this latter cannot be defined as a simple "linear" combination of components that are completely independent of each other. On the contrary, climate is a subtle system that evolves because of a complex coupling of its internal dynamics with external factors. Considering different mechanisms is important to comprehend the changes that occurs at large scales. As solar radiation powers the climate system, researchers have for a long time investigated the role of the sun's activity in changes that occurred on the Earth, such as the 11-year cycle corresponding to the inversion of the Sun's magnetic field. If activities of our star that directly follow this cycle are correlated to changes in the total solar irradiance (TSI), recent data collected by spacecrafts from 1980 only show a change of 0.1% of TSI that fails to reproduce the historical surface temperature record evolution [1]. Therefore, if not negligible, the total change of TSI would not be enough to explain the recent changes in the average temperature.

The answer might lie in another property of the Sun, its generation of an intense magnetic field that expands over the limit of our solar system, the heliosphere. The stream of plasma (ionized gas) ejected from the Sun, also called solar wind, follows the magnetic field lines and creates a large scale shielding that prevents the penetration of high energy cosmic rays, consisting of all particles coming from outer space and remnants from interaction between atmospheric and space particles [2]. Especially, high energy particles from cosmic rays can interact with the molecules of the Earth's atmosphere. Svensmark [3], by conducting an experiment simulating the effect of the Sun's rays on the typical environment of the air over the ocean, found a striking correlation between the amount of cosmic rays and cloud cover: more precisely, the electrons released from the interaction in the air act as catalysts in the chemical reactions leading to the formation of water and sulphuric acid molecules, building blocks for the formation of clouds. Clouds' top, on the other hand, has an albedo such that Sun's rays are reflected in outer space: cloudiness on average has been demonstrated as having an important cooling effect on the Earth's temperatures.

High-energy cosmic rays coming from interstellar space (also called Galactic Cosmic Rays-GCRs) are directly affected by the magnetic field of the heliosphere. Because of its shape, GCRs lose some energy and therefore some of them never reach the Earth. Of course, the flux of cosmic rays reaching our planet depends on the strength of the shielding, modulated by the Sun's magnetic activity. The stronger it is, the better the cosmic rays are streamed away from the Earth. On a more general scale, the flux of cosmic rays might also depend on the galactic environment of the solar system. However, correlation does not mean causality and as effect on cosmic rays on formation of cloud cover still lacks of an explanation for physical process, it would be unwise to certify that this phenomena is the only origin of global warming. This theory subtends amplification: small changes occurring on a solar or galactic scale could affect on a larger scale the Earth's climate. We have taken a mechanism among many others, but the correlation between this simple illustration of nonlinearities in the Earth's climate and the theory of complex physical systems is quite interesting. In the following, simulations of plasma physics show that nonlinear mechanisms evolving at different scales give good explanations of observed phenomena.

3 Nonlinearities in High Temperature Plasmas

Since the industrial revolution, the demand for energy has not stopped but increased exponentially. If fossil fuels have been extensively used as a primary source of energy, attention is now on the reduction of their use to minimize the impact of human activities on the environment. Fusion power, which birth is due to the first investigations of nuclear fusion in the early 1930s by Rutherford, has a promising future as a clean and sustainable energy as it produces no high-level radioactive waste (on the contrary to fission power) and is fueled with isotopes of hydrogen that are commonly found on the Earth. Simple estimations predict that the only use of deuterium from sea water would fuel fusion energy for 150 billion years. On the Earth, fusion reactions are made possible by creating high temperature plasmas, confined magnetically inside chambers of various shapes. Among them, toroidal chambers called "tokamaks" are under extensive investigation, especially since the launching of the international experiment ITER, first "real" model to test the feasibility of an industrial tokamak. Therefore, enhancing plasma confinements is the key issue in the achievement of fusion power.

Unfortunately, the behavior of plasmas is very complex as it results from interactions of many phenomena happening at the same time: among them, the rise of instabilities causing disruptions. The plasma inside a toroidal chamber can be described as a fluid with electric and magnetic properties (Magneto-HydroDynamics or MHD approach) transported along magnetic field lines. Arising instabilities will also travel along those lines and grow. To avoid such unwanted event, it has been demonstrated that shearing the magnetic field lines can stop the growth of instabilities. Moreover, recent studies showed that inversing the shearing enhances the confinement of the plasma through the formation of Internal Transport Barriers, separating the core from the edge and keeping it confined. However, this configuration allows magnetic field lines to break and to reconnect via resistivity, forming singular structures called magnetic islands. In a reversed shear system, those latter are created along resistive layers (also called tearing layers) and are expected to somehow enhance each other dynamics. This MHD mode is commonly called

Double Tearing Mode. The Double Tearing Mode has been proposed as a possible explanation for current penetration observed in tokamak discharges.

Previous studies have shown that the coupling between the two surfaces where reconnection occurs is weak if the distance between them is large, and becomes stronger the more little the distance is [4]. However, less attention has been proposed to the study of the middle coupling between tearing layers. Here, we conducted simulations of this specific case by numerically solving the Reduced MHD equations expressing the magnetic flux ψ and velocity flow ϕ in slab geometry as follows:

$$\frac{\partial \nabla^2 \phi}{\partial t} + \left[\phi, \nabla^2 \phi\right] = \frac{1}{\mu_0 \rho_0}\left[\psi, \nabla^2 \psi\right], \qquad \frac{\partial \psi}{\partial t} + [\phi, \psi] = \frac{\eta}{\mu_0} \nabla^2 \psi.$$

We found that the magnetic and kinetic energies of the system evolve through an exponential growth, resulting from the linear behavior of the plasma, followed by a reduced growth resulting from nonlinear effects. Generally, the growth of resistive instabilities stops at a saturation level, when the source of magnetic energy driving the instability vanishes. Here, however, the magnetic energy continues to evolve slowly while the kinetic energy, going through a quasi steady-state evolution suddenly grows abruptly (Fig. 1) [5]. We investigated the role of the magnetic structure as a trigger for this fast growth. Among the reasons for such a hypothesis is the fact that amplitudes of perturbations of the magnetic equilibrium are much higher than that of the kinetic flow. By taking two-dimensional deformed magnetic equilibriums, we found that some configurations corresponding to bended magnetic field lines between the two tearing regions of the Double Tearing Mode are linearly

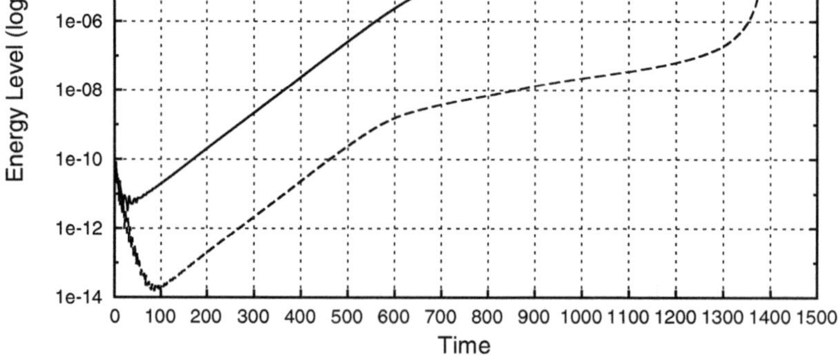

Fig. 1 Energies in a typical fast growth regime

unstable. Therefore, magnetic structures that are deformed from nonlinear interactions destabilize secondary instabilities evolving on different time scales.

4 Conclusion

Nature is full of nonlinear behavior. Understanding them is a key to predict dynamics of complex systems. Here, we presented two approaches: explaining natural nonlinearities that can count for anomalies in the Earth's climate, such as the interaction between cosmic rays and the solar wind, and simulating artificial nonlinearities via numerical tools in order to understand plasma behavior inside a tokamak, a step toward the achievement of fusion power.

References

1. Duffy PB, Santer BD, Wigley TML (2009) Phys Today 62:48–49
2. Balogh A (2006) Spatium 17
3. Svensmark H (2007) Astron Geophys 48:1.18–1.24
4. Pritchett PL, Lee YC, Drake JF (1980) Phys Fluids 23:1368–1374
5. Ishii Y, Azumi M, Kishimoto Y (2002) Phys Rev Lett 89:205002-1–205002-4

The Electric Properties of InSb Crystals for Radiation Detector

Yuki Sato, Yasunari Morita, Tomoyuki Harai, and Ikuo Kanno

Abstract Electric properties of InSb crystals were measured for the ones grown by zone melting method and liquid phase epitaxy method. It was indicated that the depletion layer thickness of radiation detectors made of epitaxially grown crystal could be nearly two times thicker than the ones fabricated with InSb which were grown by the other methods.

Keywords InSb • Photon detector • X-ray fluorescence analysis

1 Introduction

In recent years, it becomes important to detect hazardous elements such as Li, Be, and Pb contained in recycling objects. For this purpose, the X ray fluorescence analysis is a powerful and widely employed method. Therefore, the performance improvement of the X-ray fluorescence analysis contributes to the increase of the recycling rate. As a result, the X-ray fluorescence analysis with high performance contributes to the reduction of CO_2 exhaust.

In the X-ray fluorescence analysis, Si detectors are used as the X-ray detectors, generally. Si detectors are, however, not suitable for detecting heavy elements such as Pb, which emits high energy K X-rays of nearly 80 keV: the small atomic number and density of Si result in a low photon absorption rate. On the other hand, the energy of K X-rays of light elements such as Li and Be is below 100 eV.

A compound semiconductor, InSb is a promising substrate due to its high atomic numbers (In: 49, Sb: 51), high density (5.78 $g\,cm^{-3}$), and the smallest band gap energy (0.165 eV) among the developed semiconductors [1]. These features predict higher photon absorption efficiency (about 400–1,000 times higher than that of Si), and two times better energy resolution than those of Si detectors.

Y. Sato (✉), Y. Morita, T. Harai, and I. Kanno
Graduate School of Engineering, Kyoto University, Kyoto, Japan
e-mail: Y.Sato@nucleng.kyoto-u.ac.jp

T. Yao (ed.), *Zero-Carbon Energy Kyoto 2009*, Green Energy and Technology,
DOI 10.1007/978-4-431-99779-5_51, © Springer 2010

The quality of commercial InSb crystals are, however, not enough for the substrate of radiation detectors: InSb has been employed for the substrate of infrared sensors and magnetic field monitors. For the measurement of X-rays, high quality InSb crystal is necessary.

We are growing InSb crystals by zone melting method and liquid phase epitaxy method. The results of electric property measurements on these crystals are reported.

2 Experiment

The InSb crystals were grown by zone melting method and liquid phase epitaxy method. In zone melting method, 6 N (99.9999%) purity In and Sb raw materials (Osaka Asahi Co., LTD., Japan) were sealed in a quartz ampoule with Ar + H_2 (5%) at the pressure of 0.5 atm. The outer and inner diameters of the ampoule were 17 and 15 mm, respectively. In purification process, the ring heater moved horizontally along the ampoule. The heater velocity in the purification process was 5 cm h^{-1}. The purification process was repeated 20 times. In crystallization process, the heater velocity was 0.7 cm h^{-1}.

In liquid phase epitaxy (LPE) method, n-type InSb wafers (Galaxy Compound Semiconductor, Inc., USA), were used as a substrate crystal. Their diameter and thickness were 2 inches and 0.4 mm, respectively. Nearly 50 g of In–Sb (~20% in weight) was melt by the electric heater. The starting temperature of growing epitaxial crystal was 350°C. Epitaxial InSb growth started with moving the InSb substrate wafer to meet the In–Sb melt [2]. The temperature of the In–Sb melt was set to decrease 0.2°C min^{-1}. After 120 min., the substrate wafer was removed from the melt, and the epitaxial growth process was finished.

For estimation of the electric properties of the InSb crystals, the Hall measurements was used. Hall mobility, resistivity and carrier concentration were measured for zone melt crystal, LPE crystal and the substrate crystal in the temperature from 5 to 100 K.

3 Results and Discussion

The electric properties of InSb crystals are shown in Fig. 1 as a function of temperature. As a result of Hall measurements, it was indicated that the substrate crystal and crystal grown by zone melting method were n-type in the temperature range up to 100 K. On the other hand, the crystal grown by LPE method was indicated p-type in the same temperature range. In Fig. 1a, the Hall mobility of LPE crystal was smaller than those of the other crystals because the majority carrier was holes. Due to the small Hall mobility, the counting rate will become smaller than other crystals. At the temperature below 50 K, the Hall mobility of LPE crystal was, however, one order of magnitude higher than the electron mobility in Si at room temperature [3].

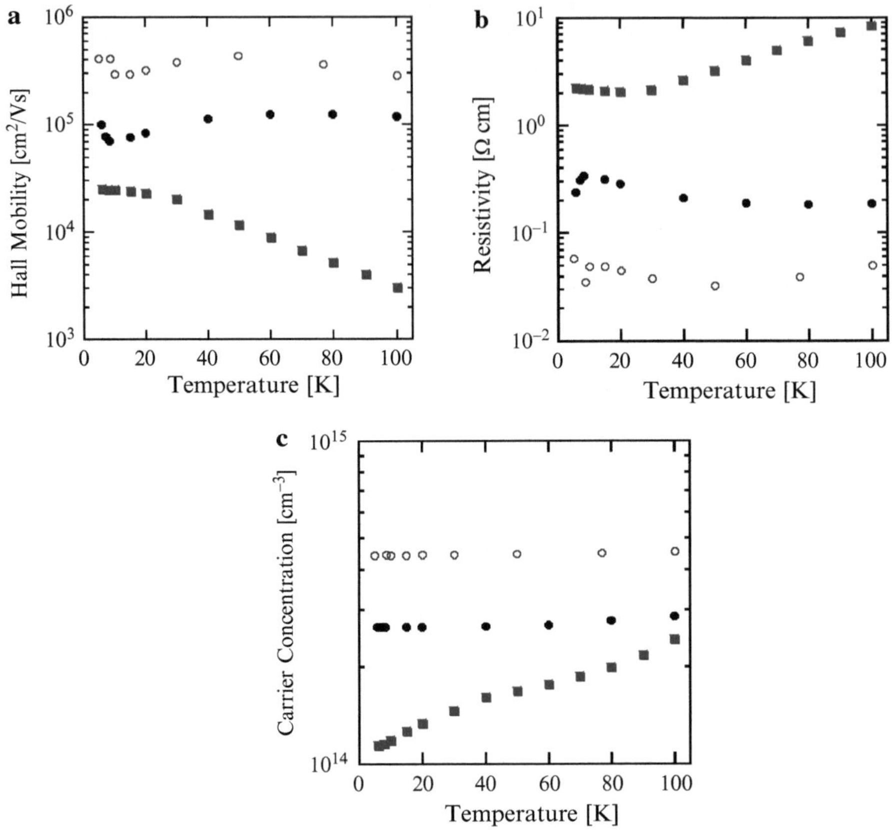

Fig. 1 The (**a**) Hall mobilities, (**b**) resistiveties and (**c**) carrier concentrations of InSb crystals as a function of temperature for zone melting method (*circles*), liquid phase epitaxy method (*squares*), substrate crystal (*open circles*)

Therefore, the counting rate of InSb detector made of LPE crystal could be higher than that of the conventional Si detector.

In Fig. 1b, at the temperature below 10 K, the resistivity of LPE crystal was several to tens of times larger than the ones of crystals grown by zone melting method and the substrate crystal, respectively. In addition, the carrier concentration of LPE crystal was smaller than the ones of the other crystals. Therefore, the leakage current of InSb detector made of LPE crystal can be smaller than the ones made of other methods.

In radiation detector, thick depletion layer is important to increase the detection efficiency. The thickness of the depletion layer is proportional to $(\mu\rho)^{1/2}$, where μ is the Hall mobility, ρ is the resistivity. The values of $(\mu\rho)^{1/2}$ as a function of temperature are shown in Fig. 2. At the temperature of 10 K, the value of $(\mu\rho)^{1/2}$ of LPE crystal was nearly two times larger than those of crystal grown by zone melting method and substrate crystal, respectively.

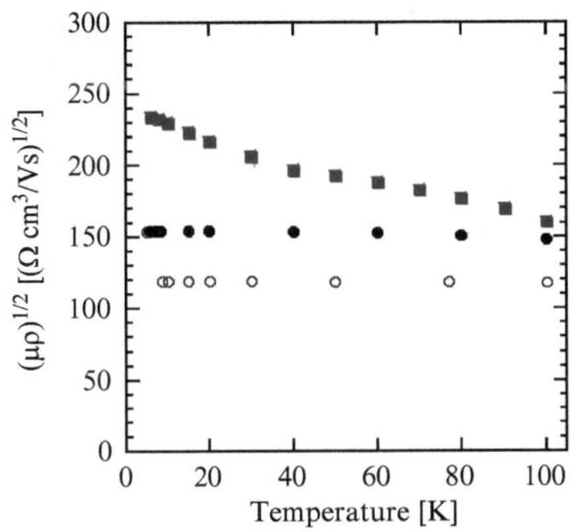

Fig. 2 The values of $(\mu\rho)^{1/2}$ as a function of temperature. *Symbols* are the same with Fig. 1

4 Conclusions

It was indicated that the depletion layer thickness of radiation detector made of LPE crystal could be nearly two times thicker than the ones fabricated with InSb crystal made by the other methods under the same bias voltage at the temperature of 10 K. In addition, the leakage current could be small. On the other hand, the Hall mobility was smaller than the other crystals. At the temperature below 50 K, the Hall mobility was one order of magnitude higher than the electron mobility in Si at room temperature. Accordingly, the LPE InSb crystal can be a promising substrate for radiation detector.

Acknowledgment This work was supported by the Kyoto University Global COE Program, "Energy Science in the Age of Global Warming".

References

1. McHarris C Wm (1986) Nucl Instrum Methods Phys Res A 242:373
2. Kanno I et al (2009) In: Proceedings, the Fifth International Symposium on Radiation Safety and Detection Technology (ISORD-5) Journal of Nuclear Science and Technology, Supplement (in press)
3. Ahmed S (2007) Physics and engineering of radiation detection. Elsevier, Oxford

Kinetic Transport Simulation of ICRF Heating in Tokamak Plasmas

Hideo Nuga and Atsushi Fukuyama

Abstract The deviation of the distribution functions from the Maxwellian due to the interaction with waves affects the propagation and absorption of the wave itself. Therefore self-consistent analysis including the modification of the momentum distribution function is required for quantitative analysis of wave heating and current drive. The modeling code TASK was updated to describe the momentum distribution function of multi-species and the wave dielectric tensor for arbitrary momentum distribution function. Numerical results of ICRF heating in tokamak plasmas which generates energetic tail of distribution function are reported.

Keywords Tokamak • Simulation • Wave heating • Momentum distribution function

1 Introduction

Plasma heating and current drive by RF waves deform the momentum distribution function of heated species. The deviation of the distribution functions from the Maxwellian affects the propagation and absorption of the wave itself. Therefore self-consistent analysis including the modification of the momentum distribution function is required for quantitative analysis of wave heating and current drive. The purpose of this study is to develop full wave analysis of ICRF simulation including the deformation of distribution function. In this paper, results of ICRF heating analysis in tokamak plasma using the integrated code TASK are reported.

The full wave component TASK/WM [1] calculates the wave electric field by solving Maxwell's equation including the plasma dielectric tensor. The bounce-averaged Fokker–Planck component TASK/FP analyzes the time evolution of the momentum distribution functions for electrons and ions by solving the Fokker–Planck equation including the quasi-linear diffusion terms calculated from the wave

H. Nuga (✉) and A. Fukuyama
Kyoto University, Sakyo-ku, Kyoto, 606-8501, Japan
e-mails: nuga@p-grp.nucleng.kyoto-u.ac.jp; fukuyama@nucleng.kyoto-u.ac.jp

T. Yao (ed.), *Zero-Carbon Energy Kyoto 2009*, Green Energy and Technology, DOI 10.1007/978-4-431-99779-5_52, © Springer 2010

electric field. The dielectric tensor component TASK/DP calculates the plasma dielectric tensor by numerically integrating the momentum distribution functions. By repeating this cycle, we can describe the time evolution or the steady state of the wave heating and current drive. We have carried out numerical analysis of ICRF heating and confirmed that the modification of momentum distribution function from Maxwellian affects the deposition to ions.

2 Model of Fokker–Planck Analysis

In this section, we briefly describe the Fokker–Planck component, TASK/FP, which solves the bounce averaged Fokker–Planck equation

$$\frac{\partial f_s}{\partial t} = E(f_s) + C(f_s) + Q(f_s) + L(f_s) + S(f_s) \tag{1}$$

where the terms $E, C, Q, L,$ and S are acceleration term by the toroidal DC electric field, collision term due to Coulomb collision, quasi-linear diffusion term due to wave–particle interaction, spatial diffusion term and source term, respectively. f_s denotes the momentum distribution function for species s, $f_s(p_{\parallel}, p_{\perp}, \rho, t)$ where $p_{\parallel}$ and $p_{\perp}$ are the parallel and perpendicular momentum at the minimum magnetic field point on the magnetic surface and ρ is the normalized minor radius of the surface. L and S terms are not included in the present analysis. FP component includes several models which describe collision and quasi-linear diffusion coefficients [2–4]. FP component includes the trapped particle effect by bounce averaging with zero banana width. This component can calculate time evolution of distribution functions not only for mainly heated species but also for the other species all together.

3 Calculation Results

We studied the case of ICRF heating of plasmas composed of two ion species, majority deuteron 95% and minority proton 5%. In this analysis, the total heating power is 0.77 MW, and the wave absorption power to electron, deuteron and proton are 0.01, 0.05, and 0.71 MW.

In this plasma, minority ion is mainly accelerated by the ion cyclotron fundamental resonance. Figure 1a shows radial profile of wave absorption by protons, and Fig. 1b shows the profile of the wave deposition power density to protons on the poloidal cross section. Most of the wave power is absorbed by the minority ions on fundamental cyclotron resonance surface.

Then FP use wave electric field which is obtained by WM to calculate time evolution of distribution function. Radial profiles of wave power absorption and

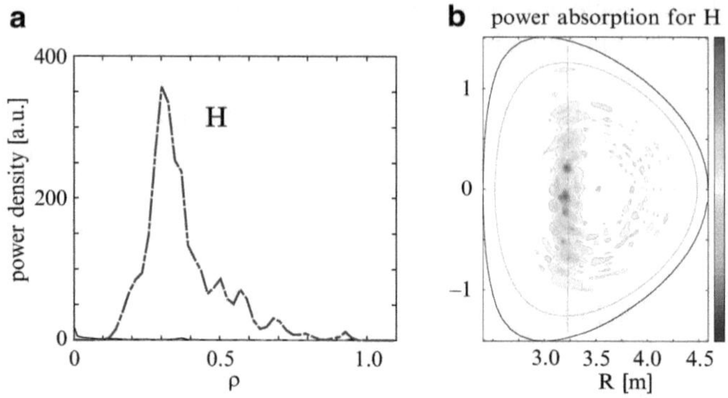

Fig. 1 (**a**) Radial profile of wave absorption calculated by WM. (**b**) Wave absorption profile for proton on the poloidal cross section

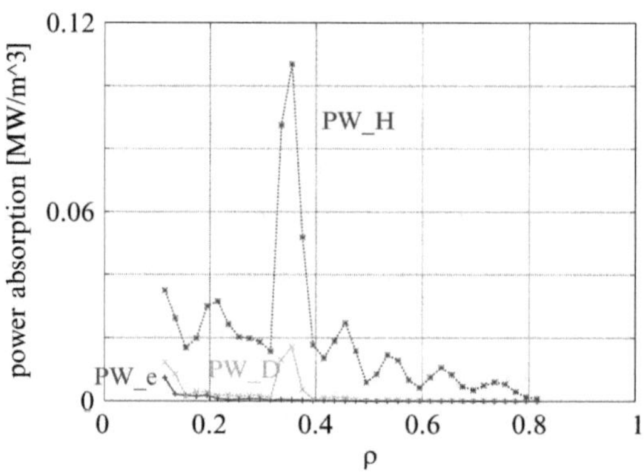

Fig. 2 Radial profile of the wave power absorption

power transfer through collision at 5 ms after the onset of wave heating are shown in Figs. 2 and 3. From these figures, we confirmed that minority ions are heated by ICRF waves, while majority ions are heated by collisions with minority ions. The collisional power gain of deuteron almost balances with the collisional power loss of proton. On the other hand, electrons are rarely heated by waves and collision except at $\rho = 0.35$ where electrons are slightly heated by collisions with minority ions. From Figs. 2 and 3, we see that the power absorption from wave is greater than the power loss due to collisional power transfer to the minority ions. The difference of the power is attributed to the increase of the tail ions because the tail formation is not saturated yet.

Figures 4 and 5 are contour of minority ion distribution function at the normalized minor radius $\rho = 0.35$ and $\rho = 0.55$ in 2D momentum space. Minority ions are heated strongly at $\rho = 0.35$ as depicted in Fig. 2. Contour of the distribution function in Fig. 4 has two tips. These strongly accelerated pitch angles are the result

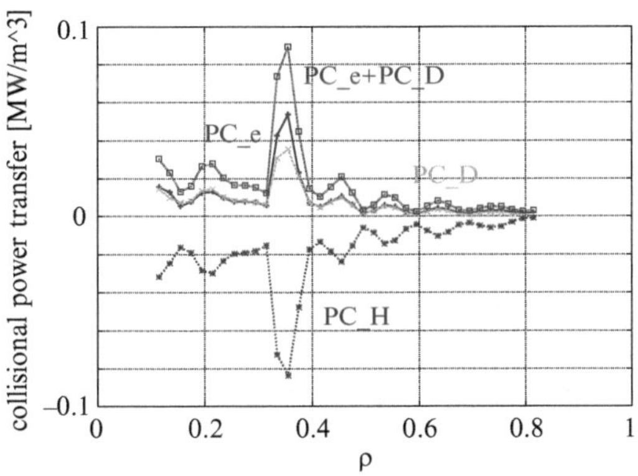

Fig. 3 Radial profile of the power transfer through collision

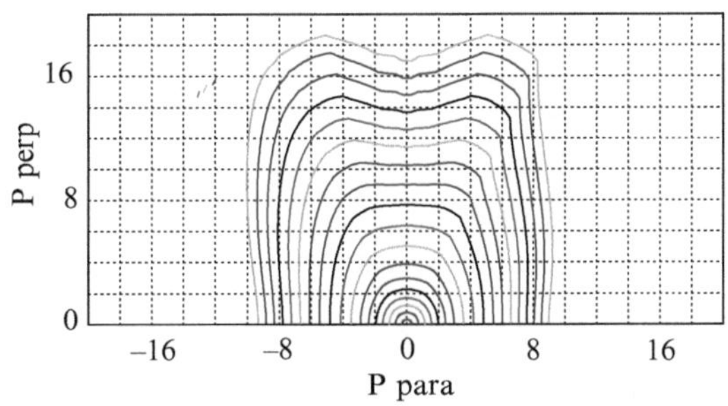

Fig. 4 Contour of distribution function of proton at $\rho = 0.35$ in 2D momentum space at 250 ms after the onset of wave

of trapped particle effect. When reflecting point of a trapped particle coincide with a cyclotron resonance surface, this strong acceleration occurs, because such trapped particles stay at the resonance surface longer than passing particles. The pitch angle is $\pi/2$ when the resonant surface is tangential to the magnetic surface ($\rho\sim 0.3$). In the present calculation, however, since the wave amplitude on the resonance surface is larger on the outer magnetic surface ($\rho = 0.35$), the accelerated pitch angle deviates from $\pi/2$.

In Figs. 6 and 7, time evolution of collision power transfer among three species at different radial position are showed. Wave accelerated protons first mainly collide with deuterons. Then as the tail of distribution function develops, however, collisional power transfer shifts from deuteron to electron. At weak absorption region $0.5 < \rho < 0.6$, distortion of distribution function is also weak (Fig. 5). Therefore, at the weak absorption region, development of collisional power transfer to electron is delayed.

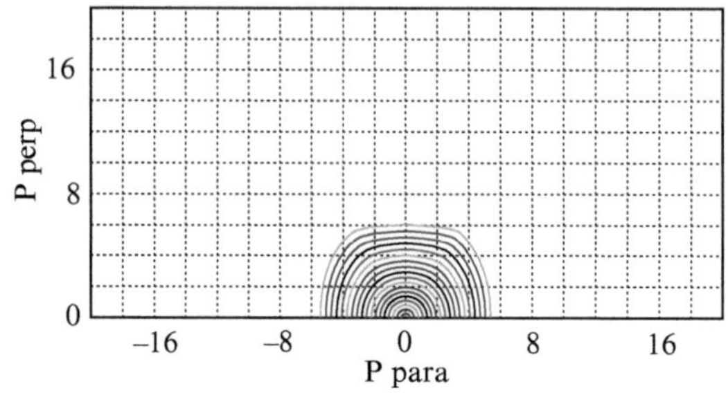

Fig. 5 Contour of distribution function of proton at $\rho = 0.55$ in 2D momentum space at 250 ms after the onset of wave heating

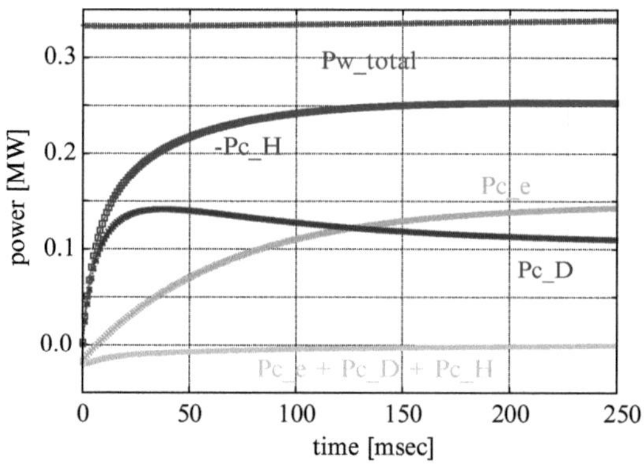

Fig. 6 Time evolution of collisional power transfer at $0.3 < \rho < 0.4$

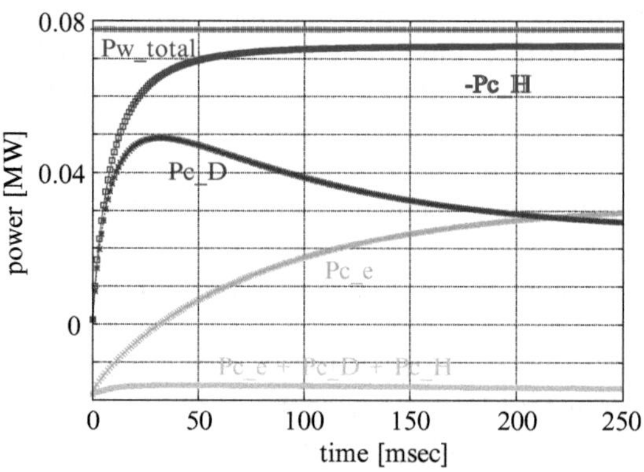

Fig. 7 Time evolution of collisional power transfer at $0.5 < \rho < 0.6$

4 Conclusion

We have updated the wave related components in the TASK code for self-consistent analysis of ICRF heating in tokamak. The present numerical analysis describes the tail formation of the resonant ion momentum distribution function. We confirmed that deformation of distribution function occurs when the plasma is heated by waves. Self-consistent analysis required for quantitative analysis of wave heating and current drive is under way.

Acknowledgements This work was supported in part by the Grant-in-Aid for scientific research, No. 20226017, from Japan Society for the Promotion of Science (JSPS).

References

1. Fukuyama A et al (2004) In: Proc. of 20th IAEA FEC, Villamoura, 2004, IAEA-CSP-25/CD/TH/P2-3
2. Karney CFF (1986) Comput Phys Rep 4:183–244
3. Killeen J, Kerbel GD, McCoy MG, Mirin AA (1986) Computational methods for kinetic models of magnetically confined plasmas. Springer, Berlin
4. Braams BJ, Karney CFF (1989) Phys Fluid B 1:1355

Electrochemical Study of Neodymium Ions in Molten Chlorides

Kazuhito Fukasawa, Akihiro Uehara, Takayuki Nagai, Toshiyuki Fujii, and Hajimu Yamana

Abstract Redox behavior of Nd in LiCl–$CaCl_2$ and LiCl–KCl was studied by cyclic voltammetry (CV). The diffusion coefficient of Nd^{3+} and formal potentials of $Nd^{3+}|Nd^{2+}$ and $Nd^{2+}|Nd^0$ couples were determined at 823 K or 923 K. The reduction wave of $Nd^{3+}|Nd^{2+}$ in LiCl–$CaCl_2$ appeared to be more positive than that in LiCl–KCl.

Keywords Neodymium • Redox potential • LiCl–$CaCl_2$ • LiCl–KCl

1 Introduction

Industrial activity using fossil fuels produces CO_2 which eventually results in global warming. Nuclear power generation has an important role in achievement of a low-carbon society with reduced CO_2 emissions. On the other hand, the successful management of nuclear spent fuels is a key concern for the future stable utilization of nuclear power. Pyrochemical reprocessing has been proposed as an alternative to the currently used aqueous process. In the pyrochemical process, spent fuel is dissolved in molten salt, and fuel materials are separated from fission products by the electro-winning method. Thus, the electrochemical properties of actinides and fission products in molten salts are essential.

Since Nd affects the fuel performance as a strong neutron absorber, it must be removed from fuel materials before re-use. However, the electrochemical behavior of Nd is complicated due to the presence of divalent state, which is unique in the lanthanide series. In order to improve the electrochemical separation of Nd, its electrochemical characteristics should be well known, but past studies on Nd have

K. Fukasawa (✉)
Graduate School of Engineering, Kyoto University, Kyoto, Japan

A. Uehara, T. Fujii, and H. Yamana
Research Reactor Institute, Kyoto University, Osaka, Japan

T. Nagai
Nuclear Fuel Cycle Engineering Lab, Japan Atomic Energy Agency, Ibaraki, Japan

T. Yao (ed.), *Zero-Carbon Energy Kyoto 2009*, Green Energy and Technology,
DOI 10.1007/978-4-431-99779-5_53, © Springer 2010

been limited to LiCl–KCl [1, 2] and NaCl–$CaCl_2$ [2] eutectic melts. In this context, we studied the electrochemical properties of Nd ions in LiCl–$CaCl_2$ to investigate the effect of Ca ions in the melt.

2 Experimental

Anhydrous chlorides used are products of AAPL (99.99% purity). All experiments were carried out under an argon atmosphere ([O_2] and [H_2O] < 1 ppm). 0.2–1.0 mol% $NdCl_3$ was dissolved in the LiCl–KCl or LiCl–$CaCl_2$ eutectic melt in a quartz tube of 13 mm inner diameter at 723 K or 823 K. A W wire (ϕ 1.0 mm), a pyrographite rod (ϕ 3.0 mm), and a Ag/Ag^+ electrode [3] were used for the working, counter and reference electrodes, respectively. In every measurement, the potential of chlorine gas evolution was determined, and this was used to calibrate the Ag/Ag^+ electrode.

3 Result and Discussion

3.1 Reduction Process of Nd^{3+}

The typical cyclic voltammograms are shown in Figs. 1 and 2. The current peak I_1 and I_1' corresponds to the $Nd^{3+}|Nd^{2+}$ redox couple while I_2 and I_2' to the Nd metal deposition/dissolution process. For the reduction process of peak I_2, there has been a discussion on whether it should be assigned to the reduction from Nd(III) to metal [1] or that from Nd(II) to metal [2].

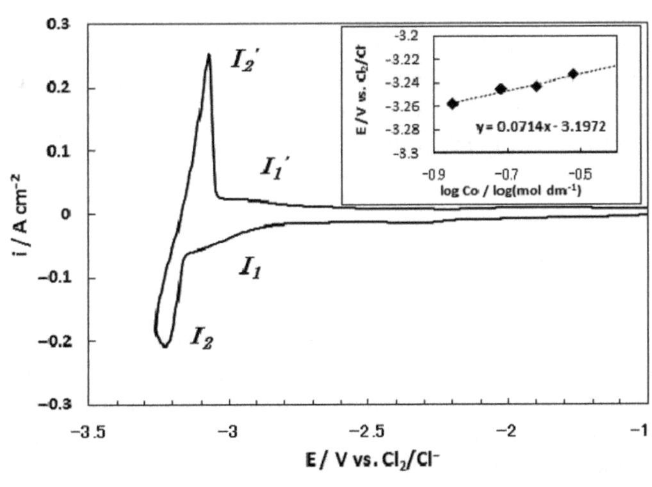

Fig. 1 Cyclic voltammogram of Nd in LiCl–KCl at 823 K. 0.47 mol% $NdCl_3$ was dissolved in the melt. Scanning rate was 0.1 mV s^{-1}

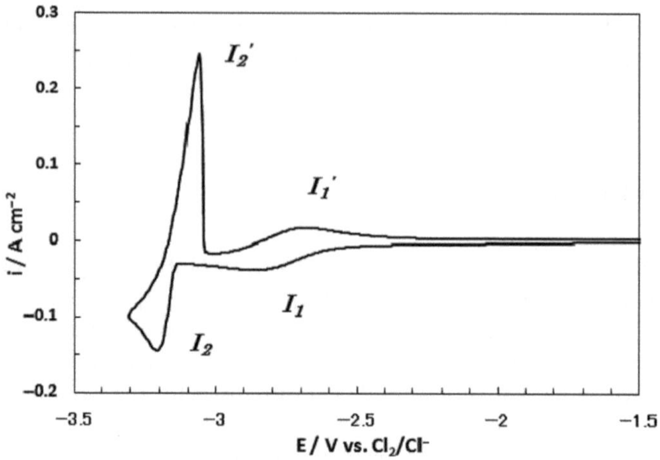

Fig. 2 Cyclic voltammogram of Nd in LiCl–$CaCl_2$ at 823 K. 0.6 mol% $NdCl_3$ was dissolved in the melt. Scanning rate was 0.1 mV s^{-1}

A theoretical description of cathode peak potential, E_{pc}, of CV for the reversible deposition of metals can be shown [4] as (1),

$$E_{pc} = E^0 + \frac{RT}{nF} \ln f_{OX} C_O - (0.9241)^2 \frac{RT}{nF}, \tag{1}$$

where R, F and T are the gas constant, the Faraday constant, the temperature, respectively. The dependence of E_{pc} on the concentration of Nd^{3+} added, C_O, was studied as shown in the box in Fig. 1. Under an assumption of activity coefficient $f_{ox} = 1$, n was evaluated to be 2.

Chronopotentiometry (CP) is an effective method to estimate n when peaks are well-separated. In the CP measurement, the ratio of the transition times, τ_1 and τ_2, gives the information on the number of the electron transfer. Our result showed seven of τ_2/τ_1 value in LiCl–$CaCl_2$, supporting $n=2$, and this agrees with the literature [5].

3.2 *Diffusion Coefficient of Nd^{3+}*

The reaction potentials of peaks I_1 and I_2 were clearly separated in the LiCl–$CaCl_2$ system. As the current peak I_1 was proportional to square root of scan rate, the reaction I_1 was recognized to be a diffusion controlled reaction. The diffusion coefficient of Nd^{3+} in this study was determined to be 1.66×10^{-5} cm^2 s^{-1} at 823 K. This was similar to the reported value in LiCl–KCl at 823 K [6], 2.30×10^{-5} cm^2 s^{-1}. We also confirmed that the temperature dependence of diffusion coefficients follow Arrhenius's law.

3.3 The Formal Potentials of Redox Couples

Determined formal potential of redox couples were listed in Table 1. The formal potential of the $Nd^{3+}|Nd^{2+}$ couple, E^{0*}(3-2), in LiCl–KCl was derived by convolution analysis of the CV curve. This value agreed with the reported values in LiCl–KCl at 723 K. For the NaCl–$CaCl_2$ and LiCl–$CaCl_2$, E^{0*}(3-2) was obtained to be the middle potential of anode and cathode peak potentials. E^{0*}(2-0) was determined by using (1). E^{0*}(3-0) was calculated from E^{0*}(3-2) and E^{0*}(2-0). E^{0*}(3-0) shifted positive in the order of LiCl–KCl, NaCl–$CaCl_2$ and LiCl–$CaCl_2$, while E^{0*}(2-0) showed similar values in all media. From the thermochemical point of view, Nd^{3+} was most unstable in LiCl–$CaCl_2$. Li^+ has the smallest ionic radius in monovalent cation series. Ca^{2+} is a divalent cation, which suggests that it has stronger polarizing power compared to K^+ and Na^+. This strong interaction among solute and solvent ions may have affected the electrochemical behavior of ions and the effect was more significant to Nd^{3+} than to Nd^{2+} in chloride melts.

Table 1 The formal potential E^{0*}/V vs. Cl_2/Cl^-

	LiCl–KCl		NaCl–$CaCl_2$	LiCl–$CaCl_2$	
T (K)	723	823	823	823	
Nd(III)/ Nd(II)	−3.082±0.010	−3.018±0.010		−2.737±0.005	[This study]
	−3.089±0.001		−2.873±0.002		[2]
	−3.098±0.010				[9]
Nd(II)/Nd(0)	−3.179±0.010	−3.127±0.010		−3.081±0.010	[This study]
	−3.206±0.003		−3.163±0.004		[21]
	−3.120±0.015				[9]
Nd(III)/ Nd(0)[a]		−3.103±0.020	−3.066±0.006	−2.966±0.015	

[a] $$E^{0*}_{Nd(\mathrm{III})/Nd(0)} = \frac{E^{0*}_{Nd(\mathrm{III})/Nd(\mathrm{II})} + 2E^{0*}_{Nd(\mathrm{II})/Nd(0)}}{3}$$

References

1. Yamana H et al (2006) J Alloys Compd 408–412:66
2. Castrillejo Y et al (2003) J Electroanal Chem 545:141
3. Nagai T, Fujii T, Shirai O, Yamana H (2004) J Nucl Sci Technol 41:690
4. Berzins T, Delahay P (1953) J Am Chem Soc 75:555
5. Bermejo R et al (2001) In: Proceedings of GLOBAL2001, Sep.13
6. Bard AJ, Faulkner LR (1980) Electrochemical methods. Wiley, New York
7. Matsumiya M, Matsumoto S, Matsuura H (2005) Electrochemistry 73(8):570
8. Kuznetsov SA et al (2000) In: Proceedings of the Workshop on Pyrochemical Separations, p295
9. Motto Y (1986) Doctoral thesis, Paris

A New Numerical Approach of Kinetic Simulation for Complex Plasma Dynamics: Application to Fusion and Astrophysical Plasmas

Kenji Imadera, Yasuaki Kishimoto, Jiquan Li, and Takayuki Utsumi

Abstract We have developed a new computational algorism based on the conservative interpolated differential operator scheme for solving the Vlasov–Poisson system. By using the developed code, we have also investigated the relation between turbulent transport and zonal flows production from the view point of entropy production. Such a self-organized system with multi-scale interaction has a generic nature for not only confined fusion plasma, but also astrophysical plasma and planet environment.

Keywords Vlasov simulation • Gyrokinetics • Entropy balance

1 Introduction

Human activity has been making an influence on the global climate through the emission of greenhouse gases, especially carbon dioxide from the burning fossil fuels such as coal, petroleum and natural gas for energy and transportation. We have been making efforts to reduce the emissions since the 1980s, however, the greenhouse gases in the atmosphere have been increasing to a level where some degree of climate change is inevitable, which lead to more serious floods and droughts. In order to keep the "sustainable development" of the world, the exploitation of alternative energy resource is crucial issue.

The nuclear fusion is one of the ideal energy resources since the fuel of the fusion is the deuterium and the lithium which can be extracted from sea water, and its reaction is safety handled without high radiation. In order to realize such a

K. Imadera (✉), Y. Kishimoto, and J. Li
Graduate School of Energy Science, Kyoto University, Gokasho, Uji, Kyoyo, 601-0011, Japan
e-mail: imadera@center.iae.kyoto-u.ac.jp; kishimoto@energy.kyoto-u.ac.jp; lijq@energy.kyoto-u.ac.jp

T. Utsumi
Department of Electronics and Computer Science, Tokyo University of Science Yamaguchi, 1-1 Daigaku-dori, Sanyou-Onoda, Yamaguchi, 756-0884, Japan
e-mail: utsumi@ed.yama.tus.ac.jp

T. Yao (ed.), *Zero-Carbon Energy Kyoto 2009*, Green Energy and Technology, DOI 10.1007/978-4-431-99779-5_54, © Springer 2010

nuclear fusion technologically and economically, a long energy confinement time for high-temperature plasma is required since the nuclei must have high energy to overcome the Coulomb barrier. Therefore, the understanding of the confinement and transport mechanism in fusion plasma is key issue and has been intensively investigated.

The study of Zonal Flows (ZFs), which mean the band-like flows such as the E \times B shear flow in fusion plasma and the zonal belt of Jupiter, has great contributions [1] for such understanding. In fact, the previous report showed the negative prospect of nuclear fusion [2], but the proper treatment of ZFs resolved such confusions, since they have strong influence on the formation of the transport barriers which can stabilize the confinement character of fusion plasma. The generation mechanism of ZFs have been widely studied and the modulational instability has been discussed as one of the plausible candidates [3]. In these previous works, the basic modulation process between ZFs and pump waves as well as their side bands were theoretically analyzed based on the Hasegawa–Mima (H–M) turbulence models. As the result, a dispersion relation of the ZFs instability was successfully derived, however, the radial profile of ZFs and its relation to global dynamics are hardly investigated in k -space.

In this paper, we have developed a new computational algorism based on the Conservative Interpolated Differential Operator (IDO-CF) scheme [4] for solving the Vlasov equation with high accuracy, which describes the complicated plasma dynamics based on the first principle. We have applied it to solve the four-dimensional (4D) gyrokinetic Vlasov–Poisson (VP) system, and investigated the relation between turbulent transport and ZFs production from the view point of entropy production. We have found that the local entropy with low velocity induces the ZFs, whereas the local entropy with high velocity triggers the heat flux and suppresses the ZFs.

The reminder of this paper is organized as follows. In Sect. 2, we briefly describe the calculation model and applied new numerical algorism. Then, we introduce the local entropy balance equation and show some numerical results focusing on the local entropy production and associated ZFs formation in Sect. 3. Finally, the concluding remarks are given in Sect. 4.

2 Calculation Model and Applied New Algorism

In this study, a physical model we have employed is based on the electrostatic global gyrokinetic model for ion in the slab configuration. The normalized basic equations are 4D gyrokinetic VP ones [5] as

$$\begin{cases} \dfrac{\partial f}{\partial t} - \dfrac{\partial \Phi}{\partial y}\dfrac{\partial f}{\partial x} + \dfrac{\partial \Phi}{\partial x}\dfrac{\partial f}{\partial y} + v_{\|}\dfrac{\partial f}{\partial z} - \dfrac{\partial \Phi}{\partial z}\dfrac{\partial f}{\partial v_{\|}} = 0 \\ -(\nabla^2 + \dfrac{\rho_{ti}^2}{\lambda_{Di}^2}\nabla_{\perp}^2)\,\Phi + \dfrac{\rho_{ti}^2}{\lambda_{De}^2}(\Phi - <\Phi>_{yz}) = \dfrac{\rho_{ti}^2}{\lambda_{Di}^2}\left(\int f dv_{\|} - 1\right) \end{cases}, \tag{1}$$

where f(t,x,y,z,{v_{||}}) is the distribution function, which independent variables describe the guiding center position, and its parallel velocity. Phi (t,x,y,z) is the electrostatic potential, ρ is the ion Larmor radius, λ and λ are the ion and electron Debye lengths, respectively. In the present case, the density is assumed to be homogeneous, and the normalized initial temperature profile is given to the *x*-direction as

$$T_i(x) = 1 - \frac{L_x}{2\pi L_T}\cos\left(\frac{2\pi x}{L_x}\right). \tag{2}$$

Here, we apply the IDO-CF scheme [4] to (1), which can capture fine-scale structures so that it can make possible to resolve the complex plasma phenomena in phase space.

3 Local Entropy Production and Associated Zonal Flows Formation

In the isolated global system without source/sink, the initial unstable profile is relaxed to a nonlinearlity saturated state. Such a transient nonlinear process is investigated from the view point of the entropy balance equation [5, 6]. Here, our concern is how the ZFs is established in such a transient phase, therefore, we have explored the local entropy balance equation from (1) as

$$\frac{\partial}{\partial t}\int\frac{\delta f^2}{2f_0}d^3Z - \int\frac{\delta f}{f_0}\frac{\partial}{\partial x}\left(\frac{\partial\Phi}{\partial y}\delta f\right)d^3Z - \int\frac{\partial\Phi}{\partial y}v_{\|}^2\delta f\frac{T'_i}{2T_i^2}d^3Z$$
$$+\int\frac{\partial\Phi}{\partial y}\delta f\frac{T_i'}{2T_i}d^3Z + \int\frac{\partial\Phi}{\partial z}\frac{v_{\|}}{T_i}\delta f d^3Z + \int\frac{\partial\Phi}{\partial z}\frac{v_{\|}}{T_i}\frac{\delta f^2}{2f_0}d^3Z = 0, \tag{3}$$

where f_0 is the Maxwellian distribution function, Δ is the perturbed one defined as Δ, T_i' is the *x*-derivative of the initial temperature profile given by (2), and $\int d^3Z = -\int_{-\infty}^{\infty}\int_0^{L_z}\int_0^{L_y} dydzdv_{\|}$. The first/second terms in (3) denote the local entropy production/convection, the third one is the heat flux, and the fifth and sixth ones are the acoustic coupling, respectively. The second one describes the vorticity flow via the gyrokinetic Poisson equation, and directly relates to the ZFs production from the H–M equation.

Here, we show some numerical results obtained by a 4D gyrokinetic full-f Vlasov simulation. In the present simulation, we consider a slab configuration with $L_x = 2L_y = 32\rho_i$, $L_z = 8{,}000\rho_i$, $L_v = 10v_{ti}$, and $L_T = 37\rho_{ti}$. The boundary condition is periodic one in real space, and fixed one to v-direction. We have employed the grid number as $(N_x, N_y, N_z, N_{v\|}) = (256,64,32,128)$.

Figure 1 shows the profile of the Lagrange derivative of local entropy perturbation (sum of first and second terms), heat flux (third term), ZFs production (fourth term) and acoustic coupling (sum of fifth and sixth terms) in (a) low velocity (ν)

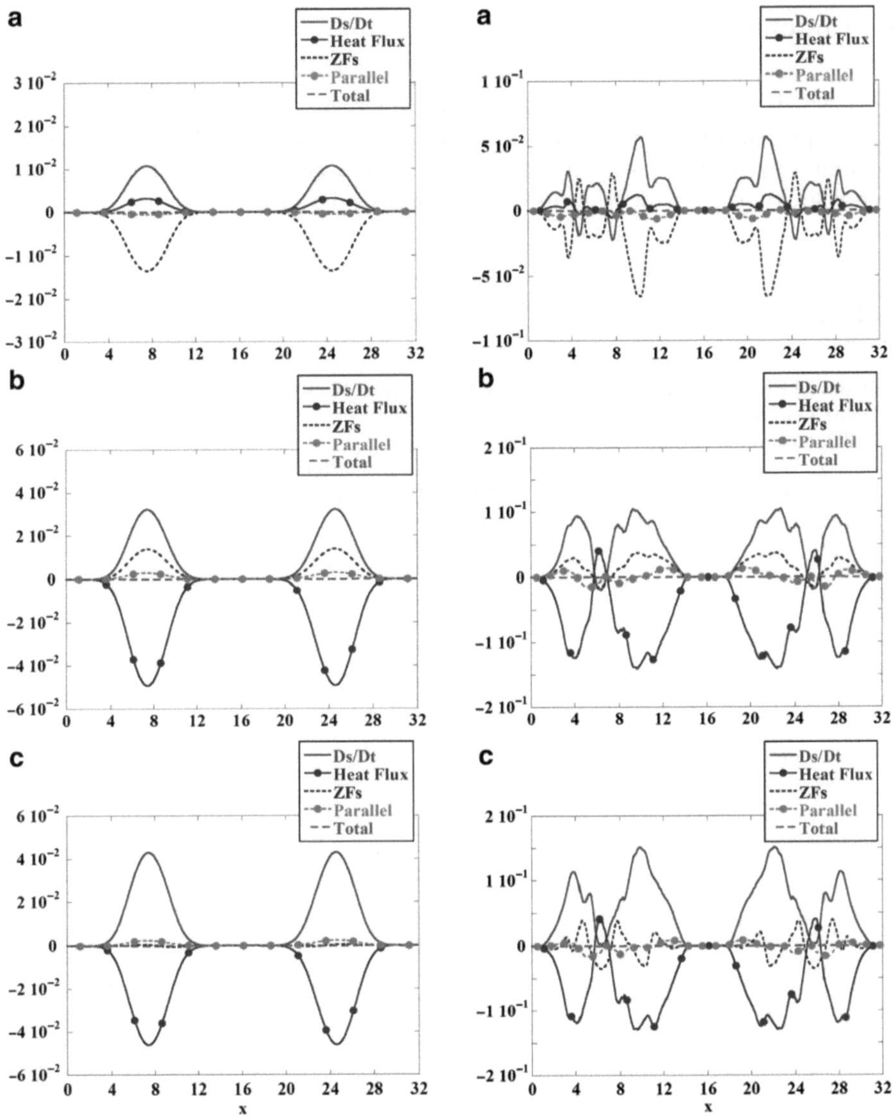

Fig. 1 The profiles of the Lagrange derivative of local entropy perturbation, heat flux, zonal flow production and acoustic coupling in (**a**) low velocity (ν), (**b**) high velocity (ν), and (**c**) whole regions are shown. The figure shows the profile at linear phase (t=20,000) (*left*) and saturation phase (t=22,000) (*right*), respectively

and (b) high velocity (ν), and (c) whole regions. It is found that the above four components balance with each other in both regions. From Fig. 1a, the local entropy with low velocity induces the ZFs, whereas the local entropy with high velocity triggers the heat flux and suppresses the ZFs, as shown in Fig. 1b.

4 Conclusion

We have developed a new computational algorism, and investigated the local entropy production and associated ZFs formation in real space. It is found that the local entropy with low velocity induces the ZFs, whereas the local entropy with high velocity triggers the heat flux and suppresses the ZFs.

References

1. Diamond PH, Itoh S-I, Itoh K, Hahm TS (2005) Plasma Phys Control Fusion 47:R35
2. Glanz J (1996) Science 74:1600
3. Li JQ, Kishimoto Y (2002) Phys Plasmas 9:1241
4. Imai Y, Aoki T, Takizawa K (2008) J Comput Phys 227:2263
5. Idomura Y, Ida M, Tokuda S, Villard L (2007) J Comput Phys 226:244
6. Watanabe T-H, Sugama H (2002) Phys Plasmas 9:3569

Relationship Between Microstructure and Mechanical Property of Transient Liquid Phase Bonded ODS Steel

Sanghoon Noh, Ryuta Kasada, and Akihiko Kimura

Abstract The oxide dispersion strengthened (ODS) Steel is one of the candidate structural materials for Gen. IV fission systems and fusion DEMO reactor because of its excellent elevated temperature strength, corrosion and radiation resistance. To fabricate these advanced nuclear systems with huge and complex structure, bonding and welding is one of unavoidable issues. Thus, suitable bonding and welding techniques need to be developed with such a process that the microstructures with very small grain size and distribution of nano-oxide particles are not remarkably changed. Transient liquid phase bonding (TLPB) is a potential joining technique to minimize the disruption of these unique microstructures.

In this study, TLPB was performed to join ODS ferritic steel blocks using an amorphous inset material based on Fe–3B–5Si compositions with high vacuum hot press. The relationship between microstructure and mechanical properties were investigated. Cross sectional microstructure in joint region were observed by electron microscopy. To evaluate mechanical properties in the joint region, tensile test was also conducted. Although the TLPB ODS steel had good tensile strength, it showed lower elongation than before the bonding.

Keywords Oxide dispersion strengthened steel • Transient liquid phase bonding

1 Introduction

Oxide dispersion strengthened (ODS) steel is one of the most candidate structural materials for advanced nuclear systems such as Gen. IV systems and fusion DEMO reactor because of its excellent elevated temperature strength, corrosion and radiation resistance [1,2]. For applications of the ODS steel to these advanced nuclear systems

S. Noh (✉)
Graduate School of Energy Science, Kyoto University, Gokasho, Uji, Kyoto, Japan

R. Kasada and A. Kimura
Institute of Advanced Energy, Kyoto University, Gokasho, Uji, Kyoto, Japan

T. Yao (ed.), *Zero-Carbon Energy Kyoto 2009*, Green Energy and Technology,
DOI 10.1007/978-4-431-99779-5_55, © Springer 2010

with a huge and complex structure, suitable bonding and welding techniques need to be developed with such a process that the microstructures with very small grain size and distribution of nano-oxide particles are not remarkably changed. [1,3,4].

The application of conventional fusion welding techniques to join ODS steels results in the disruption of fine scale-featured microstructure, and consequently a loss in the high-temperature strength [5]. Transient Liquid Phase Bonding (TLPB) is one of the potential processes which needs an insert material between two surfaces to be bonded. At the bonding temperature, melting point depressants (boron and silicon, etc.) begin to diffuse out of the liquid insert material and into the base material, resulting in isothermal solidification. Finally, the bonded region shows homogeneous concentration with the base material [6,7].

In this study, TLPB was performed to join high-Cr ODS ferritic steel blocks using an amorphous inset material based on Fe–3B–5Si compositions with high vacuum hot press. The relationship between microstructure and mechanical properties was investigated.

2 Experimental Procedure

The material used in this study is a high-Cr ferritic ODS steel (Fe(bal.)–15Cr–2W–0.2Ti–0.35Y_2O_3 in wt%) and commercial insert material (METGLAS, amorphous foil, 20 μm thickness). Their chemical compositions are listed in Table 1. The ODS steel was fabricated by mechanical alloying and hot extrusion process. Hot extrusion process was carried out at 1,150°C. Final heat treatment was 1,150°C for 1 h followed by air-cooling. The test pieces were cut into 10 mm × 10 mm × 12 mm and surfaces were wet ground with grits #2,400 and then polished with 0.25 μm diamond powders. After surface polishing, samples were stored in acetone for cleaning and degassing. Two ODS steel blocks were placed in a direction so that the joint surfaces are parallel to the extruded direction. An insert material sheet was located between two ODS steel blocks for TLP bonding. Samples were heated at 1,200°C with isotropic pressure of 25 MPa for 1h. Bonding temperature was determined based on the results of TG-DTA analysis as that showing maximum endothermic peaks. All bonds were cooled in vacuum furnace after completion of the bonding cycle.

Microstructures were observed by scanning electron microscope (SEM) with 15 kV. The sample was etched at 18 V in a solution with 5 ml perchloric acid + 95 ml ethanol for 5 min at −50°C. The wavelength dispersive X-ray spectroscopy (WDS) was also measured to analyze element distribution at the joint region. The mechanical property of the joint region was evaluated by tensile test. Figure 1 shows sampling direction and a dimension for miniaturized tensile specimen (SS-J2 type). The bonding line is located at the center of the samples. The tensile test was performed at RT at a strain rate of 0.2 mm min^{-1}. The fracture surfaces were observed by SEM.

Table 1 Composition of ODS steel and insert material

Table	Fe	C	Cr	W	Ti	Y_2O_3	B	Si
ODS steel	Bal.	< 0.02	15.0	2.0	0.20	0.35	–	–
Insert material	Bal.	–	–	–	–	–	3	5

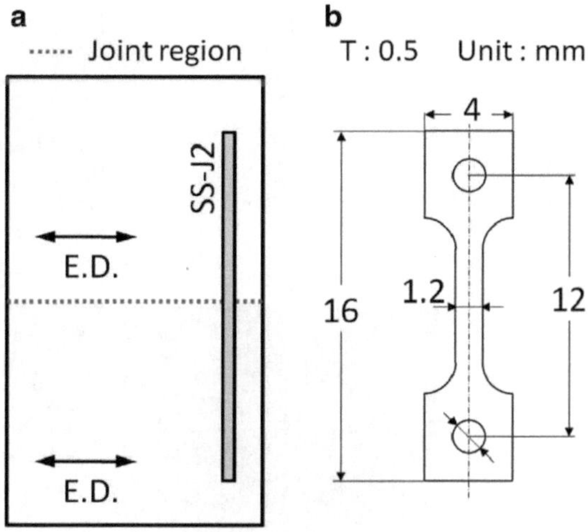

Fig. 1 Schematic of (**a**) location where tensile specimen were sampled and (**b**) its dimension

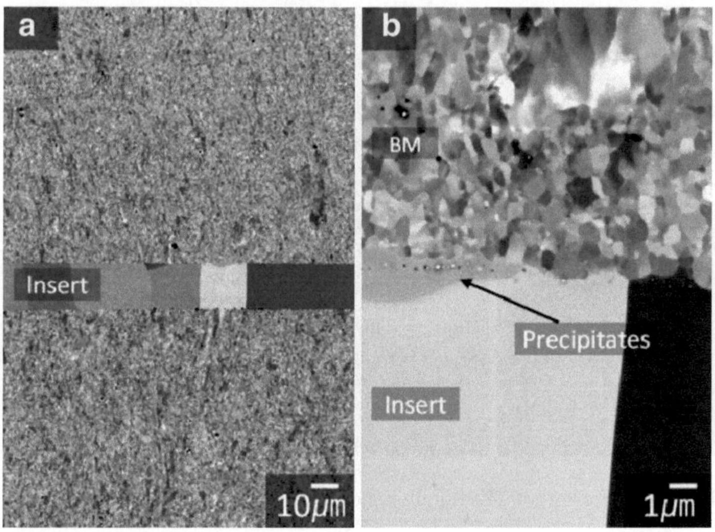

Fig. 2 SEM images of TLPB joint region on (**a**) cross section and (**b**) precipitates

3 Results and Discussion

3.1 Microstructures of the Joint Region

The cross sectional microstructures of joint region are shown in Fig. 2a,b. As shown in Fig. 2b, large grained insert up to 15 μm and some precipitates exist at the interface, although no grain growth occurred in TLPB ODS steel. WDS taken from joint

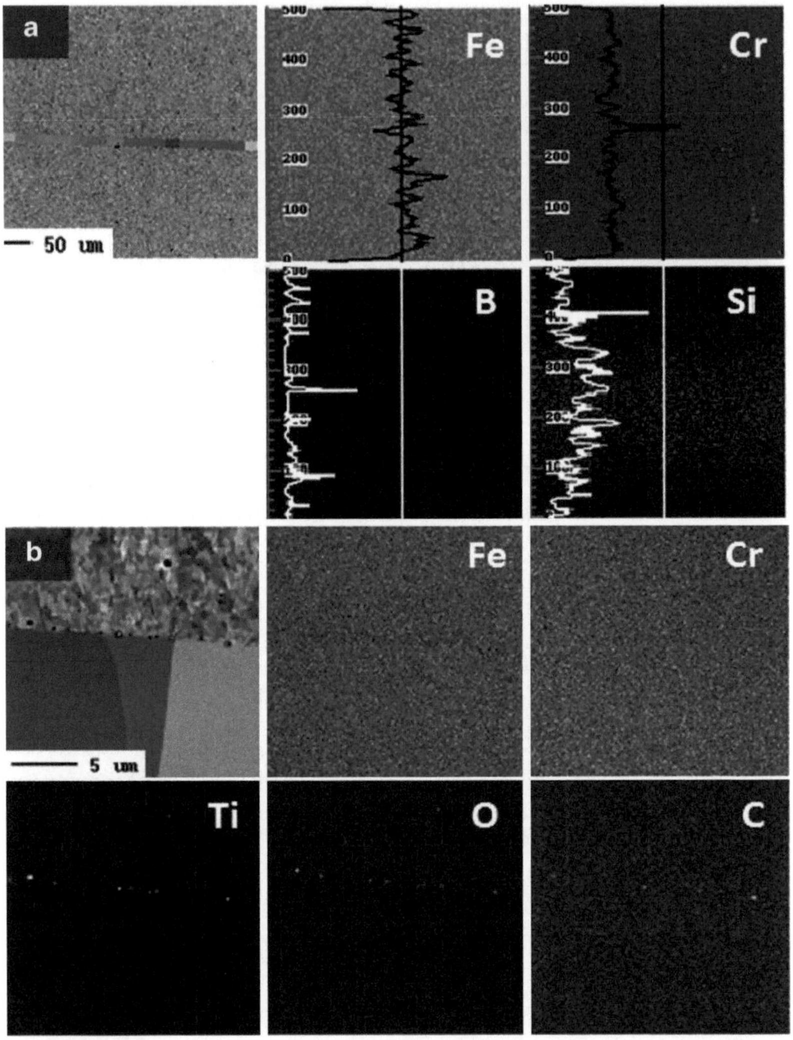

Fig. 3 WDS analysis results of TLPB joint region of (**a**) cross section and (**b**) precipitates

region shows peaks for distributions, as indicated in Fig. 3a. Remarkable element map changes were observed for Si and B, which are the melting point depressants: Si was diffused out up to 150 μm to base material during bonding process and B was distributed homogeneously because of its rapid diffusion at bonding temperature. It is interesting that chromium was also diffused in insert material. Some precipitates existed at the interface are considered to be TiC/TiO_2 based on the WDS analysis, which shown in Fig. 3b. Existence of these large grained insert material and TiC/TIO_2 precipitates may result in the negative effect on mechanical properties of the joint.

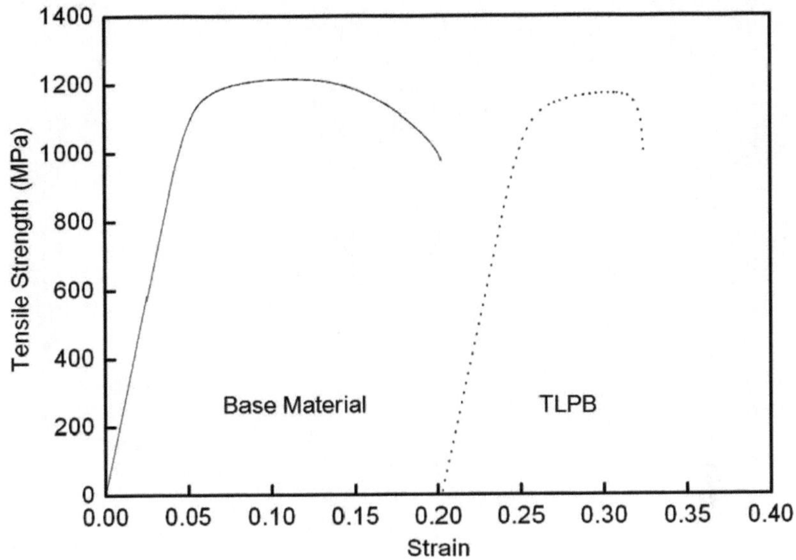

Fig. 4 Stress–strain curve of each material

Table 2 Tensile properties of joint regions

	YS (MPa)	UTS (MPa)	EL (%)
Base material	1,098	1,216	15.8
TLPB	1,008	1,170	7.8

3.2 Tensile Property of the Joint

Microstructure observation revealed that the TLPB ODS steel had TiC/TiO_2 precipitates and insert material at the bonding interface. Thus, it is necessary to investigate the effects of microstructural features developed during bonding process on the mechanical properties. Tensile tests of the joints were carried out at room temperature. Typical stress–strain curves of tensile tested base material and TLPB joint are shown in Fig. 4. The TLPB ODS steel showed a similar yield stress (YS), 1,008 MPa, which was 95% of base material. However, the reduction of total elongation (EL) was observed, which was a half of base material as summarized in Table 2. Fractographies of each tested specimen are shown in Fig. 5. In TLPB ODS steel, the failure was occurred at the interface of base material/insert material showing flat and large dimples. It seems that precipitates at bonding interface triggers micro-cracks at joint region. These micro-cracks propagate to interface boundaries of ODS steel/insert material, resulting in very flat and large dimples. This is corresponded to the result of very low area reduction in TLPB joint in Fig. 5.

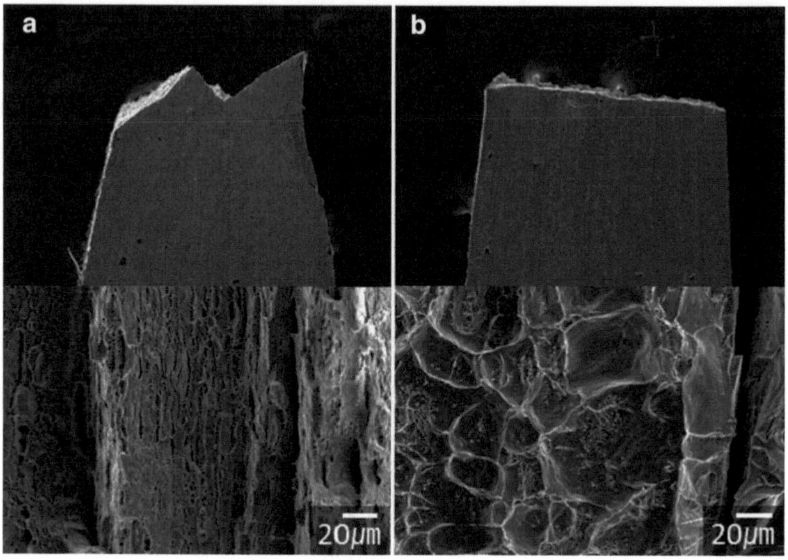

Fig. 5 Fractographs of tensile test at room temperature of (**a**) Base material, (**b**) TLPB joint region

4 Conclusion

High-Cr ODS ferritic steel was bonded by TLPB method with amorphous inset material in high vacuum isotropic hot press. Tensile tests were carried out to investigate the relationship between microstructure and mechanical property of the joints. The obtained main results are as follows:

(1) The TLPB ODS steel showed rather homogeneous distribution of the insert material elements, while at the interface TiC/TiO_2 precipitates and grain growth up to 15 μm were observed.
(2) The TLPB ODS steel had a fairly good tensile strength which was up to 95% of base material. However, it showed about half elongation of base material, which is considered to be due to the existence of TiC/TiO_2 precipitate at bonding interface.

References

1. Kimura A, Kasada R, Kohyama A, Tanigawa H, Hirose T, Shiba K, Jitsukawa S, Ohtsuka S, Ukai S, Sokolov MA, Klueh RL, Yamamoto T, Odette GR (2007) J Nucl Mater 367–370:60
2. Kimura A, Cho HS, Toda N, Kasada R, Yutani K, Kishimoto H, Iwata N, Ukai S, Fujiwara M (2007) J Nucl Sci Technol 44(3):323

3. Norajitra P, Buhler L, Fischer U, Malang S, Reimann G, Schnauder H (2002) Fusion Eng Des 61–62:449
4. Fazio C, Alamo A, Almazouzi A, De Grandis S, Gomez-Briceno D, Henry J, Malerba L, Rieth M (2009) J Nucl Mater 316–323:392
5. Shoemaker LE (1986) In: Proceedings of the International, Conference on Trends in Welding Research, Gatlinburg, ASM International, p 371
6. Khan TI, Wallach ER (1995) J Mater Sci 30:5151
7. Khan TI, Wallach ER (1996) J Mater Sci 31:2937

Nondestructive Testing of NITE-SiC Ceramics for Fusion Reactor Application

Yun-Seok Shin, Yi-Hyun Park, and Tatsuya Hinoki

Abstract Ultrasonic test (UT) method is a suitable one for examining the inherent structural defects such as pores and cracks of SiC/SiC composites. In this study, the examination sensitivity limit of UT method based on computerized tomography (C-Scan inspection) was performed by inserting an artificial defect in SiC ceramics material. The artificial defect was simulated interface defect of SiC/SiC composites. A right triangle artificial defect that has approximately 100 μm of thickness was made. The UT inspection was performed in axial and planar direction to investigate the measurement limit of width and depth. The applied transducers were focused-immersion type. The used frequencies ranges were 50 and 80 MHz. The detection limit was decided compare with results of microscope. The reflected intensity of 50% can be reliable value in detection of defects.

Keywords Nondestructive evaluation • Silicon carbide • Ultrasonic inspection • Artificial defect • Nuclear fusion

1 Introduction

The fusion energy system has been attracting as next generation power system due to global warming caused by CO_2, SO_x and NO_x gases and lack of fossil fuel supply. Silicon carbide fiber-reinforced silicon carbide matrix composites (SiC/SiC composites) have been developed one of candidate materials for fusion power device and for heat-resistance components in other advanced energy system [1], due to SiC/SiC composites have excellent high temperature properties, good corrosion resistance, low neutron absorption cross-section and stability under neutron irradiation [2–3]. It is well known that fiber reinforced brittle-matrix composites have much higher tough-

Y.-S. Shin (✉)
Graduate School of Energy Science, Kyoto University, Kyoto, Japan
e-mail: ysshin@iae.kyoto-u.ac.jp

Y.-H. Park and T. Hinoki
Institute of Advanced Energy Science, Kyoto University, Kyoto, Japan

T. Yao (ed.), *Zero-Carbon Energy Kyoto 2009*, Green Energy and Technology, 346
DOI 10.1007/978-4-431-99779-5_56, © Springer 2010

ness than brittle monolithic materials. In addition, many matrix cracks are generated prior to the final fracture of the fibrous reinforced composites [4]. The detailed analysis of failure mechanisms has shown that the fatigue lifetime of several structural components is considerably influenced by the growth rate of short cracks [5]. However, it was found that the distributed micro-damage in SiC/SiC composites was difficult to characterize by a conventional non-destructive evaluation (NDE) method, because the complex micro- and macro-structure that tends to be a masking agent for all but significant, discrete flaws [6]. Therefore, to ensure reliability and soundness of SiC/SiC composites, a NDE for them is required for both quality assurance pre-service and lifetime prediction in-service [7]. Ultrasonic test (UT) method has high sensitivity, good detecting ability and good resolution, and frequently used for NDE. C-Scan inspection as one of NDE methods is an important tool since it is able to show high resolution imaging under subsurface regions. Especially, a C-Scan image provides a two-dimensional view of a specimen, in which differences in image contrast result from object's interaction with an impinging ultrasound wave [8–9]. The purpose of this study is to employ C-Scan inspection to examine the artificial defect in monolithic SiC, to lay a foundation for application of ultrasonic C-Scan inspection in defects inspection of SiC/SiC composites.

2 Experimental Method

2.1 Specimens and Inspection Direction

The SiC specimen used in this study has been manufactured based on the Nano Infiltration Transient Eutectic Phase process (NITE process) which was developed as a processing technique for SiC/SiC composites. A ~100 μm thick carbon plate of right triangle shape was inserted in SiC to use as artificial defect. The size of specimen is $40^{L} \times 4^{W} \times 5.5^{T}\,mm^{3}$. Figure 1a,b shows the appearance of specimen and the dimension of artificial defect. The measurement directions of ultrasonic C-Scan inspection are perpendicular and vertical to the triangle as illustrated in Fig. 1c.

2.2 Apparatus and C-Scan Method

In this study, the C-Scan inspection apparatus D-View (Krautkramer Japan Co., Ltd.) was used which have high-frequency, high-resolution. It consists of computer, high frequency ultrasonic flaw detector (HIS3-HF), and driving part of scanner. Component of the C-Scan apparatus was illustrated in Fig. 2. HIS's exchangeable analog/digital modules and time resolution are 10–125 MHz probe and 5 ns, respectively. The scanning range and pitch are X: 300, Y: 300, Z: 100 mm and 0.01–9.99 mm, respectively. The transducer in this case was a longitudinal immersion 50 and 80 MHz focusing transducer, which was mounted to an *x–y* scanning

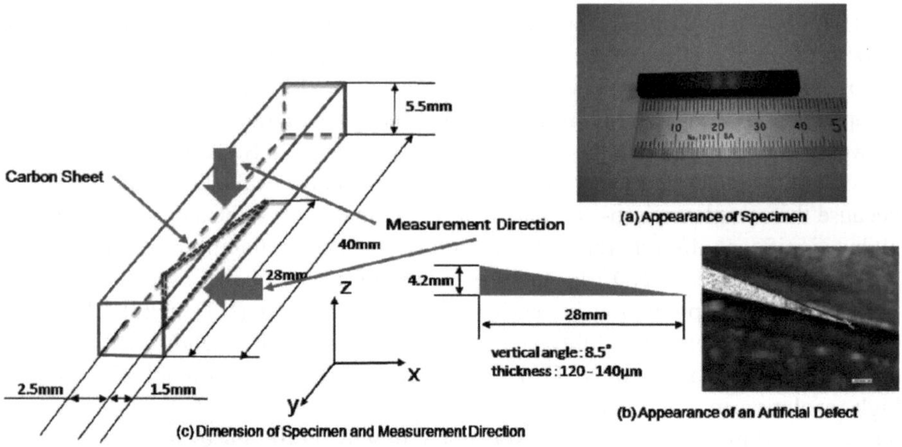

Fig. 1 Schematic illustration is shown dimension of specimen, artificial defect and measurement direction of ultrasonic C-Scan inspection. (**a**) Appearance of specimen, (**b**) appearance of artificial defect (**c**) show inspection direction by ultrasonic C-Scan method

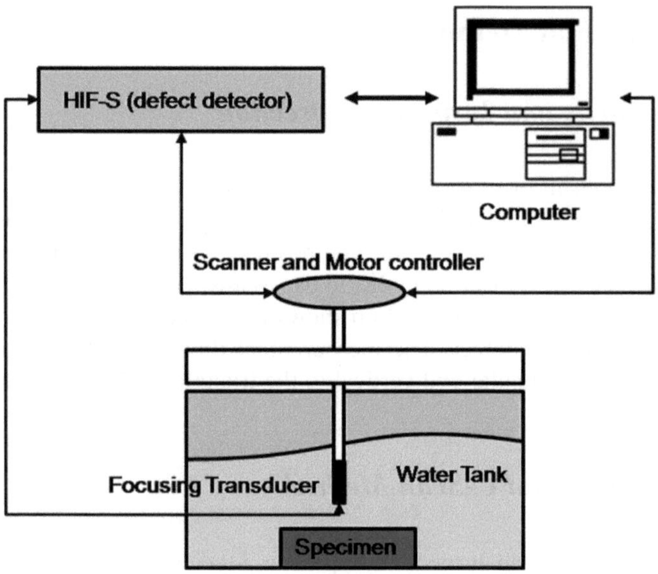

Fig. 2 A high frequency C-Scan system configured for water immersion measurements

frame, focal length are 25.0, 12.5 mm. The electronic gate is set to collect echoes corresponding to a certain depth, or perhaps an interface, the C-Scan image becomes a plan view of a slice of the specimen where the slice thickness is proportional

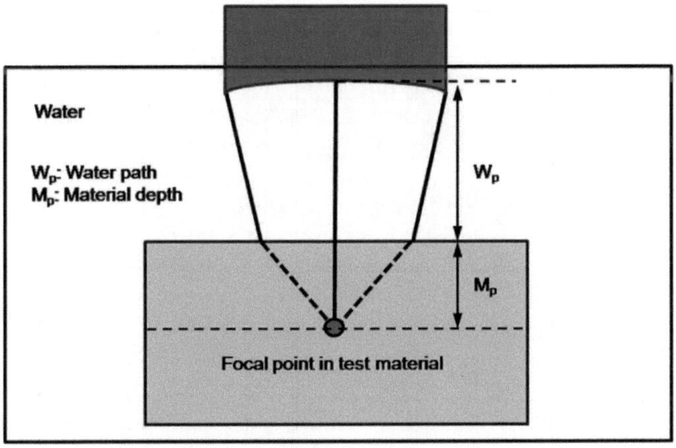

Fig. 3 Schematic illustration is shown change of focal point by Snell's Law, be caused by difference of acoustic impedance

to the width of the gate. Typically, focal point in materials is defined by Snell's law, due to difference of acoustic impedance between boundaries of media [10]. Figure 3 is the focal point by Snell's law in material which is influenced by acoustic impedance. The Snell's law is given by

$$\frac{\sin\theta_1}{c_1} = \frac{\sin\theta_2}{c_2}, \tag{1}$$

where θ_1 is angle of incidence to medium-1, θ_1 is angle of incidence to medium-2, c_1 is ultrasonic wave velocity to medium-1, c_2 is ultrasonic wave velocity to medium-2. This change in the focal length can be predicted by (2). For example, given a particular focal length and material path, this equation can be used to determine the appropriate water path to compensate for the focusing effect in the test material.

$$M_p = (F - W_p)\bullet(c_w / c_{tm}), \tag{2}$$

where, W_p, M_p, c_w, c_{tm} and F are water path, material depth, ultrasonic wave velocity in water, ultrasonic wave velocity and focal length in material, respectively. The detection capability was investigated by measurement of *x–y* plane, is displacement between the bottom side and. Also, inspection capability of depth direction was calculated by trigonometric function from the result of C-Scan on depth direction (*x–z* plane).

3 Results and Discussion

The detection limit of defect measured by C-Scan inspection method was influenced by the frequency. The C-Scan was carried out to measure artificial thickness in the *x–y* plane and width *y–z* plane of specimen as shown Fig. 1.

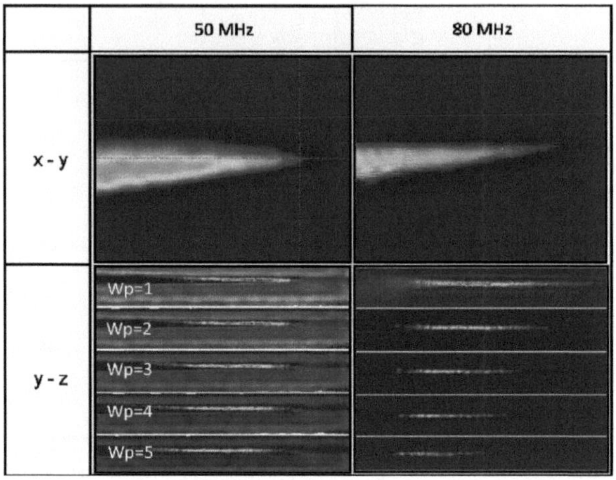

Fig. 4 Comparison of 50 and 80 MHz frequencies on *x–y* and *y–z* planes

Figure 4 shows the result of C-Scan inspection of 50 and 80 MHz that was obtained on *x–y* and *y–z* plane directions. The *x–y* plane direction was inspected the thickness of internal artificial defect according with water path (W_p) in the range of 1–5 mm. The *y–z* plane direction was investigated between a hypotenuse and a base line according to the frequencies on an artificial defect. In the case of *x–y* plane direction, by using 50 MHz, the depth detection capability was not significantly depended on water path. There is not great dependent of attenuation in water path. However, in the case of 80 MHz frequency, attenuation was greatly influenced by water path. The widths by −6 dB method (*y–z* plane) were measured with 240 and 180 μm at the frequencies of 50 and 80 MHz, respectively.

Figure 5 shows the results of the thickness of carbon plate measured by digital optical microscope and ultrasonic C-Scan inspection. The thickness measured by digital optical microscope is approximately 120–140 μm. In contrast, the thickness measured by C-Scan inspection was about 180–240 and 120–180 μm using frequencies of 50 and 80 MHz, respectively. In this result, dependence property of axial direction was not showed a sharp difference. However, there is a large difference in water path of 1 mm due to the existence of dead zone below surface, where has a overlap of reflected echo between the surface and reflecting interface.

Figure 6 shows the relationship between the measured thickness of carbon plate and reflected echo. For comparison, the thickness measured by optical microscope was also presented. As can be seen from this figure, the thickness of carbon plate measured by 50 and 80 MHz were about 140–160 and 120–160 μm, respectively, in the reflected echo intensity range of 0.5–1.0. The measured thicknesses are in the same level when the reflected echo intensity was higher than 0.5 for both cases. However, a large difference between the measured thickness and real one can be observed for the reflected echo intensity was lower than 0.5. Especially, in the case of 50 MHz with ~0.4 reflected echo intensity, the measured thickness was as high

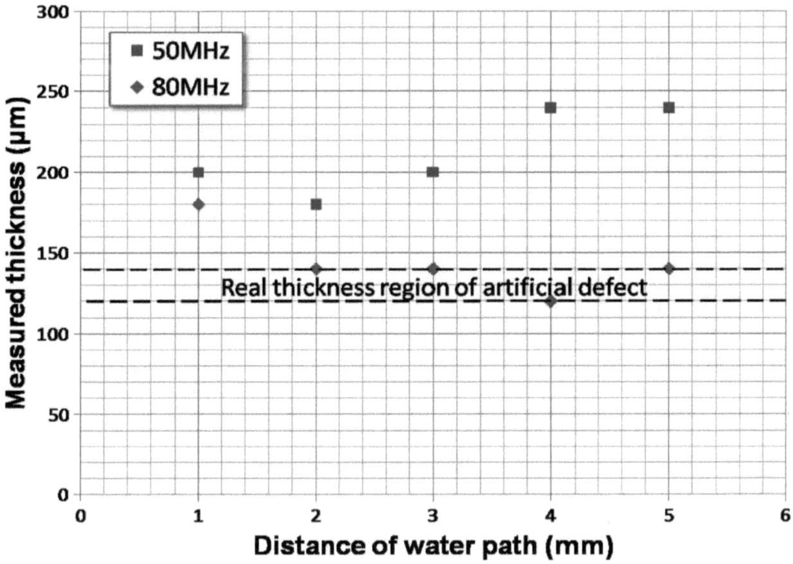

Fig. 5 The relationship between the thickness of carbon plate measured by C-Scan inspection and the distance of water path. The results were tested by 50 and 80 MHz frequencies on *x–y* plane

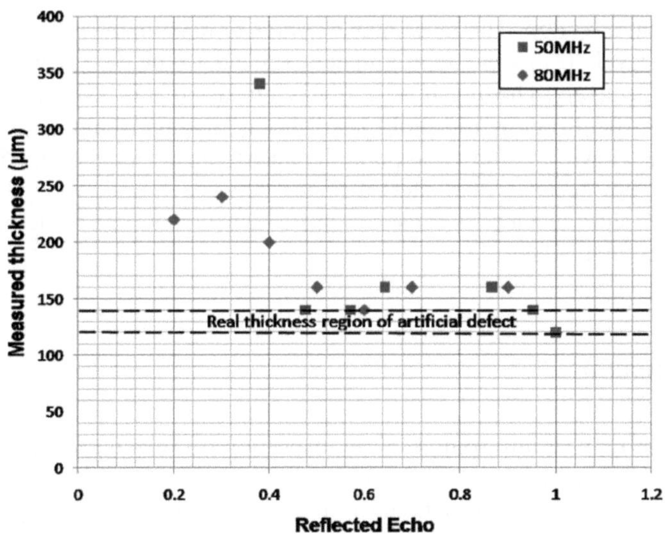

Fig. 6 The thickness of artificial defect measured by C-Scan inspection as a function of reflected echo. The results were tested by 50 and 80 MHz frequencies on *x–y* plane

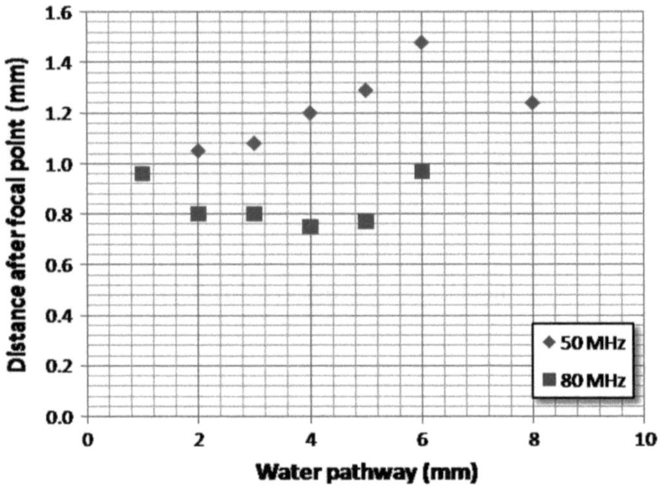

Fig. 7 The detection limit to depth direction is shown in according with water pathway

as about 340 μm. The measured large difference in attenuation can be attributed to the width of reflected echo is increased with decreasing the reflected echo amplitude by attenuation. This attenuation leads to measurement deviation on −6 dB measuring method. In conclusion, the reliable threshold level of reflecting echo amplitude could be decided to be above 0.5, approximately.

Figure 7 was shown that is relationship between water path and range of focal region in axial direction. The investigable depths of maximum (reflecting intensity of threshold level 20%) were approximately 1.5 and 1.0 on the frequencies 50 and 80 MHz, respectively. The threshold level has a means that is investigated defect or not. It was difficult below threshold level that is distinguishing defect echo and signal noises. However, after focal point the investigated depths of maximum were 0.8 and 0.6 mm because of a short distance between transducer and surface of specimens. A short distance is caused high frequency and large angle of refraction that based on Snell's law.

4 Conclusion

The detection limit of ultrasonic C-Scan inspection was evaluated by using a right triangle artificial defect. The following conclusions were drawn:

The origin of measuring deviation was the superposition between subsurface echo within dead zone and reflecting interface, and ultrasonic attenuation. The superposition brought about measuring deviation in axial direction when the subsurface was higher than 1 mm.

The relationship between the reflecting echo intensity and the measured thickness of carbon plate was investigated. On one side of the measurement precision, a reliable threshold level of reflecting echo amplitude was approximately above 0.5 to defect on direction at right angles to incident direction of ultrasonic.

The investigable depths of maximum were approximately 1.5 and 1.0 on the frequencies 50 and 80 MHz, respectively. The investigable depths of maximum were 0.8 and 0.6 mm because of a short distance between transducer and surface of specimens.

References

1. Snead et al LL (2000) Evaluation of neutron irradiated near-stoichiometric silicon carbide fiber composites. J Nucl Mater 283–287:551–555
2. Katoh Y et al (2004) SiC/SiC composites through transient eutectic-phase route for fusion applications. J Nucl Mater 329–333:587–591
3. Nozawa T et al (2004) Neutron irradiation effects on high-crystallinity and near-stoichiometry SiC fibers and their composites. J Nucl Mater 329–333:544–548
4. Okabe T et al (1999) A new fracture mechanics model for multiple matrix cracks of SiC fiber reinforced brittle-matrix composites. Acta Mater 47(17):4299–4309
5. Miller KJ (1982) The short crack problem. Fatigue Eng Mater Struct 5:223–232
6. Roth DJ, Verrilli MJ (2005) Initial attempt to characterize oxidation damage in C/Sic composite using an ultrasonic guided wave method. J Am Ceram Soc 88(8):2164–2168
7. Thompson RB et al (1976) Goals and objectives of quantitative ultrasonics. IEEE Trans Sonics Ultrasonics 23(5):292–299
8. Gordon GA et al (1993) Ultrasonic C-Scan imaging for material characterization. Ultrasonics 31(5):373–380
9. Restori M, Wright JE (1977) C-Scan ultrasonography in orbital diagnosis. Br J Ophthalmol 61:735–740
10. Schmerr LW (1998) Fundamentals of ultrasonic nondestructive evaluation a modeling approach. Springer, New York

Numerical Simulation on Subcooled Pool Boiling

Yasuo Ose and Tomoaki Kunugi

Abstract In this paper, the numerical simulations based on the MARS (Multi-interface Advection and Reconstruction Solver) with a phase-change model including the bubble growth and the condensation processes are performed. As the results, it was found that the numerical results for both bubble growth and condensation rates were very slow compared with the experimental results and the existing analytical model. In order to solve this discrepancy, the original model is improved by introducing the following models based on the quasi-thermal equilibrium state: (1) the improved phase-change model including the large density change between water and vapor; (2) a relaxation time model derived by considering the unsteady heat conduction. Resulting from the numerical simulation with the present improved model, the numerical results in both bubble growth and condensation processes show in good agreement with the experimental results and the existing analytical model.

Keywords Subcooled boiling • Phase change • Multiphase flow • Numerical simulation

1 Introduction

In order to achieve the zero CO_2 emission, the improvement of heat transfer efficiency is very important. Boiling heat transfer has the most distinguished efficiency which can be enormous heat transfer coefficient compared to the single-phase convection, and it is important phenomena in various engineering fields such as nuclear reactor, chemical plants, mechanical devices, etc. This study is to focus on a subcooled pool boiling phenomena. Since the subcooled pool boiling is occurred under the condition below saturation temperature, it is the most complex phenomena consisted of the processes of the bubble growth, the condensation and

Y. Ose (✉) and T. Kunugi
Department of Nuclear Engineering, Kyoto University, Kyoto 606-8501, Japan
e-mails: ose@nucleng.kyoto-u.ac.jp; kunugi@nucleng.kyoto-u.ac.jp

T. Yao (ed.), *Zero-Carbon Energy Kyoto 2009*, Green Energy and Technology,
DOI 10.1007/978-4-431-99779-5_57, © Springer 2010

the extinction caused by the phase-change. Since the prediction of boiling phenomena is very important, it has been proposed numerous prediction models subjected to the boiling phenomena based on the experimental data and/or theoretical considerations. However, there is no direct numerical prediction model because of the lack of the experimental databases and the theoretical considerations. In recent years, with great advances in computer, numerical simulations for directly treating the bubble dynamics regarding the nucleate boiling have been performed by several investigators [1–3]. However, they only performed the numerical simulation of the saturated pool boiling. Moreover, it is not clear whether the numerical simulations with a high accuracy interface tracking for the large bubble deformation can be applied. In this situation, Kunugi et al. [4] carried out three-dimensional pool and forced convective subcooled flow boiling phenomena by the direct numerical simulation based on the MARS [5] which is based on a high-accuracy interface volume-tracking procedure.

In this study, it is focused on the clarification of the heat transfer characteristics of the subcooled pool boiling, the discussion on its mechanism, and the establishment of a phase-change model for numerical simulation on the subcooled pool boiling phenomena. In this paper, the numerical simulations based on the MARS with an improved phase-change model for both bubble growth and condensation processes are performed, and then the results of the bubble growth and the condensation processes are compared with the experimental results and the existing analytical model.

2 Numerical Simulations

The phase-change model in the MARS is consisted of a nucleation model and the bubble growth/condensation model [4]. The bubble growth/condensation model is based on an enthalpy method (hereafter, we call the "original model"). In this study, the numerical simulations for the subcooled pool boiling are conducted as the same conditions of the visualization experiments [6]. Figure 1 shows the bubble volume variation with time in the subcooled pool boiling obtained by the experiments under

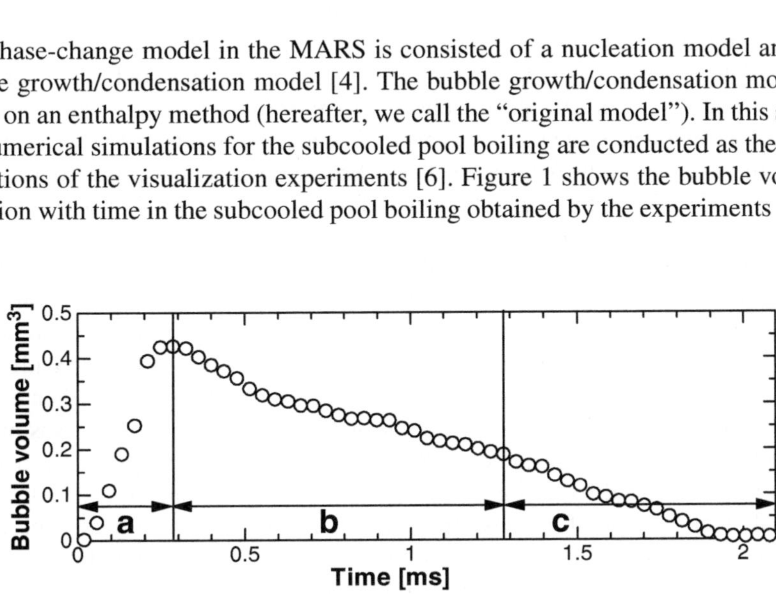

Fig. 1 Bubble volume variation with time in subcooled boiling obtained by experiments

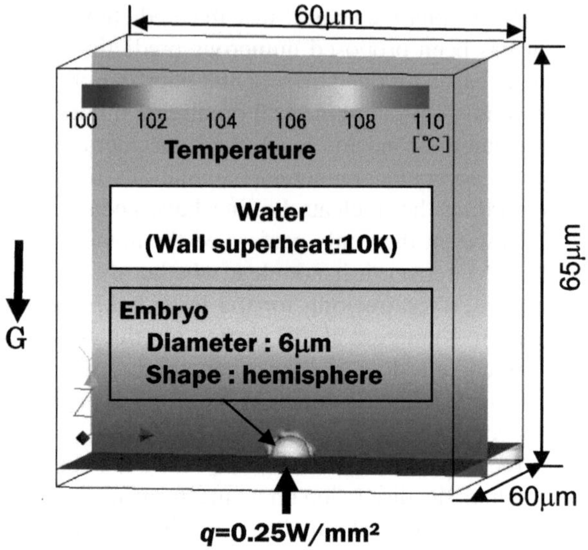

Fig. 2 Computational domain for bubble growth process

an atmospheric pressure, the degree of subcooling of 10.3 K and the wall heat flux of 0.25 W mm^{-2}. From the experimental results, it was found that the subcooled bubble behavior consisted of three processes as follows: (1) the bubble growth process with very rapid expansion of the embryo at inception boiling as shown in Fig. 1 at region a; (2) the transition process, that is that the buoyancy effect on the bubble growth process, that is the bubble expansion is balanced with the condensation bubble process as shown in Fig. 1 at region b; (3) the bubble condensation process mainly occurs after the bubble departure from the heating surface as shown in Fig. 1 at region c. Since the measures in time and space for each process are very large, the present study focuses on the bubble growth: (1) and the condensation processes: (3). The computational domain for the bubble growth process is shown in Fig. 2. In order to resolve the nucleate bubble, the grid size was chosen as 1 μm in *x*-, *y*- and *z*-directions, respectively. The periodic boundary conditions were imposed at the *x*- and *z*-directions. The non-slip wall for the velocity boundary condition was imposed to the wall, and the upper boundary condition in *y*- direction was set to a constant hydraulic pressure condition. The embryo of the vapor bubble with a critical diameter was introduced at the nearest grid of the heated surface when the wall temperature became larger than a superheated limit T_{SH}. The embryo was put as a hemisphere of shape at the center of the heated surface. The T_{SH} was set to 110°C which was estimated by using a waiting time obtained from the experiment and the analytical solution of the unsteady heat conduction, and then finally the critical diameter was obtained as about 6 μm. Figure 3 shows the computational domain and the boundary conditions for the condensation process. The uniform computational grids of 50 μm in *x*-, *y*- and *z*-directions were used. The departure bubble

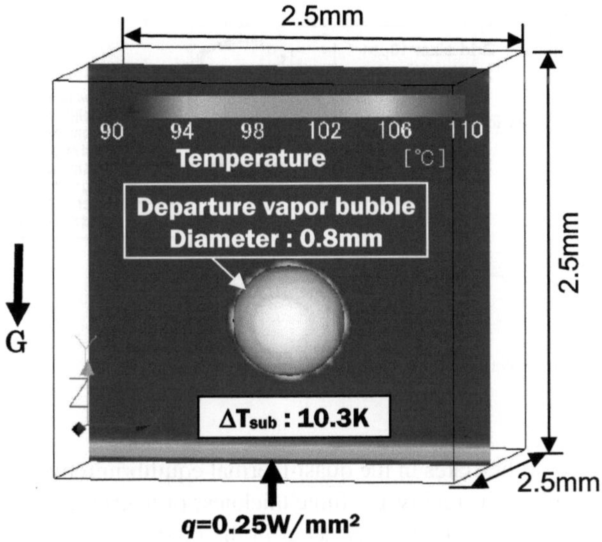

Fig. 3 Computational domain for condensation process

of 0.8 mm in diameter obtained from the experiment was used as an initial vapor bubble, and then it put as a sphere bubble in the subcooled pool. Other conditions were the same as the bubble growth process simulation.

3 Improvements of Phase-Change Model

Figures 4 and 5 show the bubble volume variation with time in the bubble growth and the condensation processes, respectively. The open circle and the broken line show the experimental results and the numerical result based on the original phase-change model, respectively. The dotted line in Fig. 4 depicts the Rayleigh's equation [7] of the existing analytical model. However, the numerical result shows very slow variation compared with the experimental results and the Rayleigh's equation. Since the original phase-change model could not treat a large volume expansion and contraction such as the water and vapor system. In this paper, the density change between water and vapor at the phase-change process is considered as the volume change by introducing the phase-change Δg_v, as follows (hereafter, we call the "improved phase-change model"):

$$\Delta g_v = (\rho_l C_{pl} \Delta T) / (\rho_g h_{lv}) \; [= \text{(Sensible heat)} / \text{(Latent heat)}]. \tag{1}$$

Here, ρ is density, C_p is specific heat at constant pressure, ΔT is degree of superheat, h_{lv} is latent heat and the suffixes of g and l denote gas phase and liquid one, respectively. Furthermore, the original model is based on the zero thickness of the interface and the rapid change of the states, i.e., a quasi-thermal equilibrium state is assumed. However,

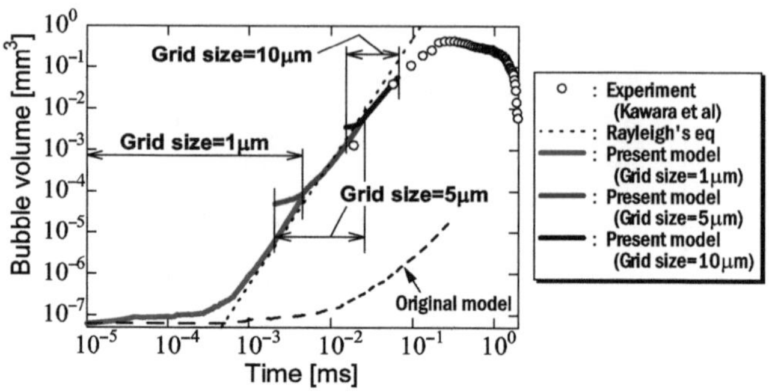

Fig. 4 Bubble volume variation with time in bubble growth process

a very slow change of the states in the quasi-thermal equilibrium hypothesis is ignored in the original model. In the reality, the finite thickness of interface exists and both slow and rapid state-changes simultaneously occur in the phase-change process. Therefore, it must consider some relaxation time for consuming the latent heat at the interface region in the phase-change process. In this sense, an effect of the unsteady heat conduction on the phase-change process in the interface region has to be considered. In this study, the relaxation time t_Δ is considered that the phase-change line/face passes through the grid width Δ, so that t_Δ can be defined by using the thermal diffusivity of the fluid α as $t_\Delta \equiv \Delta^2/\alpha$. On the other hand, in general a thermal penetration length δ is obtained: $\delta = (12\alpha t)^{1/2}$. If $t = t_\Delta$, then $\delta = (12)^{1/2}\Delta$. Therefore, the ratio of the grid width to the thermal penetration length is a constant: $\Delta/\delta = (12)^{-1/2} \cong 0.3$. This means that the phase-change volume within the relaxation time will be less than 70% of the computational cell, not 100% (*cf.* the original model is perfectly changed from one phase to the other during one time-step: very rapid process was assumed.). Therefore, the relaxation time can be introduced as a VOF limiter. For example, the relaxation time for both phase-change lines/faces is assumed to be 15%: $0.15 \leq F \leq 0.85$. In Figs. 4 and 5 for the bubble growth and the condensation processes, respectively, a solid line shows the numerical results obtained by the present improved model. In the present study, in order to further progress the numerical simulations for the bubble growth process, the patch-work calculations with changing the grid size are performed. The grid size is changed from 1 to 10 μm. The present results for both the bubble growth and the condensation processes are in good agreement with both the experimental results and the Rayleigh's equation. This suggests the present improved model may have a big potential to predict the bubble growth and the condensation processes directly.

4 Conclusions

The numerical simulations based on the MARS with the phase-change model were conducted for the bubble growth and the condensation processes. Since the numerical result based on the original phase-change model could not retrieve the experimental

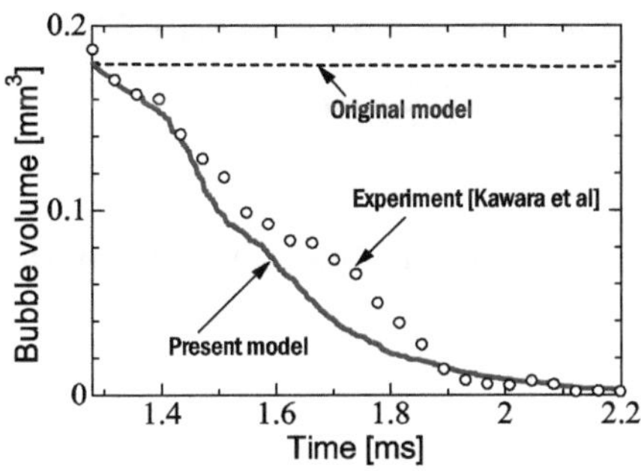

Fig. 5 Bubble volume variation with time in condensation process

data and the existing analytical models such as the Rayleigh's equation, the phase-change model was improved in this study as follows:

(1) The phase-change model was considered to the large density change between water and vapor.
(2) A relaxation time derived by considering the unsteady heat conduction is introduced as the VOF limiter for the phase-change interface.

The numerical results of both bubble growth and condensation processes based on the improved model show in good agreement with the experimental data and the Rayleigh's equation. Therefore, the present boiling and condensation model may have a big potential to predict the bubble growth and the condensation processes directly.

Acknowledgment This work was partly supported by a "Energy Science in the Age of Global Warming" of Global Center of Excellence (G-COE) program (J-051) of the Ministry of Education, Culture, Sports, Science and Technology of Japan.

References

1. Lee RC, Nydahl JE (1989) Numerical calculation of bubble growth in nucleate boiling from inception through departure. J Heat Transf 111:474–479
2. Welch SWJ (1998) Direct simulation of bubble growth. Int J Heat Mass Transf 41:1655–1666
3. Son G, Dhir VK (2008) Numerical simulation of nucleate boiling on a horizontal surface at high heat fluxes. Int J Heat Mass Transf 51:2566–2582
4. Kunugi T et al (2002) Direct numerical simulation of pool and forced convective flow boiling phenomena. Heat Transf 2002:497–502
5. Kunugi T (2001) MARS for multiphase calculation. Comput Fluid Dyn J 9:563–571
6. Kawara Z et al (2007) Visualization of behavior of subcooled boiling bubble with high time and space resolutions. PSFVIP 6:424–428
7. Rayleigh L (1917) On the pressure developed in a liquid during the collapse of a spherical cavity. Phil Mag 34:94–98

Framework of a Risk Monitor System for Nuclear Power Plant

Ming Yang, Jiande Zhang, Zhijian Zhang, Hidekazu Yoshikawa, and Morten Lind

Abstract This paper presents a design of a risk monitor system which will integrate several system modelling methods including Event Tree (ET), Multilevel Flow Models (MFM), Goal Tree and Success Tree (GTST), Fault Tree (FT)/Event Tree (ET) and GO-FLOW into one unified framework. All these modelling methods can complete each other and the proposed risk monitor system is therefore with the characters of easy modelling, modifying, validating and understanding.

Keywords Probabilistic risk assessment • Risk monitor • Operator support • Nuclear power plant

1 Introduction

Risk monitor is a plant-specific real-time analysis tool used to determine the instantaneous risk based on the actual status of the systems and components. At any given time, it reflects the current plant configuration in terms of the known status of the various systems and/or components. From the viewpoint of providing effective operator support, the methods used in a risk monitor must be efficient enough to detect or predict: (1) the occurrence of the abnormal events regarding to the operating status of the systems, equipments and even components, as a condition monitor, (2) the frequencies of occurrence of unexpected top events in plant units, as a forecasting analyzer, (3) the reliabilities of the systems, equipments and human actions completing their defined tasks within a specified time and under specified working conditions, as a reliability analyzer, and (4) the propagation of abnormal events through their safety systems, as an accident closeness analyzer. In addition, user friendly interface is also very important

M. Yang (✉), J. Zhang, Z. Zhang, H. Yoshikawa, and M. Lind
College of Nuclear Science and Technology, Harbin Engineering University, Harbin, China

M. Lind
Technical University of Denmark, Kongens Lyngby, Denmark

T. Yao (ed.), *Zero-Carbon Energy Kyoto 2009*, Green Energy and Technology, 360
DOI 10.1007/978-4-431-99779-5_58, © Springer 2010

that system information can be indicated in a clear and convincing way, which would be helpful to an overall understanding of the system problems.

This paper presents an unified framework based on Multilevel Flow Models (MFM) [1]. The proposed framework can automatically realized mapping MFM models into Fault Tree (FT), Event Tree (ET) and GO-FLOW models, therefore, the risk monitor system developed by the proposed framework can be easily understood, built, modified and verified. By utilizing the merits of each methods, the proposed framework can provide a comprehensive solution to the design and implement of risk monitor system for nuclear power plant.

2 Multilevel Flow Models

Multilevel Flow Models is a functional system modeling method based on means–end and part–whole conceptions. Along the means–end dimension MFM represents a system in terms of goals, functions and components each of which can be described at different levels of part–whole decomposition. Goal means the objectives or purpose that the system or the sub-system is designed or constructed to achieve. Component is what the system or the equipment consists of. Function is the means by which the physical component will achieve the goal. There are several kinds of relations between goal, function and component: realize relation, achieve relation, and condition relation. A realize relation affiliates physical component to function by emphasizing that a physical component is used to realize a specific function. An achieve relation connects a group of functions to a goal by emphasizing that these functions are used to obtain a specific goal. A condition relation connects a goal to a function by stressing that the goal must be achieved in order to realize the function. MFM describes and handles character and behavior of the process system with a set of interrelated flow structure, where the hierarchical structure is constructed by using both achieve relation and condition relation. Usually, there are three kinds of flow structures, i.e., mass flow structure, energy flow structure, and information flow structure. The symbols of functions and relations of MFM are shown in Fig. 1.

MFM not only can describe large and complex process in simple, plant-wide concepts, but also can develop selected parts of the process in greater detail.

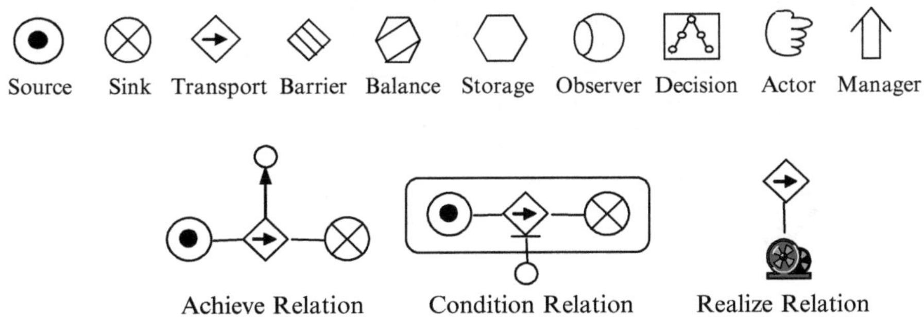

Fig. 1 Symbols of MFM relations

The explicit description of goals and functions as well as the mean–ends structure provides an easy understanding way of systems to operators. In addition, since MFM describes the relationships between physical component, function and goal, it is easier to combine with other system modeling methods such as fault tree, event tree and GO-FLOW, to provide more powerful means for the development of risk monitor.

3 Framework of Risk Monitor System

Based on above requirements for a risk monitor, this paper proposes a unified framework monitoring by integrating several basic system modeling methodologies including Fault Tree (FT)/Event Tree (ET) [2], Goal Tree-Success Tree (GTST) [3] and GO-FLOW [4] into one unified framework based on MFM. The proposed risk monitor system is shown in Fig. 2. A sequence-of-interest ET describes safety-related goals required to be achieved to prevent severe core damage following a postulated initiating event. A GTST model describes how a safety-related goal is achieved, at a high level, using a hierarchical goal–subgoal structure. MFM describe how a sub-goal is achieved, at a deeper level, by decomposing a sub-goal into several connected flow functions. Fault Tree models describe how unexpected component failures will result in the goals unavailable. GO-FLOW models are used for solving the dynamic probabilistic problems qualitatively.

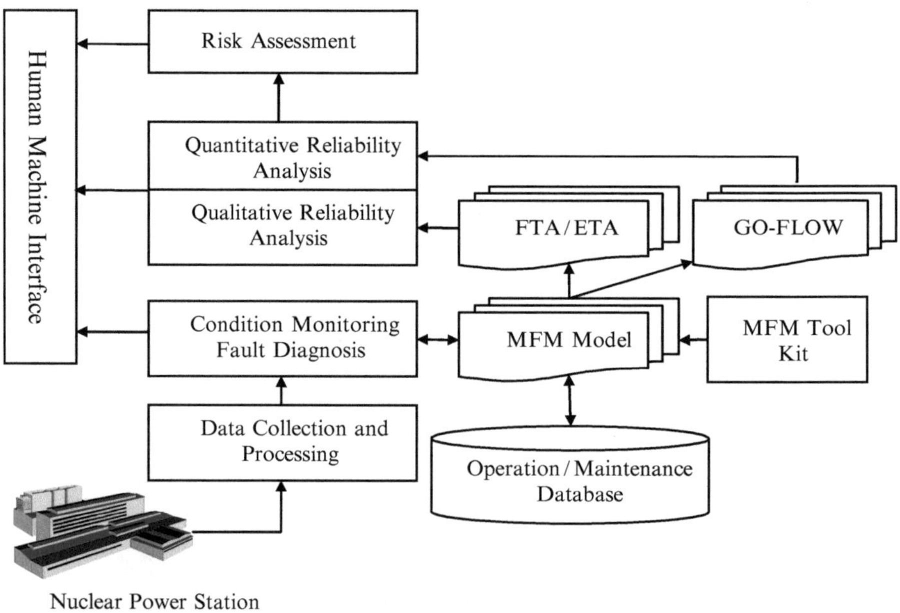

Fig. 2 Structure of risk monitor system for nuclear power plant

The authors are now studying the methods of automatically generating fault tree, event tree and GO-FLOW models from MFM models and developing MFM Model Development Tool Kit. By this way, the effort for system modeling could be greatly reduced and the proposed framework can provide a comprehensive solution to the tasks of risk monitor for nuclear power plant. The proposed risk monitor system can not only monitor the risk level changes due to component aging, failures, test, maintenance and system operation condition changes quantitatively, but also find out the weaknesses in the system, and providing an explicit understanding of how to avoid it, which would be very helpful for operator's decision-making especially in emergency cases.

4 Conclusions

This paper presents a design of a risk monitor system which will integrate several system modelling methods including Event Tree (ET), Multilevel Flow Models (MFM), Goal Tree and Success Tree (GTST), Fault Tree (FT)/Event Tree (ET) and GO-FLOW into one unified framework. All these modelling methods can complete each other and the proposed risk monitor system is therefore with the characters of easy modelling, modifying, validating and understanding.

Acknowledgments This study is supported by National Natural Science Foundation (NFSC) of China (Grant No. 60604036), Heilongjiang Provincial Foundation for Returned Scholars (Grant No. LC06C06) and the 111 project (Grant No. b08047).

References

1. Lind M (1994) Modeling goals and functions of complex industrial plants. Appl Artif Intell 8:259–283
2. Roberts NH, Vesely WE, Hassal DF et al (1981) Fault tree handbook, NUREG-0492. US NRC, Washington, DC
3. Birky GJ, McAvoy TJ et al (1988) An expert system for distillation control design. Comput Chem Eng 12:1045–1063
4. Matsuoka T, Kobayashi M. (1991) GO-FLOW: a system reliability analysis methodology. Elsevier, Amsterdam, pp 525–532

Dynamic Reliability Analysis by GO-FLOW for ECCS System of PWR Nuclear Power Plant

Ming Yang, Zhijian Zhang, Hidekazu Yoshikawa, and Shengyuan Yan

Abstract This paper applies GO-FLOW methodology to analyze the dynamic reliability characters of the Emergency Core Cooling System (ECCS) of Daya Bay Nuclear Power Plant during Loss Of Coolant Accident (LOCA). Comparing the analysis results with Fault Tree Analysis (FTA) shows a good consistency which proves that the GO-FLOW methodology has a good application future in system dynamic behaviour analysis.

Keywords System reliability analysis • GO-FLOW methodology • Emergency Core Cooling System • Nuclear power plant

1 Introduction

Emergency Core Cooling System (ECCS), an important Engineered Safeguards Features (ESF) for cooling the reactor core and maintaining the integrity of fuel shielding under abnormal conditions, shows great significance to the safe operation of nuclear power plant. The ECCS of Daya Bay Nuclear Power Station consists of three interconnected subsystems including High Head Safety Injection (HHSI) system, Middle Head Safety Injection (MHSI) system and Low Head Safety Injection (LHSI) system. If Loss Of Coolant Accident (LOCA) happens, the pressure decrease of Reactor Coolant System (RCS) will result in reactor shutdown and the three subsystems will be put into operation respectively according to different pressure situations.

The responses of ECCS after LOCA are summarized as follows [1]:

1. Coldleg Direct Injection Mode (Phase 1). LHSI and HHSI systems inject cooling water into the Coldleg of RCS.
2. Coldleg Recirculation Mode (Phase 2). If the refuelling water tank is empty, LHSI system will pump the cooling water from sump to the HHSI pump. After

M. Yang (✉), Z. Zhang, H. Yoshikawa, and S. Yan
College of Nuclear Science and Technology, Harbin Engineering University, Harbin, China

T. Yao (ed.), *Zero-Carbon Energy Kyoto 2009*, Green Energy and Technology, DOI 10.1007/978-4-431-99779-5_59, © Springer 2010

being pressurized by HHSI pump, the cooling water will be also injected into the Coldleg of RCS.

3. Coldleg and Hotleg Simultaneous Injection Mode (Phase 3). The major injection flux is provided by the HHSI system through the Hotleg. The Coldleg injection is provided by the LHSI system only through the bypass valve.
4. Long Term Recirculation Mode (Phase 4). The LHSI pump draws cooling water from sump. Both of LHSI and HHSI systems injects the water into Coldleg and Hotleg.

2 Modelling ECCS by GO-FLOW

The GO-FLOW [2] models for ECCS are shown in Fig. 1.

The system modelling is followed by some assumptions:

1. ECCS is consisted only by pumps, valves and their connections. Pipelines are considered as absolutely reliable.
2. All equipments are considered as irreparable elements.
3. All relevant systems and equipments except ECCS are in good conditions when and after accident occurs.
4. Successful state is considered as every pipeline with 100% flow rate and ignores whether other flow rates can satisfy the injection requirements.
5. Failure rate of each component is constant.

3 Reliability Analysis of ECCS Under LOCA Accident

3.1 Time Point Settings

According the responses of ECCS system under LOCA accident, 13 time points are preset, as shown in Table 1.

3.2 Reliability Analysis

The reliability data used in this study is shown in Table 2. The reliability analysis results are shown in Fig. 2.

4 Conclusions

The authors also validated above analysis results by using FTA. A good consistency between two methodology were obtained. The failure probability of every time point calculated by FTA is larger than by GO-FLOW shows that FTA is more conservative.

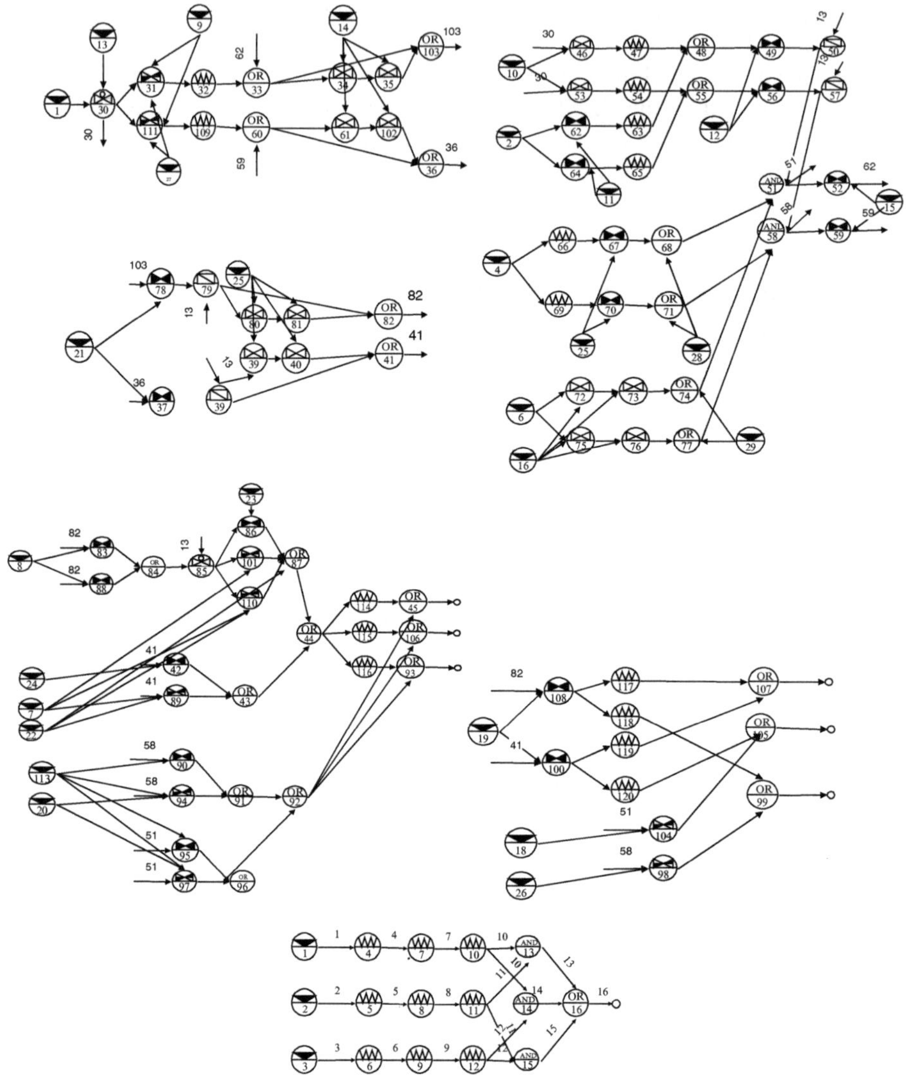

Fig. 1 GO-FLOW model for ECCS system

GO-FLOW model is quite smaller than fault tree model, suitable for solving dynamic problems, however difficult for qualitative analysis. Combining the two methodologies is therefore having a good future.

Acknowledgments The authors thank NFSC (Grant No. 60604036) and the 111 Project (Grant No. b08047). And during this study, through master Hidekazu Yoshikawa of Harbin Engineering University, Prof. Takeshi Matsuoka of Utsunomiya University and Mr. Kazuo Tamura of Itochu Technosolution Co. provided the GO-FLOW software and manual information. The authors express special thanks for their kind helps and collaborations also.

Table 1 Time point settings for ECCS under LOCA accident

Time point	Actual time	System configuration
1	0	System standby
2	0	Phase 1 begins
3	17 min	Low water level of refueling water tank
4	20 min	Phase 1 ends
5	After 20 min	Phase 2 begins
6	2 h	LHSI system in operation
7	3 h	Phase 2
8	7 h	Phase 2 ends
9	After 7 h	Phase 3 begins
10	15 h	Phase 3
11	24 h	Phase 3 ends
12	After 24 h	Phase 4 begins
13	100 h	Phase 4

Table 2 Reliability data for each component

Components	Reliability data
Electric valves	$P_g = 0.99955$
Check valves	$P_g = 0.99998$
LHSI pumps	$\lambda = 3.7 \times 10^{-5} h^{-1}$, $\mu = 0.092592\ h^{-1}$, $P_g = 0.99970$
HHSI pumps	$\lambda = 3.0 \times 10^{-5} h^{-1}$, $\mu = 0.046511\ h^{-1}$, $P_g = 0.99916$
Refueling water tank	$\lambda = 2.7 \times 10^{-8} h^{-1}$
Boron injection tank	$\lambda = 2.6\ 2.7 \times 10^{-8} h^{-1}$

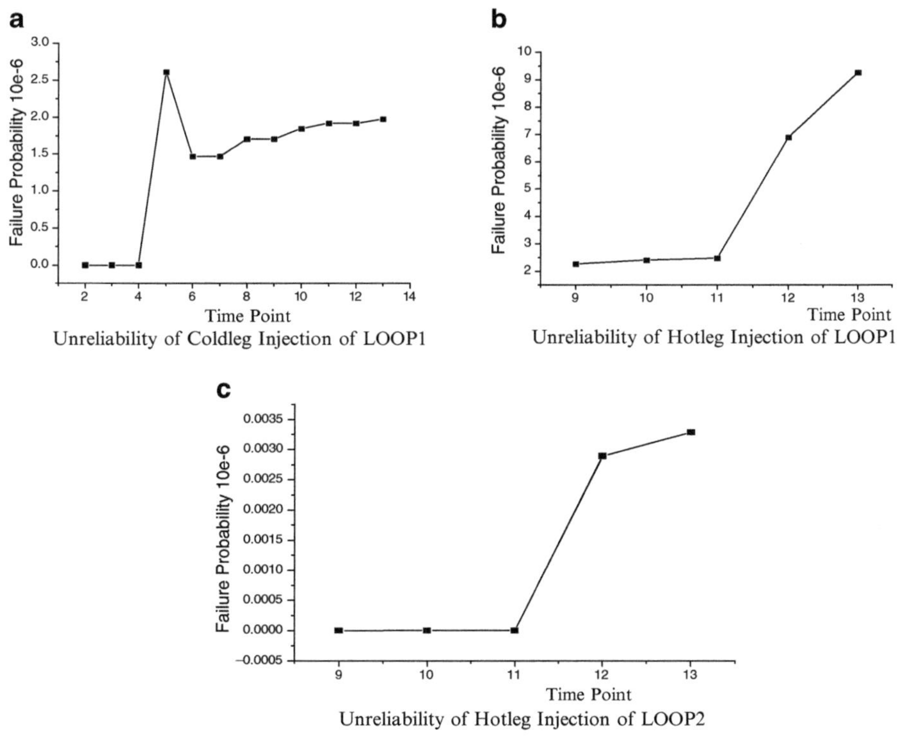

Fig. 2 Unreliability changes of HHSI and LHSI system

References

1. The Training Center of Daya Bay Nuclear Power Station (2007) Accident procedures explanation of Daya Bay Nuclear Power Station. Nuclear Power, Beijing, pp 44–89
2. Matsuoka T, Kobayashi M (1991) GO-FLOW: a system reliability analysis methodology. Elsevier, Amsterdam, pp 525–532

Prior Evaluation Method of User Interface Design

Shengyuan Yan and Kun Yu

Abstract Development of nuclear power can reduce the emissions of carbon dioxide. But nuclear power is a double-edged sword. Nuclear accidents may result more serious environment radioactive pollution. Designing and evaluating the user interface in the main control room of Nuclear Power Plants is an important approach to ensure the nuclear power safety. A prior evaluation method is proposed for user interface design in the paper. It can simulate the human–machine interaction process of Nuclear Power Plant without too much cost. The design defects of user interface design can be discovered and corrected in the design process.

Keywords Functional simulation • Evaluation • User interface design

1 Introduction

China is the second-largest contributor to energy-related carbon dioxide emissions after USA. Most of China's electricity is produced from fossil fuels, and electricity demand is growing rapidly [1]. Since growth in demand and the reliance on fossil fuels has led to much air pollution. China plans to build over 100 new nuclear power plants (NPP) by 2020, to reduce CO_2 emission. But nuclear accidents may result more serious environment pollution. Designing and optimizing user interface in main control room (MCR) of NPP has becoming an important issue.

Because most of traditional user interface evaluation method is conducted in static situation, it is difficult to reflect the relationship of displays and controls as in real system. In this paper, a novel prior evaluation method is proposed for the user interface design of NPP with applying functional simulation method. GL Studio is used to create the functional simulation models of user interface. The mentioned method will be described in detail.

S. Yan (✉) and K. Yu
College of Mechanical and Electrical Engineering, Harbin Engineering University, Harbin, China
e-mail: yanshycncn@yahoo.com.cn

T. Yao (ed.), *Zero-Carbon Energy Kyoto 2009*, Green Energy and Technology,
DOI 10.1007/978-4-431-99779-5_60, © Springer 2010

2 Functional Simulation Method

GL Studio is a tool used for developing real-time, interactive 2D and/or 3D objects [2]. This design tool provides the user with the ability to prototype any instrument or physical device, use the design in a fully immersive interactive simulation. The authors have created functional simulation models of all instruments in MRC of NPP by GL Studio. Every functional simulation model of all instruments has been built as a ".dll " file. And the ".dll " file can be used in another design as a module. Then, all the models of instruments are assembled on the user interface in MRC of NPP. Modularization notion is used in the create process to enhance the development efficiency. The final functional simulation model of user interface is shown in Fig. 1.

The functional simulation models of user interface can be displayed on the large-scale screen. It can simulate dynamic states of user interface as in operation state.

3 Evaluation Algorithm

The evaluation indexes system of user interface of NPP is proposed. It includes human–hardware interface and human–software interface. The human–hardware interface also includes displays and controls. The rock-bottom evaluation indexes of displays and controls include all the factors that may influence the final design quality of displays and controls.

The main characteristic of user interface evaluation is indeterminacy, subjective and short of information, especially for subjective evaluation. To reduce the influence of subjective judging, grey clustering theory is applied in the evaluation algorithm to enhance the accuracy of the evaluation result. Grey clustering theory method is very suitable for solving the indeterminacy problems [3]. Grey clustering method

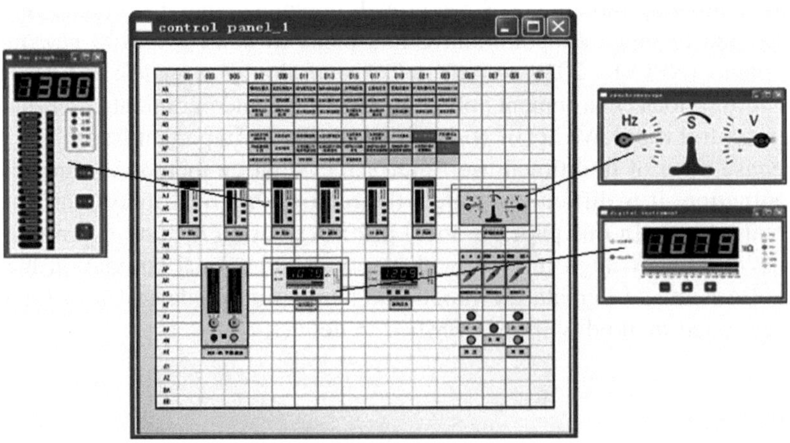

Fig. 1 Functional simulation model of user interface in MCR

divided the clustering object into grey classes by whiten function [4]. In this paper, the evaluation grade is divided into five grey classes: {excellence, preferable, fair, bad, worst}. Subjective opinions of the evaluators are turned to evaluation marks based on evaluation algorithm.

4 Instance of Prior Evaluation Method

Functional simulation based prior evaluation software of NPP is developed, which include user interface manage module, user interface display module and evaluation module. Evaluation module is used to evaluate the displays and controls of user interface in MCR. The evaluation software interface is shown as Fig. 2.

When a functional simulation based evaluation is made, first, select the console to be evaluated, the functional simulation model will emerge in work area. Second, select the evaluation menu according to which instrument to be evaluated. Third, mark each and every evaluation indexes in the evaluation interface, then calculate the total evaluation result of special instrument based on grey clustering algorithm. Figure 3 shows the case of bar-graph meter evaluation.

5 Conclusions

A prior evaluation method based on functional simulation method is proposed for user interface design of NPP. The functional simulation models of user interface have been developed based on GL Studio. The functional simulation models have

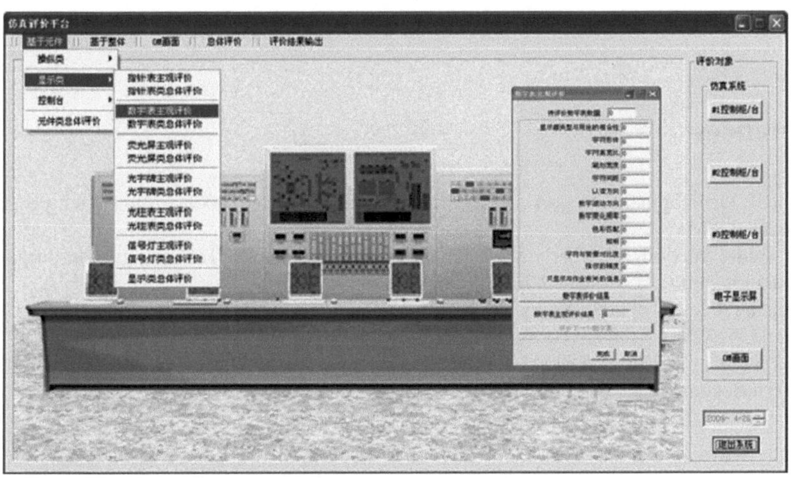

Fig. 2 Evaluate software interface of NNP

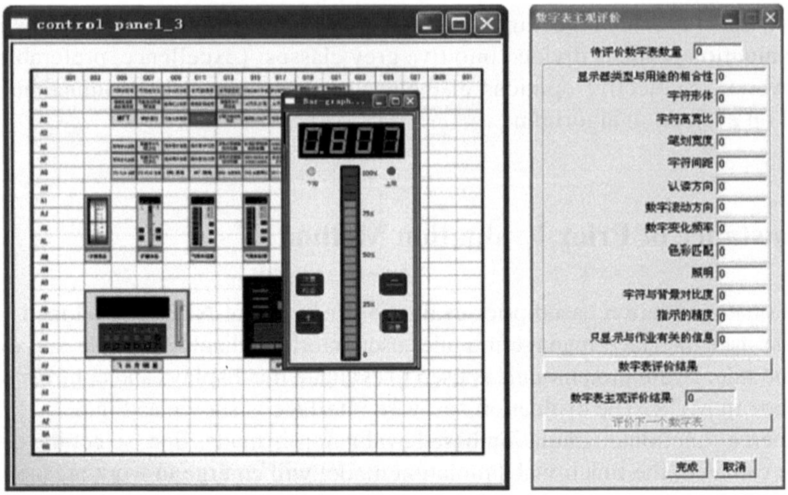

Fig. 3 Evaluation processes of bar-graph meter based on functional simulation

done a good job on simulating the dynamic states of user interface as in operation state. Then, the evaluation indexes system of user interface is proposed. Grey clustering theory is applied in evaluation algorithm to enhance the accuracy of evaluation result. Finally, the user interface prior evaluation software system for NPP was developed. The proposed evaluation method is more efficient and creditable than traditional methods, because it is more close to the real NPP MCR system.

Acknowledgments This work has been conducted under the sponsorship of the financial support from postdoctoral science-research developmental foundation of Heilongjiang Province (No. LBH-Q08111), the project supported by the National Defense Discipline Key Lab Foundation (HEUFN0803). The authors would like to express sincere thanks to the responsible persons for their supports.

References

1. Annual Development Report of China's Power Industry (2008) http://www.cec.org.cn/. Accessed 8 May 2009
2. GL Studio version 3.2 User's Manual. Distributed Simulation Technology, Inc. http://www.simulation.com/. Accessed 12 June 2006
3. Xueliang P et al (2005) Study on generation in grey system theory. J Syst Eng Electron 16(2):325–329
4. Guoping X et al (2007) A novel conflict reassignment method based on grey relation analysis. Pattern Recognit Lett 28:2080–2087

Consideration of Alumina Coating Fabricated by Sol–Gel Method for PbLi Flow

Yoshitaka Ueki, Tomoaki Kunugi, Masatoshi Kondo, Akio Sagara, Neil B. Morley, and Mohamed A. Abdou

Abstract Electrical insulation coating is extremely important for lead lithium (PbLi) magnetohydrodynamic (MHD) thermofluid research from point view of secure electric insulating wall condition. The Al_2O_3 coatings fabricated by a sol–gel method have never studied as electrical insulation coating against PbLi as a working fluid. The present paper discusses on the feasibility of the Al_2O_3 coating as an electrical insulation coating for PbLi MHD flows. It is examined the electrical insulating durability of the Al_2O_3 coating by the sol–gel method with the molten PbLi exposed. The Al_2O_3 coatings worked as electrical insulation coating for 170 h with PbLi temperature at 300°C in this study. However, the Al_2O_3 coating cured up to around 370°C seemed to have a threshold for the electric insulation break above 400°C. Nevertheless, the Al_2O_3 coating fabricated by the sol–gel method will be a potential electric insulation coating for PbLi flow under the MHD condition with the operation time and the temperature limitation.

Keywords Lead-lithium • Electrical insulation coating • Sol–gel method • Alumina

Abbreviations

APS	Air plasma spray
CVD	Chemical vapor deposition
DCLL	Dual coolant lead lithium

Y. Ueki (✉) and T. Kunugi
Department of Nuclear Engineering, Kyoto University, Kyoto, Japan
e-mails: yoshi@nucleng.kyoto-u.ac.jp; kunugi@nucleng.kyoto-u.ac.jp

M. Kondo and A. Sagara
National Institute for Fusion Science, Gifu, Japan
e-mails: kondo_masatoshi@nifs.ac.jp; sagara.akio@LHD.nifs.ac.jp

N.B. Morley and M.A. Abdou
Mechanical and Aerospace Engineering, University of California, Los Angeles, CA, USA
e-mails: morley@fusion.ucla.edu; abdou@fusion.ucla.edu

T. Yao (ed.), *Zero-Carbon Energy Kyoto 2009*, Green Energy and Technology,
DOI 10.1007/978-4-431-99779-5_61, © Springer 2010

DFLL	Dual functional lead lithium
HAD	Hot dip aluminizing
LPPS	Low pressure plasma spray
MHD	Magnetohydrodynamics
RAFM	Reduced activation ferritic/martensitic
VPS	Vacuum plasma spray

1 Introduction

Development of effective and reliable coatings is one of key fusion technologies for fusion blanket research and some fusion blanket concepts themselves [1]. Most of the efforts on electrical insulation coating development have been made for the self-cooled lithium concepts with a vanadium alloy structure (Li/V) [2]. As for PbLi electrical insulation coating, dual functional lead lithium (DFLL) adapts two optional concepts of PbLi blankets: the reduced activation ferritic/martensitic (RAFM) steel-structured He-cooled quasi-static PbLi tritium breeder (SLL) blanket and the RAFM steel structured He-gas/PbLi dual-cooled (DLL) blanket. DFLL concept considers an alumina (Al_2O_3) as a candidate for the electric insulation coating against PbLi flow for the DLL blanket [3].

There is another purpose of coatings in fusion blankets, depending on fusion blanket concepts. The water-cooled PbLi breeder (WCLL) concept with ferritic steel structures [4] adapts a coating for tritium permeation barrier (TPB) in order to reduce tritium permeation into water coolant caused by relatively high tritium partial pressure produced in PbLi breeder to an acceptable level. Most of the efforts on the TPB coating development are focused on the Al_2O_3 coatings fabricated on aluminized ferritic/martensitic steels by means of a hot dip aluminizing (HDA) [5], a chemical vapor deposition (CVD) [6], a vacuum plasma spray (VPS) [7], a detonation jet, a low pressure plasma spray (LPPS) and an air plasma spray (APS) [8]. These investigations are originally not oriented for electric insulating Al_2O_3 coating against the molten PbLi, but for the coating as TPB. Nevertheless, the investigations are also useful for the Al_2O_3 coating as the electric insulation coating especially in terms of the coating compatibility with the molten PbLi. There are considerable requirements that must be satisfied in the coating development for the fusion application. General requirements of coatings used for all fusion systems are summarized as follows [1]:

1. Potential for coating large complex geometry or configuration
2. Potential for in site self-healing of defects that might occur
3. Processing parameters compatible with material and capabilities, e.g. temperatures and times
4. Bonding/thermal expansion match with substrate
5. Acceptable neutronic properties
6. Material availability/cost
7. Safety/environmental characteristics
8. Radiation damage resistance

The Al_2O_3 coatings fabricated on aluminized ferritic/martensitic steels by means of HDA and CVD are reported to have the favorable bonding with its substrate against the molten PbLi exposed. However, the fabrication techniques are, in general, high cost and difficult to fabricate on a large complex configuration. The sol–gel method of Al_2O_3 coating has some advantages as follows:

1. Easy material availability
2. Easy process to fabricate Al_2O_3 coating on a substrate
3. Applicable to large complex configuration
4. Low cost fabricating process

On the contrary, the Al_2O_3 coating fabricated by the sol–gel method has poor bonding to its substrate compared to the above-mentioned process, e.g. HDA and CVD. The Al_2O_3 coatings fabricated by the sol–gel method have never studied as electrical insulation coating against PbLi flows. The present research focuses on the Al_2O_3 coatings fabricated by the sol–gel method and discusses the feasibility of the Al_2O_3 coating as the electrical insulation coating for PbLi flows under the MHD condition.

2 Al_2O_3 Coating by Sol–Gel Method

The sol–gel method fabrication of Al_2O_3 coating has been commercialized already. Aremco Products, Inc. commercially provides an Al_2O_3 coating material (Ceramacoat™ 503-VFG-C), which can be fabricated by the sol–gel method. The Ceramacoat™ is a single-component, Al_2O_3-filled, high-temperature (maximum durable temperature: 1,650°C), and electrical insulation coating material. The Al_2O_3 coating material contains an aluminum oxide and a mono aluminum phosphate suspended in an inorganic binder system. The cured Al_2O_3 coating is 10^7 m at room temperature in volume resistivity.

3 Electrical Insulating Test of Al_2O_3 Coating

With the molten PbLi exposed, the electrical insulating performance of Al_2O_3 coating fabricated by the sol–gel method was examined. The Al_2O_3 coating is fabricated by the sol–gel method on stainless steel cups in the procedure mentioned above. The stainless cup is made of SUS304. Figure 1 shows the Al_2O_3 coating electrical insulation test set-up. The stainless cups with the Al_2O_3 coating were filled with the molten PbLi in an electric furnace. Electrodes were attached to one for the cup substrate and the other for immersing into the molten PbLi. The electrodes were connected to a power supply to be applied a few volts, and to a Multiplexer/Voltmeter to measure the current passing through the Al_2O_3 coating with time variation.

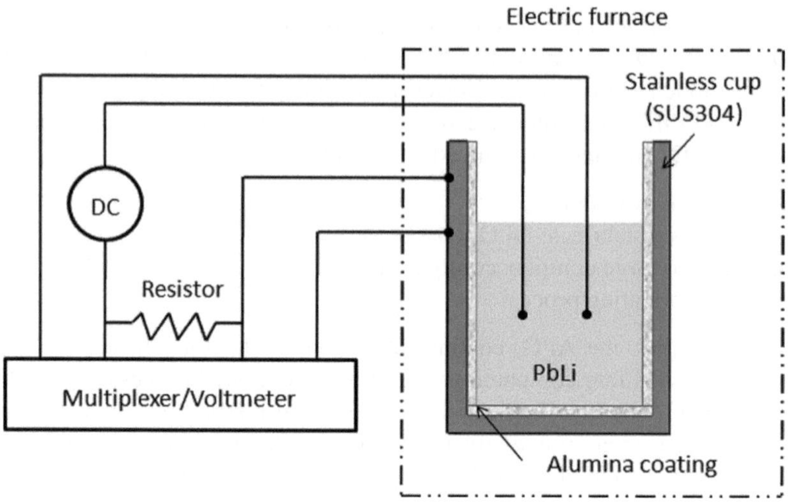

Fig. 1 Electrical insulating test experimental setup

The following three test runs were conducted with the change of PbLi temperature as the experimental parameter:

- Run #1: PbLi temperature was controlled at 300°C for around 170 h.
- Run #2: PbLi temperature was controlled to increase up to 500°C.
- Run #3: same as Run #2.

The PbLi temperature, air temperature in the electric furnace and the current passing through the Al_2O_3 coating were sampled every 10 min by the multiplexer.

4 Results and Discussion

The electrical insulation test results are shown in Figs. 2–4. Figure 2 shows the results of Run #1. The Al_2O_3 coating insulated current path between the molten PbLi and its substrate with exposed to the molten PbLi at 300°C for around 170 h. Figure 3 shows the results of Run #2. With increase in temperature, the electrical insulation of the Al_2O_3 coating was broken to let the current pass through the coating at around 430°C of PbLi. Figure 4 shows the results of Run #3. It indicates the similar result to Run #2. With increase in temperature, the electrical insulation of the Al_2O_3 coating was broken around 410°C in PbLi.

Run #1 result indicated the Al_2O_3 coating fabricated by the sol–gel method has the electrical insulating durability for approximately 170 h in this study. Run #2 and #3 results indicated the Al_2O_3 coating cured up to around 370°C seemed to have a threshold for the electric insulation break above 400°C. The curing process and temperature are dominant parameters for the coating structure or durability by the

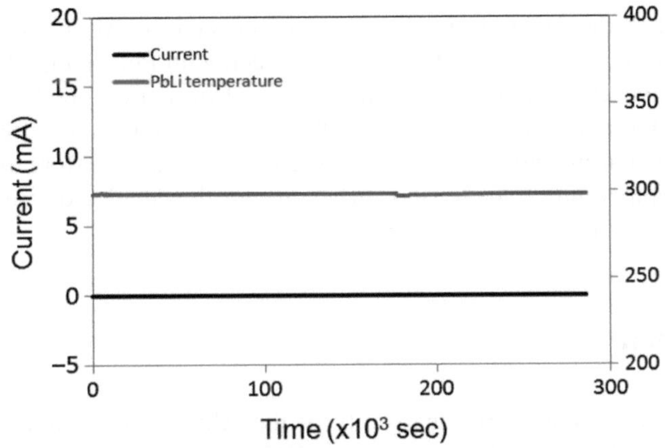

Fig. 2 Electrical insulation test result (Run #1)

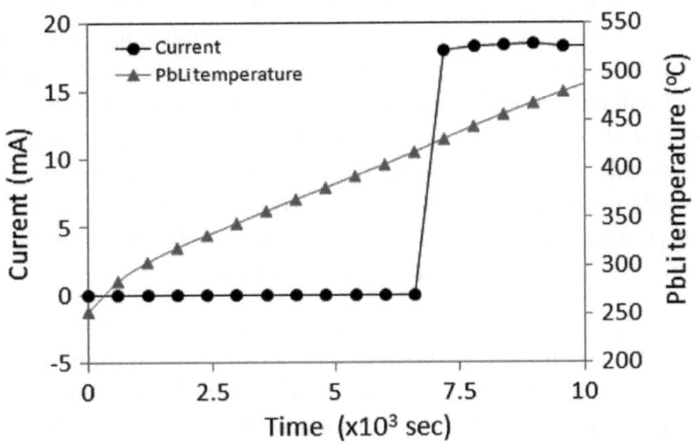

Fig. 3 Electrical insulation test result (Run #2)

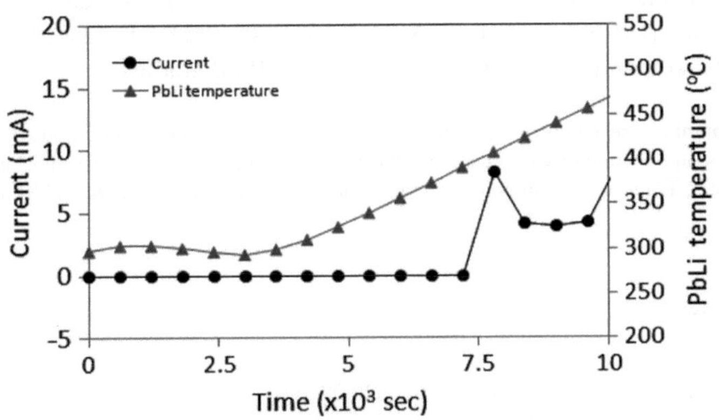

Fig. 4 Electrical insulation test result (Run #3)

sol–gel method. A thermal expansion mismatch between the Al_2O_3 coating and its substrate caused the coating detachment or the crack with PbLi temperature above the curing temperature, and eventually the electrical insulation breaks. The Al_2O_3 coating with cured at higher temperature will have a potential as the effective electrical insulation coating endurable up to the higher temperature. More experiments on the curing temperature issue must be necessary for the detailed discussion of the Al_2O_3 coating performance.

5 Conclusion

The present paper discusses the feasibility of the Al_2O_3 coating as the electrical insulation coating for PbLi flow under the MHD condition. The present conclusions are summarized as follows:

1. The Al_2O_3 coatings worked as electrical insulation coating for 170 h with PbLi at 300°C in this study. However, the Al_2O_3 coatings performance in longer operation still remains to be evaluated. Nevertheless, the Al_2O_3 coating fabricated by the sol–gel method will have a potential electric insulation coating enough for PbLi MHD thermofluid study with the operation time and the temperature limitation.
2. The Al_2O_3 coating cured up to around 370°C seemed to have a threshold for the electric insulation break above 400°C. It is discussed that the Al_2O_3 coating with cured at higher temperature will have a potential as the effective electrical insulation coating endurable up to the higher temperature.

Further investigations on the Al_2O_3 coating must be necessary for the detailed discussion of the coating performance. The insulation coating thickness is an important parameter for electrical insulation efficiency. In addition, the curing process and temperature are dominant parameters for the sol–gel method coating structure or durability. From the view point of evaluation of the coating durability, the material analysis of the Al_2O_3 coating cross-section must be necessary. Material interaction and chemistry control issues associated with the fabrication process, the stability and the performance of the Al_2O_3 coating should be discussed for the usage in fusion blanket, and even in PbLi MHD thermofluid studies.

Acknowledgments The authors were grateful for the great financial support the Ministry of Education, Culture, Sports and Technology of Japan and the US Department of Energy via the Japan–US Joint Research Project, Tritium, Irradiation and Thermofluid for America and Nippon (TITAN).

References

1. Smith DL et al (2002) Fusion Eng Des 61–62:629–641
2. Muroga T et al (2007) J Nucl Mater 367–370:780–787
3. Wu Y et al (2006) Fusion Eng Des 81:2713–2718

4. Giancarli L et al (1994) Fusion Tech 26:1079–1085
5. Glassbrenner H et al (2000) J Nucl Mater 283–287:1302–1305
6. Chabrol C et al (1997) Report CEA Grenoble No. 73/97
7. Perujo A et al (1995) Fusion Tech 28:1256–1261
8. Benamati G et al (1999) J Nucl Mater 271–272:391–395

Feasibility Study on Introducing Building Integrated Photovoltaic System in China and Analysis of the Promotion Policies

Hongbo Ren, Weisheng Zhou, and Ken'ichi Nakagami

Abstract In this study, after investigating the current status of PV system and corresponding policies, a customer aided evaluation model is developed for the analysis of PV adoption in China. Using the developed model, an investigation is conducted of feasible PV adoption for four typical buildings located in a pilot model area in China. According to the simulation results, although with reasonable environmental merits, the introduction of PV system results in increased energy cost for all the assumed buildings. Furthermore, through the sensitivity analysis, the positive effects of some promotion policies, including subsidies and electricity buy-back, on the introduction of PV system have been recognized.

Keywords BIPV system • China • Economic analysis • Government subsidies • Electricity buy-back

1 Introduction

In the past few years, response to the continuing global warming, in order to develop a sustainable and low carbon society, the Chinese government is considering to issue some new policies to promote the adoption of photovoltaic (PV) system, especially the building integrated PV (BIPV) system. At present, BIPV is mainly applied on exposed walls, sun-shading shelters, patios, tiles, roofs, sound-proof walls, as well as in fields of private apartments, schools, hospital buildings, airports, platforms of subway stations and large workshops, in the developed countries [1]. It is also expected to be an excellent option for building energy conservation and low-carbon society construction in China.

In this study, by integrating the local climate conditions and energy market information, as well as the policy considerations, a customer aided evaluation model is

H. Ren (✉)
Global Innovation Research Organization, Ritsumeikan University, Kyoto, 603-8577, Japan
e-mail: tjrhb@fc.ritsumei.ac.jp

W. Zhou and K. Nakagami
College of Policy Sciences, Ritsumeikan University, Kyoto, 603-8577, Japan

T. Yao (ed.), *Zero-Carbon Energy Kyoto 2009*, Green Energy and Technology, 380
DOI 10.1007/978-4-431-99779-5_62, © Springer 2010

developed for the analysis of PV adoption in China. Using the developed model, an investigation is conducted of feasible BIPV adoption for four typical buildings (government office, hospital, school and hotel) in a pilot model area of China.

2 Current Status of PV System Adoption in China

By the year 2006, the total PV capacity in China is about 80 MW, which is far below some pioneer countries, such as Spain, Germany, and Japan. Among all the applications, about 43% is for remote areas power supply, 40% is for communication and industrial applications [2,3]. BIPV, which takes up the dominant share in the PV market of developed countries, is rarely employed. Currently, it is mainly adopted in some model buildings.

Facing with such an unsatisfied situation, in order to cope with the economic depression, as well as to develop a sustainable and low-carbon society, the Chinese government issued the "Method of financial subsides for BIPV application" in March 2009 to boost the BIPV application in China. According to the method, the capacity of a single project should be not less than 50 kW. Currently, the subsides are set to be about 300 Yen/W for the initial cost, and is priority for the public buildings, including school, hospital, government office, etc.

3 Evaluation Method and Case Study

This model attempts to solve the optimal sizing problem of PV system utilizing the method of mathematical programming. The annual total cost is minimized from the viewpoint of long-term economics. As constraints of the evaluation model, it mainly considers the performance characteristic of PV system and the energy balance relationships of each energy flow to satisfy the load demands.

In this study, in order to research the economic feasibility of BIPV system in China, a model area in Huzhou of Zhejiang province is selected for case study [4]. In the following, four typical buildings (government office, school, hospital, and hotel), with a floor area of 10,000 m^2, are chosen as the representative categories of the public and commercial buildings in the model area. As a demonstrative example, a 150 kW grid integrated PV system is assumed for all buildings.

4 Results and Discussions

4.1 Economic Analysis of PV Adoption

Figure 1 illustrates the annual energy costs of various buildings for both conventional system (supply all electric demand with utility grid) and PV system. Generally, it can be found that for all the studied buildings, the introduction of PV

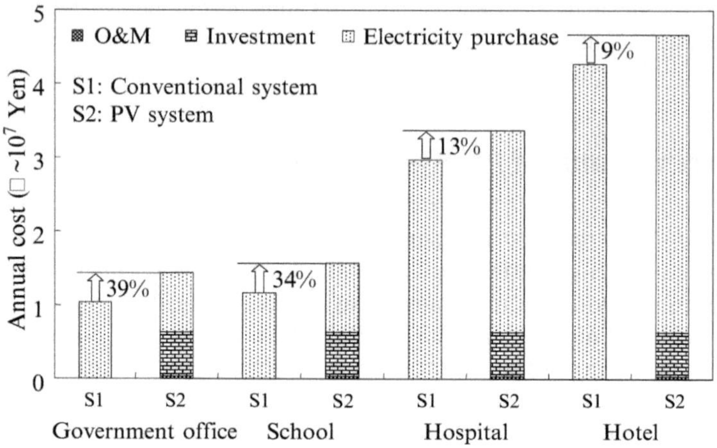

Fig. 1 Annual energy cost for various buildings

system results in increased annual cost. Compared with the conventional system, the annual total cost is increased by 39%, 34%, 13% and 9% for government office, school, hospital and hotel, respectively. It can be deduced that, unless enough economic promotion policies are introduced, it is hard to promote PV system in the studied public and commercial buildings.

4.2 *Effect of Government Subsidies*

Financial subsidy is the most conventional economic encouragement practice, which is also employed by the Chinese government to promote the PV adoption. As shown in Fig. 2, for all studied building types, the rise of subsides leads to increased cost reduction ratio. As subsides are set to about 68% of the initial cost (440 Yen/W), the cost reduction ratio increases from a negative value to a positive one. Therefore, it can be recognized that the current subsides introduced in China is not large enough to make PV system popular for the public and commercial buildings.

4.3 *Effect of Electricity Buy-back*

In the following, as a stimulative policy for the mass spread of PV systems, the excess electricity out of PV system is assumed to be sold back to the grid. Figure 3 shows the results of electricity buy-back price sensitivity. It is not surprising that increases in electricity buy-back price result in corresponding increased economic merit of PV system. Before the buy-back price is rose above 10 Yen/kWh, the introduction of electricity buy-back has no effect on the economics of the PV system. This is because of the relatively low price, which makes the electricity from

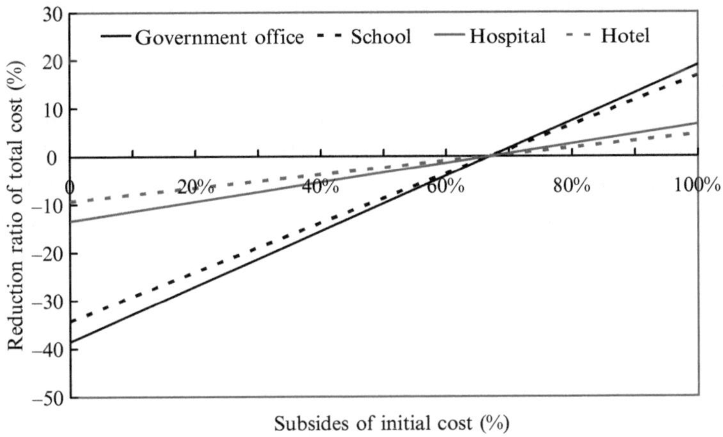

Fig. 2 Effect of subsides on the economic results

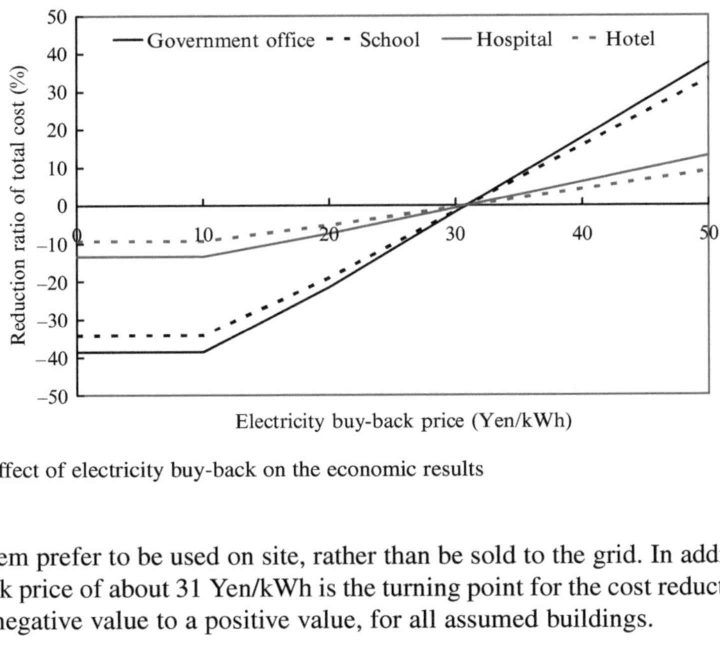

Fig. 3 Effect of electricity buy-back on the economic results

PV system prefer to be used on site, rather than be sold to the grid. In addition, the buy-back price of about 31 Yen/kWh is the turning point for the cost reduction ratio from a negative value to a positive value, for all assumed buildings.

5 Conclusions

In this study, the hypothetical introduction of a gird connected PV system for four typical buildings in a model area of China has been analyzed. A mathematical model was developed for determining the economic energy system installation at given electricity demand for the customer, and calculating the optimal installation capacities of the PV system under several conditions. According to the analysis in this study, we can get the following conclusions:

1. At current situation, although with reasonable environment benefits, it is not economic to introduce PV system in the public and commercial buildings in China.
2. The introduction of subsides and electricity buy-back can more or less promote the adoption of PV system. However, the effect of a single policy is always limited. It may be a good consideration to integrate various policies to reach a bright solar future in China.

References

1. Zhengming Z, Qingyi W, Xing Z, Hamrin J, Baruch S (2006) Renewable energy development in China: the potential and the challenges. Center for Resource Solutions, San Francisco, CA
2. National Development and Reform Commission (2007) Chinese renewable energy industry development report, 2007
3. Tsinghua University (2005) Present situation and prediction on photovoltaic in china. In: Proceeding of China Renewable Energy Development Strategy Workshop, 2005
4. The city of Huzhou. Available at http://www.huzhou.gov.cn/

Author Index

Keyword Index

MIX
Papier aus verantwortungsvollen Quellen
Paper from responsible sources
FSC® C105338

If you have any concerns about our products,
you can contact us on
ProductSafety@springernature.com

In case Publisher is established outside the EU,
the EU authorized representative is:
Springer Nature Customer Service Center GmbH
Europaplatz 3, 69115 Heidelberg, Germany

Printed by Libri Plureos GmbH
in Hamburg, Germany